Mathematics for Today

G. D. B

A. D. Bal

Macmillan Education

First published 1978
Reprinted 1979
Second edition 1980

Published by
Macmillan Education Limited
Houndmills Basingstoke Hampshire RG21 2XS
and London
Associated companies in Delhi Dublin
Hong Kong Johannesburg Lagos Melbourne
New York Singapore and Tokyo

Printed in Hong Kong

Contents

Preface

The considerable diversity of courses and teaching methods used in large comprehensive schools often leads to the point where, at the beginning of the fourth year, students are finally put into GCE O–level sets. It is at this stage that the authors feel it desirable for a single book to be used which covers fully the material for the examination, while also enabling work previously covered to be quickly revised and developed. This book has been written from a course used in such a situation and covers all of the work needed for most 'modern' examinations.

With the trend away from the more traditional syllabuses, the authors feel that there is a need for a book which covers the work for the new examinations without the restrictions on methods of teaching imposed by many of the new course textbooks. Consequently, this book has been written along traditional lines with the emphasis on examples and exercises, keeping description to a minimum. Each chapter covers a separate topic and is concluded by miscellaneous examples divided into two types. The harder examples are mainly for use in the final year as revision and are generally of the standard which students would find in an examination paper; in fact many of them are from past examination papers.

Although some mathematical rigour has been included, the level has been kept at a standard which can be coped with by even the weakest candidate. For example, the signs $\Rightarrow$ and $\therefore$ have been used exclusively instead of the logical equivalence sign $\Leftrightarrow$. With the advent of the pocket calculator, work involving the slide rule has been entirely omitted. However, a full treatment of logarithms and associated table work has been included. A chapter on the calculus has been included in this new edition at the request of many teachers, to meet the demands of a number of O-level syllabuses.

A selection of test papers is included at the end of the book, each of which is designed to occupy a single lesson of forty minutes. These papers should be used throughout the two-year course as they form an important part of the work.

The authors would like to thank the following Examination Boards for permission to include questions from their past examination papers in Mathematics at GCE O–level:

The Associated Examining Board [AEB]
University of Cambridge Local Examinations Syndicate [C]
Joint Matriculation Board [JMB]
University of London University Entrance and School Examinations Council [L]
Northern Ireland Schools Examinations Council [NI]
Oxford Delegacy of Local Examinations [O]
Oxford and Cambridge Schools Examination Board [O & C]
(Questions from School Mathematics Project papers are coded [O & C (SMP)])
Southern Universities' Joint Board for School Examinations [SUJB]
Welsh Joint Education Ccommittee [W]

Also, Josephine Buckwell for her careful typing and checking of the manuscript.

Symbols used in the text

$\therefore$	therefore
$\Rightarrow$	implies
$A \cup B$	the union of A and B
$A \cap B$	the intersection of A and B
$\mathscr{E}$	the universal set
A'	the set of elements not in A
$\varnothing$	the empty set
$A \subset B$	A is a subset of B
$\in$	belongs to
$n(A)$	the number of elements in A
$f(x)$ or $f: x \rightarrow$	function of x
$\mathbf{a}$ or $\overrightarrow{AB}$	vector or translation
$\angle AOB$	angle AOB
$\lvert\mathbf{a}\rvert$	the length of $\mathbf{a}$
$P(X)$	the probability that event X happens
$a \leftrightarrow b$	one-to-one correspondence between a and b
I	the identity matrix

1 The language of sets

1.1 Definition of a set

Throughout this book the language of sets will be used wherever possible to clarify ideas. A *set* is a collection of objects.

e.g. A = {red, orange, yellow, green, blue, indigo, violet} is the set of colours in the rainbow.

You will see that we use { } to enclose the objects. The *order* in which we write down the objects is immaterial. The objects in the set are called *members* or *elements* of the set.

We write

red $\in$ A to mean red 'belongs to' A.

We also write

black $\notin$ A to mean black does 'not belong to' A.

1.2 Sets of numbers

In mathematics, we are most likely to be talking about sets of numbers:

e.g. B = {1, 2, 3, 4, 5, 6, 7, 8}. It is important to realise that it is incorrect to write the set as B = {1, 2, . . . 8} because this tells us nothing about what happens between 2 and 8.

1.3 Operations on sets

(i) Union: Look at the two sets

A = {1, 2, 4, 7, 8} and B = {3, 4, 7}.

If we put the elements of these two sets together, we form the *union* of A and B, written $A \cup B$ (read as A union B):

i.e. $A \cup B = \{1, 2, 3, 4, 7, 8\}$.

You will notice that although 4 and 7 occur in each set, they are only written down once in the union.

(ii) Intersection: If we find all those elements which are in both sets we form the *intersection* of A and B, written $A \cap B$ (read as A intersection B):

i.e. $A \cap B = \{4, 7\}$.

1.4 Null set

If $A = \{1, 3, 7\}$ and $B = \{2, 4)$, then clearly A and B have no elements in common.

i.e. $A \cap B = \{\ \}$.

The symbol used to denote $\{\ \}$, the empty set (null set), is $\emptyset$.

$\therefore A \cap B = \emptyset$.

1.5 Universal set

In any mathematical problem, we need to restrict the objects that we are investigating. This limited set is called the *universal set*, denoted by $\mathscr{E}$. The importance of this set should become more obvious when we talk about the complement of a set in the next section.

1.6 Complement

It should be clear that if you wanted to talk about all elements which do not belong to a set you would have difficulty in describing all of them unless you had limited your interest.

e.g. If the universal set $= \{1, 2, 3, 4, 5, 6, 7, 8, 9, 10\}$ and $A = \{1, 3, 5, 7, 9, 10\}$, then the set of elements which does not belong to A, called the *complement* of A and written A', will be

$A' = \{2, 4, 6, 8\}$.

1.7 Subsets

In the sets $A = \{1, 3, 4\}$ and $B = \{1, 2, 3, 4, 6\}$ it can be seen that every element of A belongs to B. We say that A is a *subset* of B, written

$$A \subset B.$$

Notice that A is a subset of itself, because every element of A belongs to itself, and so we can write

$$A \subset A.$$

If A is not a subset of B then we write

$$A \not\subset B.$$

1.8 Alternative notation

If there are a large number of elements in the set then it becomes tedious listing them individually. We use the following alternative notation:

$$A = \{y : y \text{ is an integer and } 1 \leq y \leq 100\}.$$

This is read as: A = the set of all possible values of y such that y is an integer and lies between 1 and 100 inclusive. In other words A is the set of the first 100 natural numbers.

If we remove the restriction that y should be an integer, and write

$$B = \{y : 1 \leq y \leq 100\},$$

then B is the set of all possible numbers between 1 and 100. This is an *infinite set.*

1.9 Venn diagrams

Problems involving sets can often be considerably simplified by the use of diagrams, called *Venn* diagrams. In a Venn diagram, each set is represented by a closed curve. The universal set is represented by a rectangle.

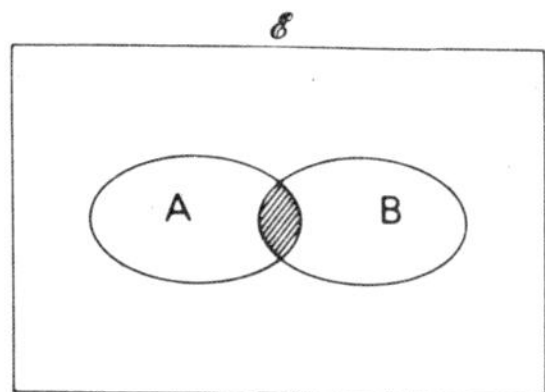

Fig. 1.1

In Fig. 1.1 we see two sets A and B. The sets overlap, showing that their intersection is not empty. Therefore the shaded region must represent $A \cap B$.

In Fig. 1.2 we see a number of examples, with the shaded area described underneath.

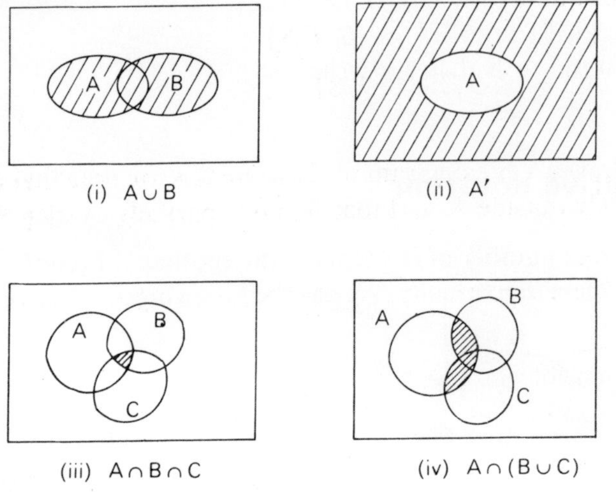

(i) $A \cup B$ (ii) A'

(iii) $A \cap B \cap C$ (iv) $A \cap (B \cup C)$

Fig. 1.2

1.10 Number of elements in a set

n (A) denotes the number of elements in set A.

e.g. if $A = \{1, 3, 9, 15, 16\}$, then $n(A) = 5$
if $B = \{x : x \text{ is an even integer and } 1 \leqslant x \leqslant 30\}$,
then $n(B) = 15$.

By looking at the Venn diagram in Fig. 1.1 it should be clear that

$$n(A \cup B) \neq n(A) + n(B)$$

because the shaded region, which is $A \cap B$, will have been included twice, once for A and once for B.

$$\therefore \; n(A \cup B) = n(A) + n(B) - n(A \cap B)$$

EXAMPLE 1

If the universal set $\mathscr{E} = \{5, 6, 7, 8, 9, 10\}$, $A = \{8, 9\}$, $B = \{5, 6, 9, 10\}$ and $C = \{6, 7\}$, find (i) $A \cap B \cap C$, (ii) $A \cap (B' \cup C)$. Draw a Venn diagram to illustrate the relationship between A, B and C.

(i) $A \cap B \cap C$ is the set of numbers common to all three sets. Since there is no number which is a member of all three sets, $A \cap B \cap C$ is empty and we write

$$A \cap B \cap C = \emptyset.$$

(ii) In order to find $A \cap (B' \cup C)$, we must first work out $B' \cup C$. B' is the set of numbers not in B, i.e. $\{7, 8\}$,
so $B' \cup C = \{7, 8\} \cup \{6, 7\} = \{6, 7, 8\}$

$$\therefore\ A \cap (B' \cup C) = \{8, 9\} \cap \{6, 7, 8\}$$
$$= \{8\}$$

In order to draw a Venn diagram to illustrate this, we note that C must lie entirely outside A, and that A and C partially overlap B. See Fig. 1.3.

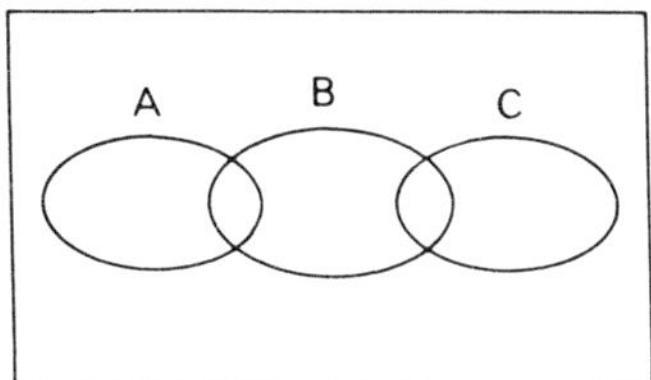

Fig. 1.3

EXAMPLE 2

If $n(X) = 21$, $n(X \cup Y) = 37$ and $n(X \cap Y) = 5$, find $n(Y)$. Illustrate with a Venn diagram.

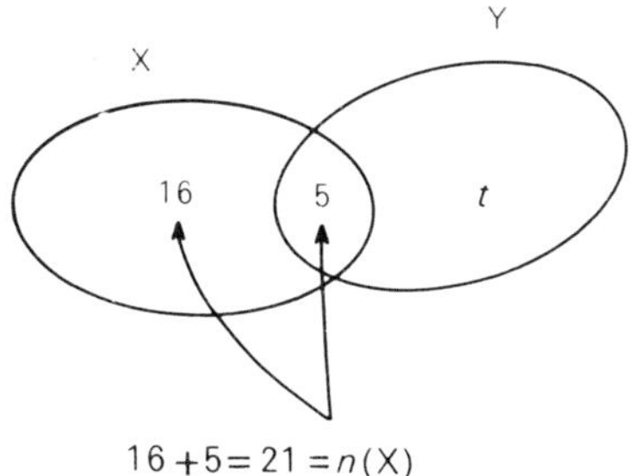

Fig. 1.4

To find $n(Y)$ we need to know t. But $n(X \cup Y) = 37$.

$$\therefore\ 16 + 5 + t = 37$$
$$\therefore\ t = 16$$
$$\therefore\ n(Y) = 5 + 16 = 21.$$

EXAMPLE 3

The universal set $\mathscr{E}$ is the set of points inside a square ABCD of side 6 cm. R is the set of points P such that PA > PB. S is the set of points Q such that DQ > 4 cm. Draw a diagram and shade $R \cap S$.

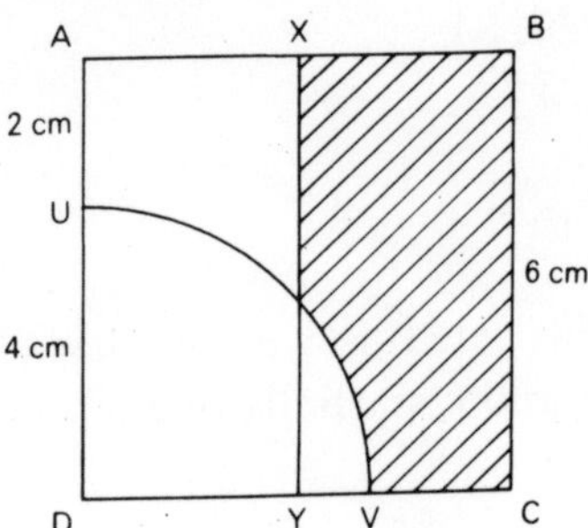

Fig. 1.5

XY is the line joining the mid-points of AB and DC, and along it PA = PB.

$\therefore$ R is the set of points to the right of XY.

UV is an arc of a circle of radius 4 cm.

$\therefore$ Q is the set of points outside the circle.

Hence $R \cap S$ is the shaded region shown in Fig. 1.5.

EXAMPLE 4

$A = \{x : x \text{ is an integer and } 1 \leq x \leq 20\}$.
$B = \{z : z \text{ is a prime number}\}$

Find $A \cap B'$.

First note that A = {1, 2, 3, 4, 5, 6, 7, 8, 9, 10, 11, 12, 13, 14, 15, 16, 17, 18, 19, 20}. $A \cap B'$ is the set of numbers which belong to A and do not belong to B—that is, the elements of A which are not prime numbers.

$\therefore A \cap B' = \{4, 6, 8, 9, 10, 12, 14, 15, 16, 18, 20\}$.

EXAMPLE 5

A, B and C are three sets which have at least one element in common. Draw a Venn diagram and shade the set represented by $(A' \cup B)' \cap (B \cup C')$.

In more complicated examples it may be necessary to build up the final diagram from a number of simpler ones. In this case we need one for $A' \cup B$, then one for $(A' \cup B)'$, and finally one for $(B \cup C')$.

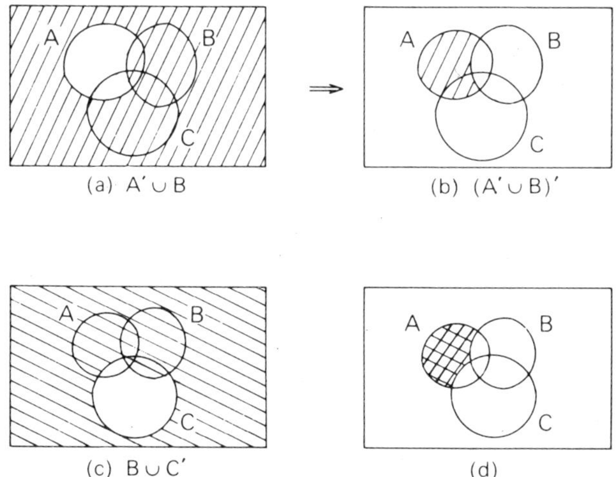

(a) $A' \cup B$ (b) $(A' \cup B)'$ (c) $B \cup C'$ (d)

Our answer is obtained by comparing diagrams (b) and (c) and seeing where the shading overlaps. Diagram (d) shows us in fact that $(A' \cup B)' \cap (B \cup C') = A \cap (B \cup C)'$.

EXAMPLE 6

Three sets, A, B and C satisfy the following conditions:

(i) $A \cap C = \emptyset$ (ii) $n(B \cap C) > 0$ (iii) $B' \cap C = \emptyset$.
(iv) $A \subset B$

Draw a Venn diagram to illustrate the relationship between them.

(i) tells us that A and C do not overlap. (ii) tells us that B and C have elements in common. (iii) tells us that $C \subset B$ (i.e. no elements of C are outside B) (iv) tells us that $A \subset B$. The Venn diagram must be as in Fig. 1.6.

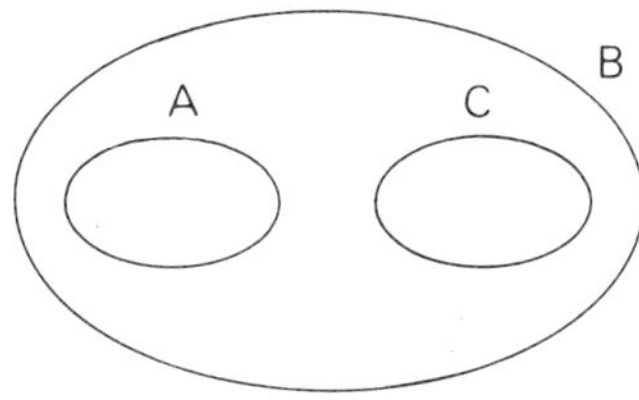

Fig. 1.6

EXAMPLE 7

Some rich men own horses. Some people who own horses also ride them. Is it true, then, that some rich people ride horses?

In reality, of course, the answer to this question is yes, but in fact we are not given enough information to answer the question. Let O be the set of people who own horses, H be the set of people who ride, and R be the set of rich men.

(i) The first statement tells us $O \cap R \neq \emptyset$.

(ii) The second statement tells us $O \cap H \neq \emptyset$.

However, it does not tell us anything about H and R. Hence, the Venn diagram could look like that in Fig. 1.7, in which case no rich man rides a horse.

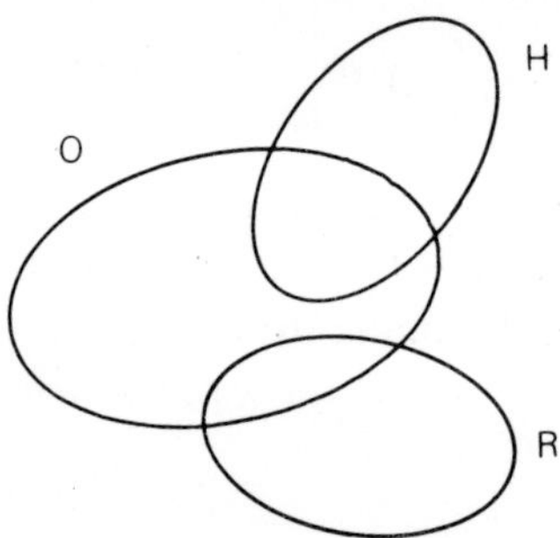

Fig. 1.7

EXERCISE 1

1 If $A = \{1, 2, 4\}$, $B = \{3, 4\}$, $C = \{1, 2, 5\}$, find (i) $A \cup B$; (ii) $B \cap C$; (iii) $A \cup (B \cap C)$; (iv) $(A \cap B) \cup C$.

2 If $X = \{a, b, c, e\}$, $Y = \{d, f, e, h, a\}$ and $Z = \{a, c, f, g\}$, find (i) $X \cap Y$; (ii) $X \cup Y$; (iii) $(Z \cap X) \cup Y$; (iv) $(Y \cap Z) \cup (X \cap Z)$.

3 If the universal set $\mathscr{E} = \{1, 2, 3, 4, 5, 6\}$, using the sets in question 1 find (i) A'; (ii) $A' \cap B'$; (iii) $(A \cup B)'$; (iv) $A' \cap B' \cap C'$.

4 If $\mathscr{E} = \{$letters of the alphabet$\}$, using the sets in question 2, find (i) $X' \cap Y$; (ii) $Z' \cap X$; (iii) $X' \cap Y \cap Z'$.

5 Draw a Venn diagram for each of the following with the set required shaded: (i) $(A \cap B)'$; (ii) $A' \cap B'$; (iii) $(A' \cap B)$.

6 Draw a Venn diagram to show the relationship between A, B, C and $\mathscr{E}$ from questions 1 and 3.

7 List the elements of the following sets:

(i) $A = \{x : 1 < x < 10, x \text{ is a prime number}\}$

(ii) $B = \{x : 1 < 2x < 10, x \text{ is an odd number}\}$
(iii) $C = \{x : 1 < 3x < 10, x \text{ is a multiple of } 3\}$
(iv) $D = \{x : 3 < 2x + 1 < 12, x \text{ is a square number}\}$.

8 Find the largest and smallest elements of the following sets:
(i) $A = \{y : 1 \leq y \leq 10, y \text{ an even number}\}$
(ii) $B = \{t : t \text{ is a positive prime number between } 50 \text{ and } 100\}$
(iii) $C = \{m : \text{the digits of } m \text{ add up to } 6, m \text{ is positive, and no digit is zero}\}$
(iv) $D = \{n : 3 \leq 2n \leq 40 \text{ and leaves a remainder of } 3 \text{ when } n \text{ is divided by } 7\}$.

9 In the following question x is an integer.
If $X = \{x : 3 < 5x + 1 < 18\}$ and $Y = \{x : 2 < 4x - 1 < 16\}$, find (i) $X \cap Y$; (ii) $X \cup Y$.

10 If $D = \{y : 1 < y < 20 \text{ and } y \text{ is a square number}\}$, what is $n(D)$?

11 $P = \{p : p \text{ is a positive prime number, and } 10 \leqslant p \leqslant 100\}$; find $n(P)$.

12 In Fig. 1.8, $\mathscr{E}$ is the set of points inside the circle, R is the set of points inside the square ABCD, and T is the set of points inside the semi-circle to the right of XY. Shade the following sets: (i) $T' \cap R$; (ii) $R' \cap T$.

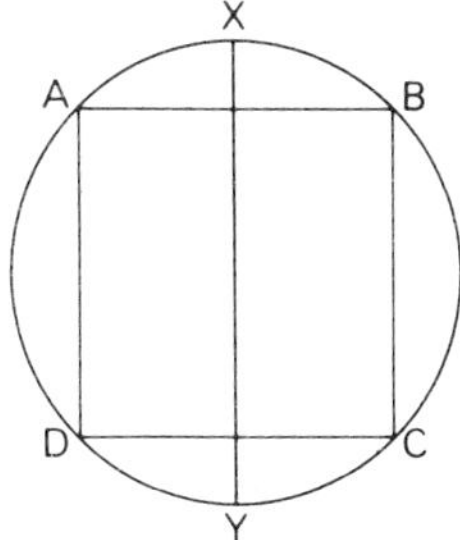

Fig. 1.8

13 If $n(A) = 15$, $n(A \cap B) = 8$ and $n(A \cup B) = 19$, what is $n(B)$? Illustrate with a Venn diagram.

14 If $n(X \cup Y) = n(X) + n(Y)$, what is $n(X \cap Y)$?

15 If $n(M) = 20$ and $n(N) = 17$, what are the smallest and largest possible values of $n(M \cup N)$?

16 If $n(A) = 8$ and $n(B) = 7$, what are the smallest and largest possible values of $n(A \cup B)$? If $n(A \cup B) = x$ and $n(A \cap B) = y$, what is the largest value of $x - y$?

17 A, B and C are three sets which have at least one element in common. Draw Venn diagrams to illustrate the following sets: (i) $(A \cap B') \cup C$; (ii) $(A' \cap B') \cup C'$.

R and S are two sets which have at least one element in common. Draw Venn diagrams to illustrate the following sets: (i) $R \cap S'$; (ii) $R' \cup S$; (iii) $(R' \cup S)'$.

19 List all the subsets of the set $A = \{a, b, c\}$

20 If $A \subset B \subset D$, which of the following statements are always true: (i) $A \cap D = \emptyset$; (ii) $A \cup B = D$; (iii) $n((A' \cap B')') \leq n(D)$; (iv) $B' \subset D$?

21 In the following questions one of the sets may be a subset of the other. Insert the appropriate sign from $\supset$, $\subset$, $\not\supset$.

(i) {positive multiples of 3} . . . {positive multiples of 9}
(ii) {positive odd numbers} . . . {positive prime numbers}
(iii) {all integers} . . . {all fractions}
(iv) {all isosceles triangles} . . . {all equilateral triangles}
(v) {triangular numbers} . . . {square numbers}.

HARDER EXAMPLES 1

1 If $\mathscr{E}$ = {buses}, R = {red buses} and D = {double-decker buses}, write in terms of R and D an expression for the set of all single-decker buses which are not red. Illustrate with a Venn diagram, showing this set shaded.

2 $A + B$ denotes the set $(A \cap B') \cup (B \cap A')$. Draw a Venn diagram and shade the set $A + B$.

3 $A - B$ denotes the set $A \cap B'$. Draw a Venn diagram and shade the set $A - B$.

4 If $A = \{1, 2, 3, 4\}$, $B = \{1, 3, 6, 7\}$, $C = \{4, 5, 7, 8\}$ and $\mathscr{E}$ = {integers from 1 to 10 inclusive}, use the definitions of questions 2 and 3 to find (i) $A + B$; (ii) $B - C$; (iii) $A + (B + C)$; (iv) $A - (B - C)$; (v) $(A - B) + C$; (vi) $(A + B) - (B + C)$; (vii) $A - C$.

5 How many subsets (including $\emptyset$ and the set itself) are there in a set containing (i) 2 elements; (ii) 3 elements; (iii) 4 elements; (iv) n elements?

6 A, B and C are three sets which satisfy the following conditions: $n(A \cap B \cap C) = 4$, $n(A \cap B) = n(A \cap C) = 10$, $B \cap C \cap A' = \emptyset$, $n(C) = 30$, and $n(A) = n(B)$. If $n(A \cup B \cup C) = 50$, find $n(A)$.

7 Three sets, A, B and C, satisfy the following conditions: (i) $A \subset B$; (ii) $A \cap C \subset B$; (iii) $A \cap C \neq \emptyset$. State which of the following statements must be true and which must be false. For those which could be true or false, give examples to indicate how this is possible. (i) $C \subset B$; (ii) $A \subset C$; (iii) $n(B \cap C) > 0$; (iv) $n(A) + n(C) < n(B)$; (v) $C \cup A \subset B$.

8 The universal set $\mathscr{E}$ consists of the positive integers less than

25. A and B are subsets, A containing those members of $\mathscr{E}$ which are multiples of 4 and B those members of $\mathscr{E}$ which are multiples of 3. List the members of the sets A, B, $A \cap B'$, $A' \cap B$, $A \cup B$, $A' \cup B'$, and $(A \cap B') \cup (A' \cap B)$. Represent the sets $\mathscr{E}$, A and B by a Venn diagram and shade in the region or regions representing $(A \cap B') \cup (A' \cap B)$. Show that this set can also be written as $(A \cup B) \cap (A' \cup B')$.

[O]

9 AB is a line of length 4 cm.
(i) Shade horizontally the set X where $X = \{\text{points } P : \angle PAB \leq 40°\}$.
(ii) Shade vertically the set Y where $Y = \{\text{points } Q : BQ \geq AQ\}$.
(iii) Measure the distance between the two points of $X \cap Y$ which are furthest apart.

[O]

10 In a school where 36 study foreign languages in the sixth form, 26 study French, 18 study German, 11 study French and German, 10 study French and Latin, 6 study German and Latin, and 4 study French, German and Latin. By drawing a Venn diagram and by letting

F = {those who study French}
G = {those who study German}
L = {those who study Latin}

find: (i) the number who study French and German but not Latin;
(ii) the number who study Latin;
(iii) the number who study at least two foreign languages.

[W]

11 (a) The universal set $\mathscr{E}$ comprises the 16 positive integers $\{5, 6, 7 \ldots, 20\}$, and P, Q, R are subsets of $\mathscr{E}$ defined as follows:

$P = \{x : x \text{ is a multiple of } 5\}$
$Q = \{x : x \text{ is a factor of } 20\}$
$R = \{x : x \text{ is a prime number}\}$.

(i) List the elements of each of the sets P, Q, R.
(ii) List the elements of each of the sets $P \cup R$, $P \cap Q \cap R$, $(P \cup Q) \cap R'$.
(b) Of the 100 members of a sports club, 48 play squash (S), 45 play badminton (B) and 52 play table tennis (T); 15 members play squash and badminton, 18 play squash and table tennis and 21 play badminton and table tennis. Each

club member plays at least one of these games and x members play all three games. Illustrate the sets S, B and T on a clearly labelled Venn diagram, and in your diagram show, in terms of x, the number of members in each region. Hence or otherwise, calculate (i) the value of x; (ii) the number of members who play only squash.

[NI]

12 In a certain examination there are four papers, named p, q, r and s. Every candidate must take two papers, which must be p and q, or p and r, or q and s: no other combination is allowed. If

$\mathscr{E} = \{\text{candidates}\}$
$P = \{\text{candidates taking paper p}\}$
$Q = \{\text{candidates taking paper q}\}$
$R = \{\text{candidates taking paper r}\}$
$S = \{\text{candidates taking paper s}\}$

explain why (a) $P' = S$; (b) $R \subset P$; (c) $P \cap S = \emptyset$, and write down three similar relations concerning Q, R and S or their complements. In a year when 500 candidates took paper p, 400 took paper q and 120 took paper r, find the total number taking the examination and the number taking paper s.

[L]

2 The number system

2.1 Equivalent fractions

Figure 2.1 shows two ways of representing the same fraction.

$$\therefore \frac{2}{6} = \frac{1}{3}.$$

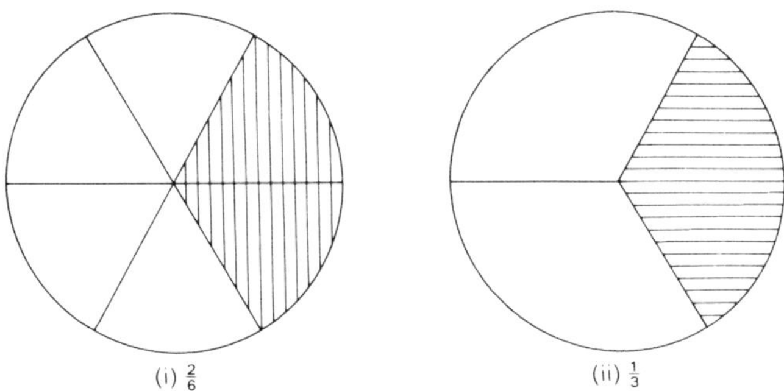

Fig. 2.1

We say that these are *equivalent fractions.* $\frac{2}{6}$ is obtained from $\frac{1}{3}$ by multiplying the top and bottom lines by 2. The top line is called the numerator, and the bottom line the denominator. The fraction $\frac{1}{3}$ is said to be in its lowest terms as there is no number which divides exactly into the numerator and denominator.

EXAMPLE 1

Reduce $\frac{45}{60}$ to its lowest terms.

$$\frac{{}^{9}\cancel{45}}{\cancel{60}_{12}} = \frac{9}{12} \qquad \text{(cancelling by 5)}$$

$$\frac{\cancel{9}^{3}}{\cancel{12}_{4}} = \frac{3}{4} \qquad \text{(cancelling by 3)}$$

$$\therefore \frac{45}{60} = \frac{3}{4}$$

[*Note:* it is possible to cancel straight away by 15.]

2.2 Improper fractions and mixed numbers

A fraction such as $\frac{18}{5}$, where the numerator is bigger than the denominator, is called an *improper fraction.* $18 \div 5 = 3$ with a remainder of 3.

$$\therefore \frac{18}{5} = 3\tfrac{3}{5}.$$

A number such as $3\frac{3}{5}$, which consists of an integer part 3 and a fractional part $\frac{3}{5}$, is called a *mixed number.*

EXERCISE 2a

Express the following fractions in their lowest terms:

1 $\frac{9}{81}$ **2** $\frac{12}{64}$ **3** $\frac{9}{48}$ **4** $\frac{220}{770}$ **5** $\frac{13}{65}$ **6** $\frac{45}{100}$ **7** $\frac{200}{550}$ **8** $\frac{65}{117}$ **9** $\frac{48}{328}$ **10** $\frac{91}{182}$ **11** $\frac{121}{990}$ **12** $\frac{105}{189}$ **13** $\frac{168}{264}$ **14** $\frac{81}{189}$ **15** $\frac{625}{750}$ **16** $\frac{192}{240}$ **17** $\frac{143}{169}$ **18** $\frac{335}{402}$ **19** $\frac{315}{420}$ **20** $\frac{77}{117}$

Express the following improper fractions as mixed numbers, with the fractional part in its lowest terms:

21 $\frac{13}{4}$ **22** $\frac{19}{8}$ **23** $\frac{37}{7}$ **24** $\frac{64}{11}$ **25** $\frac{51}{6}$ **26** $\frac{34}{10}$ **27** $\frac{78}{8}$ **28** $\frac{243}{21}$ **29** $\frac{335}{45}$ **30** $\frac{330}{20}$.

Express the following mixed numbers as improper fractions:

31 $5\frac{5}{8}$ **32** $2\frac{3}{7}$ **33** $4\frac{1}{6}$ **34** $3\frac{9}{11}$ **35** $15\frac{1}{8}$.

2.3 Lowest common multiple (LCM)

LCM stands for *lowest common multiple*, which is the smallest number that is exactly divisible by two given numbers. The LCM of 3 and 4 is the smallest number that is divisible by 3 and 4, i.e. 12. We shall write this:

LCM (3, 4) = 12.

It is often easier to express each of the two numbers in its prime factors in order to find the LCM.

EXAMPLE 2

Find LCM(60, 45). First express each number as a product of its prime factors:

$$60 = 2 \times 2 \times 3 \times 5$$
$$45 = 3 \times 3 \times 5.$$

To obtain the LCM, take the largest number of each prime number which occurs in either 60 or 45 and multiply them together. There are two 2's in 60, two 3's in 45 and one 5 in each.

$$\therefore \text{LCM}(60, 45) = 2 \times 2 \times 3 \times 3 \times 5 = 180.$$

EXERCISE 2b

Find the LCM of the following pairs of numbers:

1	15, 20	**3**	9, 12	**5**	9, 7	**7**	12, 15	**9**	30, 48
2	18, 24	**4**	20, 30	**6**	18, 36	**8**	16, 28	**10**	210, 1155.

2.4 Addition and subtraction of fractions

In order to add $\frac{1}{4}$ and $\frac{1}{3}$, each fraction must be written with the same denominator. The denominator is chosen as the LCM of the two denominators.

$$\therefore \tfrac{1}{4} + \tfrac{1}{3} = \tfrac{3}{12} + \tfrac{4}{12} = \tfrac{7}{12}.$$

Similarly

$$\tfrac{3}{5} + \tfrac{2}{7} = \tfrac{21}{35} + \tfrac{10}{35} = \tfrac{31}{35}.$$

Subtraction is carried out in the same manner:

$$\tfrac{9}{11} - \tfrac{2}{5} = \tfrac{45}{55} - \tfrac{22}{55} = \tfrac{23}{55}.$$

Problems involving mixed numbers have to be carried out in a slightly different manner.

EXAMPLE 3

$$\text{Evaluate } 5\tfrac{6}{7} + 8\tfrac{2}{3} = 13 + \tfrac{6}{7} + \tfrac{2}{3} \qquad \text{[add the integral parts first]}$$
$$= 13 + \tfrac{18}{21} + \tfrac{14}{21} = 13 + \tfrac{32}{21}$$
$$= 14\tfrac{11}{21}.$$

EXAMPLE 4

Evaluate $81\frac{2}{7}-63\frac{5}{9}$. This problem can be simplified by splitting the first number into $80+1\frac{2}{7}$. The problem then becomes

$$\begin{aligned} 80+1\tfrac{2}{7}-63-\tfrac{5}{9} &= 17+\tfrac{9}{7}-\tfrac{5}{9} \\ &= 17+\tfrac{81}{63}-\tfrac{35}{63} \\ &= 17\tfrac{46}{63}. \end{aligned}$$

EXERCISE 2c

Evaluate:

1 $\frac{3}{4}+\frac{2}{3}$

2 $\frac{5}{8}+\frac{3}{5}$

3 $\frac{2}{9}+\frac{1}{12}$

4 $\frac{4}{11}+\frac{3}{22}$

5 $\frac{1}{10}+\frac{1}{100}$

6 $\frac{8}{9}-\frac{1}{3}$

7 $\frac{4}{5}-\frac{3}{10}$

8 $\frac{7}{9}-\frac{5}{12}$

9 $\frac{3}{8}-\frac{1}{16}$

10 $\frac{5}{48}-\frac{3}{64}$

11 $1\frac{1}{2}+2\frac{3}{4}$

12 $2\frac{2}{3}+3\frac{1}{2}$

13 $2\frac{3}{7}+1\frac{1}{3}$

14 $5\frac{6}{7}+3\frac{1}{14}$

15 $8\frac{3}{8}+5\frac{1}{10}$

16 $12\frac{3}{4}-8\frac{1}{8}$

17 $19\frac{5}{8}-7\frac{3}{4}$

18 $8\frac{1}{3}-6\frac{5}{6}$

19 $12\frac{3}{5}-5\frac{2}{7}$

20 $5\frac{1}{11}-4\frac{1}{3}$

21 $7\frac{1}{5}-2\frac{3}{4}$

22 $21\frac{1}{2}-5\frac{1}{8}$

23 $3\frac{7}{8}+4\frac{2}{5}$

24 $103\frac{1}{2}-97\frac{5}{8}$

25 $219\frac{1}{5}-198\frac{1}{15}$

26 $30\frac{3}{7}+9\frac{5}{8}$

27 $2\frac{7}{8}-1\frac{1}{17}$

28 $3\frac{3}{5}-2\frac{9}{11}$

29 $81\frac{4}{5}+21\frac{5}{9}$

30 $1000\frac{1}{4}-996\frac{5}{18}$.

2.5 Multiplication of fractions

Multiplication of fractions is performed by multiplying the two numerators together, and the two denominators:

e.g. $\frac{3}{8}\times\frac{5}{7}=\frac{15}{56}$.

An example is often simplified if the method of cancelling is applied to the fractions to be multiplied:

e.g. $$\frac{{}^{2}\cancel{22}}{\cancel{25}_{5}}\times\frac{{}^{1}\cancel{3}}{\cancel{7}_{1}}\times\frac{{}^{1}\cancel{7}}{\cancel{9}_{3}}\times\frac{{}^{1}\cancel{5}}{\cancel{11}_{1}}=\frac{2}{15}.$$

If this is found without cancelling, the answer might be difficult to reduce to the simplest equivalent fraction.

Multiplication of mixed numbers is best carried out by expressing each fraction as an improper fraction:

e.g. $$4\tfrac{2}{5}\times 1\tfrac{4}{11}=\frac{{}^{2}\cancel{22}}{\cancel{5}_{1}}\times\frac{{}^{3}\cancel{15}}{\cancel{11}_{1}}=6.$$

2.6 Division of fractions

In order to find $\frac{3}{5} \div \frac{7}{10}$ it is necessary to invert the second fraction and change the operation of division to that of multiplication.

$$\therefore \frac{3}{5} \div \frac{7}{10} = \frac{3}{\not{5}_1} \times \frac{\not{10}^2}{7} = \frac{6}{7}.$$

EXAMPLE 5

Evaluate $(\frac{2}{9} + \frac{3}{5} \times 1\frac{3}{4}) \div (-\frac{3}{8} \times \frac{2}{5} + 1\frac{2}{3} \times \frac{2}{3})$.

It is necessary to work out each bracket separately first.

$$(\tfrac{2}{9} + \tfrac{3}{5} \times 1\tfrac{3}{4}) = \tfrac{2}{9} + \tfrac{3}{5} \times \tfrac{7}{4} = \tfrac{2}{9} + \tfrac{21}{20}$$

$$= \tfrac{40}{180} + \tfrac{189}{180} = \tfrac{229}{180}$$

$$(-\tfrac{3}{8} \times \tfrac{2}{5} + 1\tfrac{2}{3} \times \tfrac{2}{3}) = -\tfrac{6}{40} + \tfrac{5}{3} \times \tfrac{2}{3} = \tfrac{10}{9} - \tfrac{6}{40}$$

$$= \tfrac{400}{360} - \tfrac{54}{360} = \tfrac{346}{360}.$$

We can now carry out the division:

$$\frac{229}{180} \div \frac{346}{360} = \frac{229}{\not{180}_1} \times \frac{{}^{\not{2}}\not{360}^{\,1}}{\not{346}_{173}} = \frac{229}{173} = 1\tfrac{56}{173}.$$

EXAMPLE 6

Evaluate $\dfrac{11\frac{3}{4} - 6\frac{7}{8}}{5\frac{1}{2} - 4\frac{1}{8}}$.

This is simplified if top and bottom lines of the fraction are multiplied by the LCM of the denominators occurring. In this case LCM = 8.

$$\therefore \frac{11\frac{3}{4} - 6\frac{7}{8}}{5\frac{1}{2} - 4\frac{1}{8}} = \frac{94 - 55}{44 - 33} = \frac{39}{11} = 3\tfrac{6}{11}.$$

EXERCISE 2d

Evaluate:

1 $\frac{2}{3} + \frac{4}{7}$

2 $\frac{3}{4} + \frac{5}{9}$

3 $\frac{3}{7} + \frac{2}{3} + \frac{4}{9}$

4 $\frac{2}{5} - \frac{1}{10}$

5 $\frac{6}{7} - \frac{3}{14}$

6 $1\frac{1}{2} + 3\frac{1}{4}$

7 $2\frac{3}{8} + 4\frac{2}{5}$

8 $1\frac{3}{4} + 1\frac{1}{8} + 1\frac{2}{5}$

9 $(1\frac{1}{4} + \frac{3}{8}) \times (6\frac{1}{2} + 2\frac{1}{3})$

10 $\frac{5}{7} \div (\frac{3}{4} + \frac{2}{7})$

11 $(\frac{2}{5} - \frac{1}{10}) \div 1\frac{2}{7}$

12 $\frac{1}{18} + \frac{7}{9} + \frac{3}{5} - \frac{7}{15}$

13 $1\frac{3}{4} \times 2\frac{7}{8}$

14 $4\frac{3}{4} \div \frac{2}{9}$

15 $(1\frac{1}{2}+\frac{3}{4})\times(\frac{2}{5}+\frac{1}{10})$
16 $2\frac{5}{8}-3\frac{1}{4}+1\frac{3}{5}$
17 $(\frac{1}{2}-\frac{1}{3})\div(\frac{1}{4}-\frac{1}{5})$
18 $(\frac{3}{4}-\frac{5}{8})\div(\frac{1}{4}+\frac{1}{2})$
19 $(\frac{1}{2}+\frac{1}{3})\times(\frac{1}{4}+\frac{1}{5})$
20 $1\frac{2}{5}\times(\frac{3}{8}+\frac{2}{3})$
21 $3\frac{3}{4}\times 1\frac{2}{7}$
22 $(2\frac{3}{4}-1\frac{5}{8})\div 1\frac{1}{2}$
23 $(\frac{5}{11}\div\frac{7}{9})\times 38\frac{1}{2}$
24 $3\frac{1}{2}\times 4\frac{1}{2}\times 5\frac{1}{2}$
25 $10\frac{1}{4}\times 10\frac{1}{2}$
26 $(1\frac{1}{2}-1\frac{1}{3})\times(\frac{1}{4}+\frac{1}{3})$
27 $(5\frac{1}{2}\div 4\frac{1}{2})\div 3\frac{1}{2}$
28 $(1\frac{1}{2}-1\frac{1}{3})\div(\frac{1}{4}+\frac{1}{3})$
29 $7\frac{1}{2}\times(1\frac{3}{4}+\frac{2}{5})$
30 $(1\frac{1}{3}+1\frac{1}{4}+1\frac{1}{2})\div(\frac{1}{2}+\frac{3}{4}+\frac{2}{3})$.

2.7 The number line

If we try to subtract 8 from 5 this is not possible if the universal set is the set of positive integers $\{0, {}^{+}1, {}^{+}2, {}^{+}3, \ldots\}$. In order to carry out the subtraction, we introduce a new set of integers called negative integers, i.e. $\{{}^{-}1, {}^{-}2, {}^{-}3, {}^{-}4, \ldots\}$. We define ${}^{-}1 = 3-4$, and using this value of negative one, we get ${}^{-}2 = {}^{-}1 + {}^{-}1$, ${}^{-}3 = {}^{-}2 + {}^{-}1$ and so on. We can now represent the positive and negative integers on a number line.

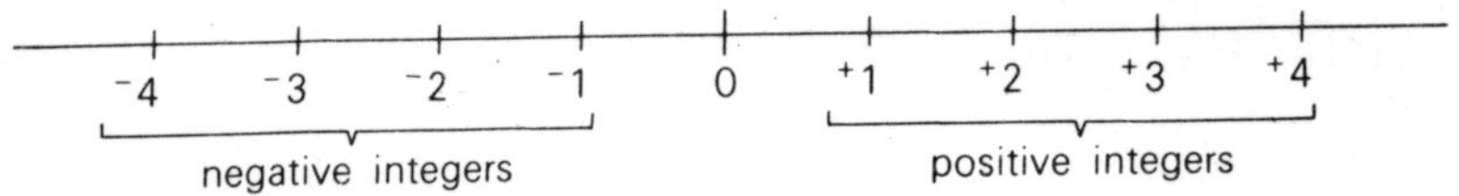

Fig. 2.2

In future, these will be referred to as directed numbers.

2.8 Addition and subtraction of directed numbers

In the statement ${}^{+}3 + {}^{+}5$ the + sign is used in two ways. The + sign above the number tells us what sort of number it is (in this case positive), and the + sign between the two numbers is a mathematical operation, in this case addition.

We will use the following definitions:

(i) Addition of a positive number ${}^{+}n$ increases the number by n.
(ii) Addition of a negative number ${}^{-}n$ decreases the number by n.
(iii) Subtraction of a positive number ${}^{+}n$ decreases the number by n.
(iv) Subtraction of a negative number ${}^{-}n$ increases the number by n.

[*Note:* great care should be taken with rule (iv).]

$$^{-}5 + {}^{+}6 = {}^{+}1 \quad \text{rule (i)}$$
$$^{-}2 + {}^{-}4 = {}^{-}6 \quad \text{rule (ii)}$$
$$^{+}7 - {}^{+}2 = {}^{+}5 \quad \text{rule (iii)}$$
$$^{-}2 - {}^{-}4 = {}^{+}2 \quad \text{rule (iv)}$$

It is customary to omit the + sign in front of a positive number, and to write the − sign in front of a negative number at the same level as that of the subtraction sign. The above examples become: $-5+6=1$, $-2+-4=-6$, $7-2=5$, $-2--4=2$.

EXERCISE 2e

Evaluate:

1 $-5+-11$
2 $-8--2$
3 $8--5$
4 $-3+7$
5 $-11+-5--2$
6 $-14--3+-5$
7 $28-14--15$
8 $-31+-26--11$
9 $18+-18--2$
10 $-41--100+-37$
11 $-100--87$
12 $-93+-41$
13 $97+-43$
14 $121--21-37$
15 $19--36-43$
16 $-16--13-17$
17 $-11--5-18$
18 $-27-36+57$
19 $-191--192-193$
20 $99-100--101$
21 $1\frac{3}{4}-2\frac{7}{8}$
22 $-3\frac{1}{2}--2\frac{1}{2}$
23 $-6\frac{1}{4}--5\frac{1}{4}$
24 $3\frac{3}{4}-4\frac{5}{8}$
25 $\frac{3}{4}--\frac{2}{3}$
26 $1\frac{1}{12}-1\frac{3}{4}$
27 $9\frac{7}{8}--\frac{3}{4}-1\frac{1}{2}$
28 $1\frac{1}{5}-2\frac{3}{7}$
29 $-7\frac{1}{2}--3\frac{2}{5}$
30 $3\frac{1}{2}+-2\frac{1}{5}$
31 $4\frac{1}{6}--3\frac{1}{5}$
32 $11\frac{1}{2}--10\frac{1}{2}+-3\frac{1}{2}$
33 $-6\frac{2}{5}+-3\frac{1}{7}$
34 $-2\frac{1}{2}-3\frac{1}{4}+5\frac{1}{8}$
35 $16\frac{1}{2}+-3\frac{1}{4}--3\frac{1}{8}$
36 $2\frac{7}{8}--3\frac{3}{7}$
37 $10\frac{1}{2}--8\frac{1}{4}--3\frac{1}{2}$
38 $5\frac{1}{6}+-6\frac{1}{11}$
39 $-7\frac{1}{8}-6\frac{3}{4}$
40 $-8\frac{3}{8}--3\frac{7}{8}+-2\frac{1}{2}$.

2.9 Multiplication and division of directed numbers

In multiplying and dividing directed numbers, the operations are carried out assuming both numbers to be positive and the appropriate sign is then added according to the following rules:

(i) *positive* multiplied by *positive* = *positive*
(ii) *negative* multiplied by *positive* = *negative*
(iii) *positive* multiplied by *negative* = *negative*

(iv) *negative* multiplied by *negative* = *positive*
(v) *positive* divided by *positive* = *positive*
(vi) *positive* divided by *negative* = *negative*
(vii) *negative* divided by *positive* = *negative*
(viii) *negative* divided by *negative* = *positive.*

Using these rules, we have:

(i) $^{+}7 \times {}^{+}5 = {}^{+}35$ (ii) $^{-}3 \times {}^{+}7 = {}^{-}21$
(iii) $^{+}2 \times {}^{-}11 = {}^{-}22$ (iv) $^{-}8 \times {}^{-}5 = {}^{+}40$
(v) $^{+}24 \div {}^{+}8 = {}^{+}3$ (vi) $^{+}27 \div {}^{-}9 = {}^{-}3$
(vii) $^{-}32 \div {}^{+}4 = {}^{-}8$ (viii) $^{-}25 \div {}^{-}5 = {}^{+}5$

EXERCISE 2f

Evaluate:

1 $-3\frac{1}{2} \times -2\frac{1}{4}$
2 $-\frac{3}{8} \div 2\frac{1}{2}$
3 $3\frac{1}{2} \times -5\frac{1}{4}$
4 $11\frac{1}{4} \div -3\frac{3}{4}$
5 $6\frac{1}{2} \times -2\frac{3}{4}$
6 $7\frac{1}{2} \div -3\frac{1}{2}$
7 $-1\frac{1}{4} \times -1\frac{1}{2} \times -1\frac{3}{4}$
8 $-2\frac{1}{2} \div -5\frac{1}{2}$
9 $1\frac{3}{4} \div -\frac{3}{8}$
10 $1\frac{1}{4} \times -1\frac{3}{4} \times -\frac{1}{2}$
11 $-\frac{1}{2} \times (\frac{3}{8} - 1\frac{1}{2})$
12 $-\frac{1}{4} \times (\frac{1}{2} + -\frac{3}{8})$
13 $\frac{3}{7} \times (\frac{1}{2} - \frac{5}{9})$
14 $1\frac{3}{4} \times (-\frac{3}{8} - \frac{2}{5})$
15 $\frac{3}{4} \times (-1\frac{1}{2} + -3\frac{1}{2})$
16 $-\frac{2}{5} \times (-17 - -5)$
17 $1\frac{1}{7} \times (-3 + -4)$
18 $\frac{2}{3} \times (-\frac{1}{8} - -\frac{3}{4})$
19 $(-\frac{1}{4} \times -\frac{3}{4}) \div -\frac{1}{8}$
20 $(-\frac{3}{4} \div -\frac{3}{5}) \div -\frac{1}{8}$.

2.10 Surds

$\sqrt{36}$ means the square root of 36, i.e. the number which when multiplied by itself gives 36.

$$\therefore \sqrt{36} = 6.$$

If we try to find $\sqrt{18}$, there is no integer which when multiplied by itself gives the answer 18. However, we can find a decimal which when multiplied by itself gives an answer very close to 18:

$$4.243 \times 4.243 = 18.003049.$$

We can say that $\sqrt{18} = 4.24$ (correct to 2 decimal places). Tables of square roots are usually included in logarithm tables. Examples using these can be found in chapter 3.

It is sometimes convenient to simplify square roots without working them out. The procedure is as follows:

since $\quad 18 = 9 \times 2$

then $\quad \sqrt{18} = \sqrt{9} \times \sqrt{2} = 3 \times \sqrt{2} = 3\sqrt{2}$

$$\sqrt{108} = \sqrt{36} \times \sqrt{3} = 6 \times \sqrt{3} = 6\sqrt{3}$$

Answers which contain the square root sign are said to be in surd form. A surd is an example of an *irrational* number, i.e. a number which cannot be expressed as a fraction.

2.11 Rationalising the denominator

$1/\sqrt{3}$ is rather an awkward number to work out, and can be simplified as follows. [*Note:* $\sqrt{3} \times \sqrt{3} = 3$.] To get a fraction equivalent to $1/\sqrt{3}$, multiply numerator and denominator by $\sqrt{3}$.

$$\therefore \frac{1}{\sqrt{3}} = \frac{1}{\sqrt{3}} \times \frac{\sqrt{3}}{\sqrt{3}} = \frac{\sqrt{3}}{3}.$$

This is considerably easier to work out.

EXAMPLE 7

Simplify $\sqrt{75} + \sqrt{12}$.

$$\sqrt{75} = \sqrt{3 \times 25} = 5\sqrt{3}$$

$$\sqrt{12} = \sqrt{3 \times 4} = 2\sqrt{3}.$$

Adding these results

$$\sqrt{75} + \sqrt{12} = 5\sqrt{3} + 2\sqrt{3} = 7\sqrt{3}.$$

2.12 Another look at the number line

In section 2.7, we drew the number line, containing the integers. On that line must now be added the rational numbers, and the irrational numbers. It would seem that things get a little crowded, as we can see in Fig. 2.3 overleaf.

The full theory of the number line is very complicated, and will not be developed further here. Figure 2.4 shows the relationship between the three types of numbers.

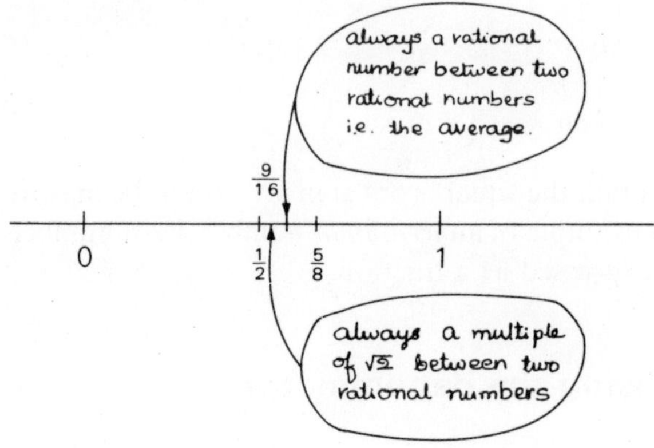

Fig. 2.3

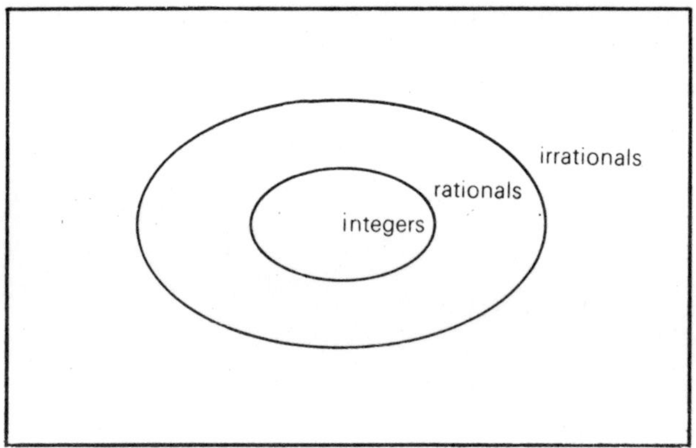

Fig. 2.4

EXERCISE 2g

Simplify the following:

1 $\sqrt{18}+3\sqrt{2}$

2 $\sqrt{24}+\sqrt{150}$

3 $\sqrt{18}-\sqrt{8}$

4 $\sqrt{90}+\sqrt{40}$

5 $2\sqrt{32}-\sqrt{128}$

6 $\sqrt{63}-\sqrt{28}$

7 $\sqrt{45}+\sqrt{20}$

8 $\sqrt{99}-\sqrt{176}$

9 $\sqrt{28}+\sqrt{7}+\sqrt{63}$

10 $\sqrt{12}+\sqrt{18}-\sqrt{48}+\sqrt{98}$.

Rationalise the denominator in the following questions, and simplify your answers where possible:

11 $\frac{1}{\sqrt{2}}$ **14** $\frac{1}{\sqrt{6}}$ **17** $\frac{1}{\sqrt{125}}$ **20** $\frac{1}{\sqrt{2}-\sqrt{8}}$.

12 $\frac{2}{\sqrt{8}}$ **15** $\frac{12}{\sqrt{18}}$ **18** $\frac{5}{\sqrt{10}}$

13 $\frac{3}{\sqrt{3}}$ **16** $\frac{3}{\sqrt{45}}$ **19** $\frac{1}{\sqrt{2}+\sqrt{8}}$

2.13 Number bases

In everyday arithmetic, when we write down a number such as 1247, we really mean $1000+200+40+7$. Writing this in tabular form:

thousands	hundreds	tens	units
$10 \times 10 \times 10$ $= 1000$	10×10 $= 100$	10	1
1	2	4	7

We say that the number 10 is the base.

When a number is written 2651_8, the suffix 8 tells us that the base is 8. In an exactly similar way to the above we have

$8 \times 8 \times 8$ $= 512$	8×8 $= 64$	8	1
2	6	5	1

$$\begin{aligned} \text{i.e.} \quad 2 \times 512 &= 1024 \\ 6 \times 64 &= 384 \\ 5 \times 8 &= 40 \\ 1 \times 1 &= 1 \\ \text{total} \quad & 1449 \end{aligned}$$

$$2651_8 = 1449_{10}$$

[*Note:* a digit greater than the base being used cannot appear in the number.]

A number in base 10 is called a denary number.
A number in base 8 is called an octal number.
A number in base 2 is called a binary number.

EXERCISE 2h

Convert the following numbers to denary numbers:

1	125_8	**6**	6250_8	**11**	2147_8	**16**	4554_6
2	317_8	**7**	3140_6	**12**	3451_6	**17**	203020_4
3	1061_8	**8**	50001_6	**13**	7770_8	**18**	11111_2
4	100111_2	**9**	101010_2	**14**	6413_8	**19**	6767_8
5	11101_2	**10**	110011_2	**15**	1000111_2	**20**	44444_6.

2.14 Change of base

To change a denary number into a different base, proceed as in the following example.

EXAMPLE 8

Change the denary number 2893 into an octal number.

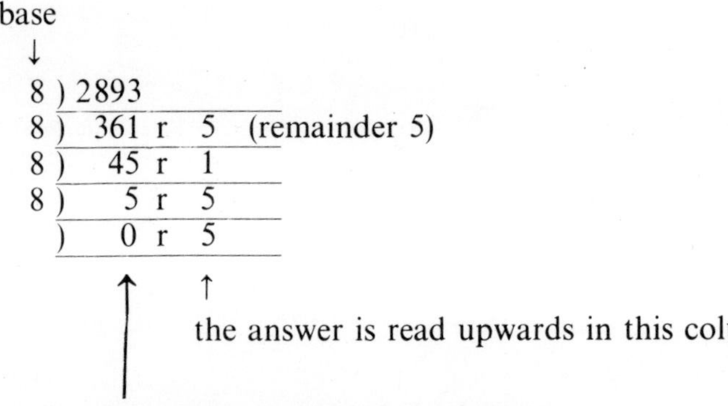

$\therefore\ 2893 = 5515_8$.

EXAMPLE 9

Change 55_6 into an octal number. The simplest method for this sort of change is to convert to a denary number first.

	units
6	1
5	5

$$\begin{aligned} 5 \times 6 &= 30 \\ 5 \times 1 &= \underline{\ 5} \\ \text{total} & \ \ 35 \end{aligned}$$

$\therefore 55_6 = 35_{10}$.

Now convert 35 into the octal scale

$$\begin{array}{r|l} 8 & 35 \\ \hline 8 & 4 \ \text{r}\ 3 \\ \hline & 0 \ \text{r}\ 4 \\ \hline \end{array}$$

$\therefore 55_6 = 43_8$.

2.15 Addition and subtraction

EXAMPLE 10

(i) $457_8 + 325_8$ (ii) $431_6 - 255_6$.

(i)

$$\begin{array}{r} 457 \\ +325 \\ \hline 1004 \\ \hline {\scriptstyle 11} \end{array}$$

Since the base is 8, a 1 is carried into the next column each time 8 is reached.

(ii)

$$\begin{array}{r} {\scriptstyle 10} \\ {\scriptstyle 3\ \not{2}\ 9} \\ \not{4}\ \not{3}\ \not{1} \\ -2\ 5\ 5 \\ \hline 1\ 5\ 4 \end{array}$$

Remember we borrow 8 each time from the next column.
$\therefore$ the 1 becomes a 9, the 2 becomes 10.

EXERCISE 2i

Evaluate in the base stated:

1 $25_8 + 61_8$
2 $36_8 + 77_8$
3 $1011_2 + 1111_2$
4 $51_6 + 121_6$
5 $73_8 + 444_8$
6 $1111_2 + 1111_2$
7 $1111_3 + 1111_3$
8 $653_8 + 431_8$
9 $521_8 + 432_8$
10 $510_6 + 454_6$
11 $77_8 - 63_8$
12 $127_8 - 75_8$
13 $1111_2 - 1011_2$
14 $321_6 - 55_6$
15 $431_8 - 76_8$
16 $10001_2 - 1111_2$
17 $2201_3 - 222_3$
18 $316_8 - 277_8$
19 $413_8 - 337_8$
20 $412_6 - 354_6$.

2.16 Multiplication and division

EXAMPLE 11

Evaluate (i) $235_6 \times 44_6$ (ii) $10100001_2 \div 111_2$.

(i)
```
  235
 × 44
14320
 1432
20152
11
```

(ii)
```
           10111
111 ) 10100001
       111
        1100
         111
        1010
         111
          111
```

It can be seen that the layout is identical to the normal processes of long multiplication and division.

EXERCISE 2j

Evaluate in the given base:

1 $33_8 \times 25_8$ **5** $(47_8)^2 \times 12_8$ **9** $100011_2 \div 111_2$

2 $27_8 \times 63_8$ **6** $11011_2 \times 1011_2$ **10** $24415_6 \div 35_6$

3 $(1011_2)^2$ **7** $651_8 \div 21_8$ **11** $2057_8 \div 21_8$

4 $35_6 \times 21_6$ **8** $1331_8 \div 33_8$ **12** $1000110_2 \div 1010_2$.

2.17 Decimals

When we write down a number such as 0.285, we mean

$\frac{2}{10} + \frac{8}{100} + \frac{5}{1000}$.

2.18 Conversion of a fraction to a decimal

EXAMPLE 12

Convert (i) $\frac{3}{8}$; (ii) $\frac{2}{7}$ into a decimal.

(i)
```
     0.375
8 ) 3.0000
    24
     60
     56
      40
```

The process of long division is used.

```
(ii)      0.2857142857142 . . .
     7 ) 2.0000000
         14
         --
          60
          56
          --
           40
           35
           --
            50
            49
            --
             10
              7
             --
              30
              28
              --
               20
```

In this example, the decimal does not terminate, but when the number 2 is reached it repeats itself. We write $\frac{2}{7} = 0.\dot{2}8571\dot{4}$.

A decimal which does not repeat itself, and is non-terminating, cannot be expressed as a fraction. An example of such a decimal is π (see chapter 7).

2.19 Decimal places and significant figures

It would be impossible for a human being to measure accurately a length such as 4.38627 cm. What we would have to do is to state a length as near as possible to this length that we could measure. If we assume that we can measure to within 0.1 of a centimetre, what length is 4.38627 nearest to? Is it 4.3 cm, or 4.4 cm? The answer, of course, is 4.4 cm.

We say that 4.38627 = 4.4 correct to one decimal place. We shall abbreviate this to 4.4 (1 dec. pl.).

EXAMPLE 13

What are the following numbers correct to two decimal places: (i) 4.874; (ii) 3.125; (iii) 5.001; (iv) 9.999?

(i) 4.874 = 4.87 (2 dec. pl.). The 4 after the 7 does not affect it.
(ii) 3.125 = 3.12 or 3.13 (2 dec. pl.), since the 2 is followed by a 5. In this book, we will always give the latter answer, i.e. 3.125 = 3.13 (2 dec. pl.).
(iii) 5.001 = 5.00 (2 dec. pl.). Note that the two 0's must be left.
(iv) 9.999 = 10.00 (2 dec. pl.). The last 9 increases every other 9 by 1.

A statement such as 'The distance between the spacecraft and the space station was 23658231.2 km' is really absurd, in that when measuring in millions of kilometres a few hundred either way is not important. In this case, a more sensible answer would be 23700000 km.

We say that 23 658 231.2 = 23 700 000 correct to three significant figures, abbreviated to 23 700 000 (3 SF).

EXAMPLE 14

Evaluate correct to 3 SF (i) 99890; (ii) 48.3759; (iii) 0.002578.

(i) 99890 = 99900 (3 SF). The last 9 increases the 8 by 1.
(ii) 48.3759 = 48.4 (3 SF). We are allowed to enter decimal places.
(iii) 0.002578 = 0.00258 (3 SF). The significant figures are counted from the first non-zero digit.

EXERCISE 2k

Evaluate:

1 99.394 (1 dec. pl.)
2 9.099 (2 dec. pl.)
3 0.09401 (3 dec. pl.)
4 5.9906 (1 dec. pl.)
5 6.009 (2 dec. pl.)
6 3.9999 (2 dec. pl.)
7 9325 (3 SF)
8 4 985 000 (2 SF)
9 5.865 (3 SF)
10 0.00991 (1 SF)
11 4.0002 (2 SF)
12 99.999 (3 SF)

Convert the following fractions into decimals:

13 $\frac{3}{8}$ **15** $\frac{4}{7}$ **17** $\frac{5}{12}$ **19** $\frac{7}{16}$
14 $\frac{5}{9}$ **16** $\frac{3}{11}$ **18** $\frac{7}{8}$ **20** $\frac{13}{32}$.

HARDER EXAMPLES 2

1 Simplify (i) $\sqrt{13^2 - 5^2}$; (ii) $\dfrac{\sqrt{63} + \sqrt{28}}{\sqrt{175} - \sqrt{63}}$.

2 State which of the following is the closest approximation to $\dfrac{\pi\sqrt{10}}{(0.1)^2}$:
(a) 0.1; (b) 50; (c) 1000; (d) 3000; (e) 100000.
[JMB]

3 State two irrational numbers which lie between $\frac{1}{2}$ and $\frac{1}{4}$.
4 If $a\sqrt{3} + b\sqrt{2} = m\sqrt{3}$, what can you say about a, b and m?
5 Given that $a^2 = b^2 + c^2$
(i) find a when $b = \sqrt{3}$ and $c = \sqrt{6}$
(ii) find b when $a = 1$ and $c = \sqrt{\frac{1}{2}}$.
[C]

6 Simplify $\left(\sqrt{50}+\frac{1}{\sqrt{8}}\right)^2$.

7 If n is a natural number, 234_n means $2n^2+3n+4$.
Write down the meaning of 144_n.
Show that, for all values of the natural number n,
(i) 144_n is a perfect square; (ii) 132_n is not a prime number. [L]

8 What base has been used in the following calculations:
(i) $235+121=400$; (ii) $(21)^2=1211$; (iii) $1227\div 21=47$?

9 Find $\sqrt{42+\sqrt{42+\sqrt{49}}}$.

10 The denary integer x is such that $10\,000_2<x<10100_2$.
(i) Write down all the possible values of x.
(ii) If the units digit is 3 when x is written in the octal scale, write down the denary value of x. [C]

11 Find the arithmetic mean of $\sqrt{5}$, $5\sqrt{5}$, $\sqrt{75}$, $\sqrt{45}$.

12 Without using tables and showing your working clearly, arrange in order of increasing magnitude: $\sqrt{2}$, $1\frac{2}{5}$, 1.42.

13 Simplify $(1\frac{1}{2}+1\frac{1}{3})\div(1\frac{1}{2}-1\frac{1}{7})$.

14 Simplify $(\sqrt{2}+\sqrt{8})^2$. Do you think that the product of two irrational numbers is always a rational number? Give reasons.

15 The following table is to indicate to which of the sets—{integers}, {rationals}—the numbers listed do or do not belong. The first line of the table has been filled in. Complete the table.

	{integers}	{rationals}
$\sqrt{\frac{49}{4}}$	no	yes
$\sqrt{121}$		
$\sqrt{1.44}$		
$\sqrt{9}+\sqrt{7}$		

[NI]

16 If t is a very large positive integer, between what two integers does the number $2t+1+\dfrac{4t}{2t+\frac{1}{2t}}$ lie?

17 Prove that the number n^2+n+1, where n is an integer is always odd.

3 The metric system

3.1 Units of length

In the metric system, the unit of length is the metre.

1 kilometre (km) = 1000 metres (m)
1 centimetre (cm) = 0.01 m
1 millimetre (mm) = 0.001 m

3.2 Units of mass

The standard unit of mass is the kilogram.

1 gram (g) = 0.001 kilogram (kg)
1 milligram (mg) = 0.001 g

3.3 Area and volume

(i) The unit of area is denoted by a 2, written as a power after the unit. For example, if the lengths are measured in cm, then the area is written in cm^2.
(ii) Volume. If the lengths are measured in m, then the volume is in m^3. However, 1000 cm^3 = 1 litre (l).

EXAMPLE 1

If 500 cm^3 of a substance weigh 250 g, find the mass in kg of 1 litre.

The mass of 500 cm^3 is 250 g.

$\therefore$ the mass of 1000 cm^3 is 500 g.

$$1 \text{ g} = 0.001 \text{ kg}$$
$$500 \text{ g} = 500 \times 0.001 \text{ kg} = 0.5 \text{ kg}$$

$\therefore$ 1 litre has a mass of 0.5 kg.

EXERCISE 3a

1 Express the following lengths in cm:
(i) 2.1 m (ii) 586 mm (iii) 200 m (iv) 0.06 m
(v) 1.3 mm (vi) 956 km (vii) 0.02 km (viii) 985 mm.

2 Express the following lengths in km:
(i) 856 cm (ii) 9000 cm (iii) 8900 mm (iv) 86.5 cm
(v) 12 mm (vi) 20.3 cm.

3 Express the following masses in g:
(i) 5 kg (ii) 0.08 kg (iii) 560 mg (iv) 25 kg
(v) 6.3 mg (vi) 850 kg.

4 Change the following areas into cm^2 (remember that since 100 cm = 1 m, 10000 cm^2 = 1 m^2):
(i) 8.2 m^2 (ii) 6870 mm^2 (iii) 0.026 km^2.

5 Change the following volumes into litres:
(i) 800 cm^3 (ii) 40 m^3 (iii) 1 km^3.

6 Express 12 cm as a fraction and a decimal of:
(i) 1 m (ii) 2.4 km (iii) 120 mm.

3.4 Density

It is common knowledge that 1 cm^3 of lead has a mass considerably greater than 1 cm^3 of polystyrene. To express this fact, we use a quantity called the density of a substance, which is simply the mass of a given volume of that substance:

$$\text{density} = \frac{\text{mass}}{\text{volume}} \qquad \text{(A)}$$

From this equation, by multiplying both sides by the volume, we obtain:

$$\text{mass} = \text{density} \times \text{volume} \qquad \text{(B)}$$

On dividing (B) by the density, we obtain:

$$\frac{\text{mass}}{\text{density}} = \text{volume} \qquad \text{(C)}$$

Equations (A), (B) and (C) are very useful.

EXAMPLE 2

A block of zinc was found to have a mass of 142 g. When immersed in a beaker of water, originally full to the brim, a volume of 20 cm^3 was spilt. Find the density of the zinc.

$$\text{Density} = \frac{\text{mass}}{\text{volume}} = \frac{142}{20} = 7.1 \text{ g/cm}^3.$$

EXAMPLE 3

A bottle is being designed to hold 17.4 g of turpentine. What must be the minimum volume of the bottle if the density of the turpentine is 0.87 g/cm^3?

For the minimum volume:

$$\text{volume} = \frac{\text{mass}}{\text{density}} = \frac{17.4}{0.87} = \frac{1740}{87} = 20 \text{ cm}^3.$$

EXERCISE 3b

Complete the following table:

substance	volume (cm^3)	density (g/cm^3)	mass (g)
aluminium	41	2.7	
gold	100		1932
lead		11.37	45.48
sand	30	2.63	
glass	30		75
ice		0.92	73.6
olive oil	24	0.92	
air	200		0.258
hydrogen		0.00009	11.34

3.5 Monetary systems

Although the metric system of measurement is now widely used throughout the world, each country has its own monetary system, and it is necessary to be able to change from one system to another.

EXAMPLE 4

If the exchange rate for American dollars is \$2.103 = £1 (sterling), find, to the nearest cent, how much you would get in exchange for £240.

Since

£1 = \$2.103

then

£240 = \$240 × 2.103
= \$504.72.

Since there are 100 cents in \$1, the answer is already correct to the nearest cent.

EXAMPLE 5

Given that the exchange rate for the French franc is £1 = 9.10 francs, find how many American dollars you would get in exchange for 1000 francs.

This problem is not a simple change, as we are given the necessary information in two parts.

$$\pounds 1 = 9.10 \text{ francs}$$

$$\therefore \frac{\pounds 1}{9.10} = 1 \text{ franc}$$

$$\therefore \frac{\pounds 1000}{9.10} = 1000 \text{ francs}$$

but

$$\pounds 1 = \$2.103$$

$$\therefore \frac{\pounds 1000}{9.10} = \frac{\$1000}{9.10} \times 2.103$$

$$\therefore 1000 \text{ francs} = \$231.098$$

$$= \$231.10 \quad \text{correct to the nearest cent.}$$

EXERCISE 3c

Suppose the exchange rates for £1 are:

Denmark	= 12.38 kroner
Germany	= 5.36 marks
Italy	= 1465 lire
Spain	= 120 pesetas
United States	= 2.10 dollars.

1 Change into pounds (give your answer to the nearest penny):
(i) 400 kroner (ii) 60 marks (iii) 500 lire
(iv) 2000 pesetas (v) 40 dollars (vi) 250 marks
(vii) 2700 lire (viii) 800 dollars (ix) 58 kroner
(x) 24 marks.

Make the following changes, giving your answer correct to 2 decimal places:

2 (i) £12 into marks (ii) £15 into dollars
(iii) £8.53 into pesetas (iv) £95 into lire
(v) £6.50 into kroner (vi) £800 into kroner.

3 (i) 500 dollars into marks (ii) 80 kroner into pesetas
(iii) 2000 lire into pesetas (iv) 400 pesetas into lire
(v) 50 marks into kroner

3.6 Large and small numbers

A number such as 5000000000 is clearly far too cumbersome to write out in full. It is in fact 5×1000000000 or 5×10^9.

[10^9 means $10 \times 10 \times 10 \times 10 \times 10 \times 10 \times 10 \times 10 \times 10$. See chapter 4.]

The expression 5×10^9 is said to be the *standard form* (note that the first number is between 1 and 10) Similarly for small numbers, 0.0000046 is really $4.6 \div 10^6$. We can write this in another way, i.e. 4.6×10^{-6}. We have changed the 6 to -6, and $\div$ to $\times$. Notice that, in each case, the number at the front is between 1 and 10.

EXAMPLE 6

Express the numbers 34600 and 2800 in standard form. What is their sum?

$$34\,600 = 3.46 \times 10\,000 = 3.46 \times 10^4$$
$$2800 = 2.8 \times 1000 = 2.8 \times 10^3.$$

It is impossible to add two numbers in standard form unless the power of 10 is the same. We choose the first number, and write $3.46 \times 10^4 = 34.6 \times 10^3$.

$$\therefore\ 34.6 \times 10^3 + 2.8 \times 10^3 = 37.4 \times 10^3 = 3.74 \times 10^4.$$

EXERCISE 3d

Express the following numbers in standard form:

1	2800	**8**	0.165	**15**	$2.8 \times 10^5 + 1000$
2	6 000 000	**9**	0.000 001	**16**	$3.64 \times 10^9 + 10^8$
3	3028	**10**	0.000 12	**17**	$6.4 \times 10^3 + 3.8 \times 10^2$
4	294	**11**	28 830	**18**	$10^{-1} + 10^{-2}$
5	37 230 000	**12**	0.000 000 4	**19**	$3.1 \times 10^{-1} + 2.6 \times 10^{-2}$
6	0.0031	**13**	19 000 000	**20**	$4.1 \times 10^{-5} + 3 \times 10^{-6}$.
7	0.027	**14**	$10^4 + 10^3$		

3.7 Speed problems

Units: metres per second (m/s), kilometres per hour (km/h). Average speed is defined by:

$$\text{average speed } (v) = \frac{\text{distance } (s)}{\text{time } (t)}$$

$$\text{i.e. } v = \frac{s}{t}$$

giving also $\quad s = vt \quad$ [distance = speed × time]

and $\quad t = \frac{s}{v}.$

EXAMPLE 7

A car travels for 30 minutes at 50 km/h and then travels a distance of 10 km at 40 km/h. Find the average speed for the whole journey.

[*Note:* the answer is not $\frac{50+40}{2} = 45.$]

Since average speed $= \frac{\text{total distance}}{\text{total time}}$ we need to find the time that the second part of the journey takes, and the distance covered in the first part of the journey.

$$t = \frac{s}{v}$$

$$\therefore t = \frac{10}{40} = \tfrac{1}{4}\text{ h}$$

$$\text{average speed} = \frac{25+10}{\frac{1}{2}+\frac{1}{4}} \quad \text{(distance in first part} = 25\text{ km)}$$

$$= 35 \div \tfrac{3}{4} = 35 \times \tfrac{4}{3} = \tfrac{140}{3}$$

$$= 46\tfrac{2}{3}\text{ km/h.}$$

3.8 Distance-time graphs

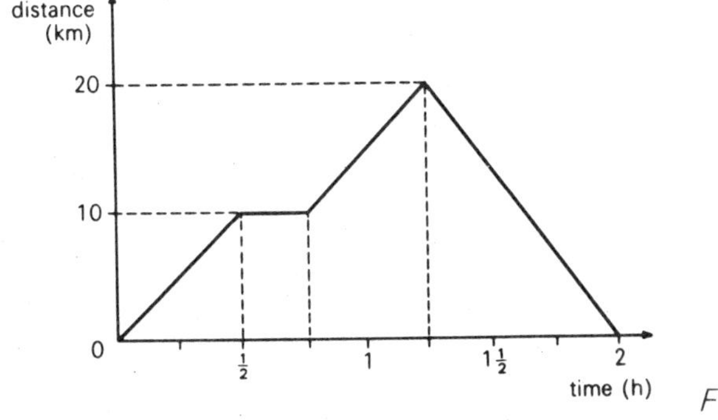

Fig. 3.1

The graph in Fig. 3.1 shows a journey travelled by a car.

This type of graph is called a distance – time graph.

first $\frac{1}{2}$hr : car travels 10 km at an average speed of 20 km/h
next $\frac{1}{4}$hr : car is at rest
next $\frac{1}{2}$hr: car travels 10 km at an average speed of 20 km/h.
final $\frac{3}{4}$hr : car returns to its starting point at an average speed of
$20 \div \frac{3}{4} = 26\frac{2}{3}$ km/h.

3.9 Speed-time graphs

Figure 3.2 shows the speed-time graph of an automatic lift. Its speed increases from 0 to 40 cm/s for the first 30 seconds. It then travels at a constant speed of 40 cm/s for 20 seconds. The lift then comes to rest in the final 20 seconds. We can find the *distance* travelled from a *speed–time* graph by finding the total shaded area. In this case:

$$\text{Area} = \underbrace{\tfrac{1}{2} \times 30 \times 40}_{\text{first triangle}} + \underbrace{40 \times 20}_{\text{rectangle}} + \underbrace{\tfrac{1}{2} \times 20 \times 40}_{\text{second triangle}}$$

$$= 600 + 800 + 400$$
$$= 1800 \text{ cm.}$$

Alternatively, the shape is a trapezium. Its area is given in chapter 7 as

$\frac{1}{2}(70 + 20) \times 40 = 1800$ cm.

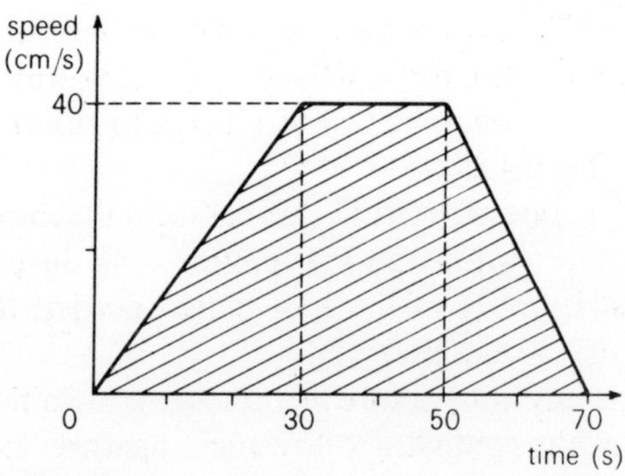

Fig. 3.2

EXERCISE 3e

1 A car travels for 30 mins at 50 km/h. It then stops for 10 minutes before continuing its journey. It covers the last 25 km in 40 minutes.
 (i) What is the total distance travelled?
 (ii) What is the average speed for the whole journey?
 Draw the distance – time graph for the journey.

2 Figure 3.3 shows the speed – time graph for part of a train's journey. What is the average speed for this part of its journey, and what distance did it cover?

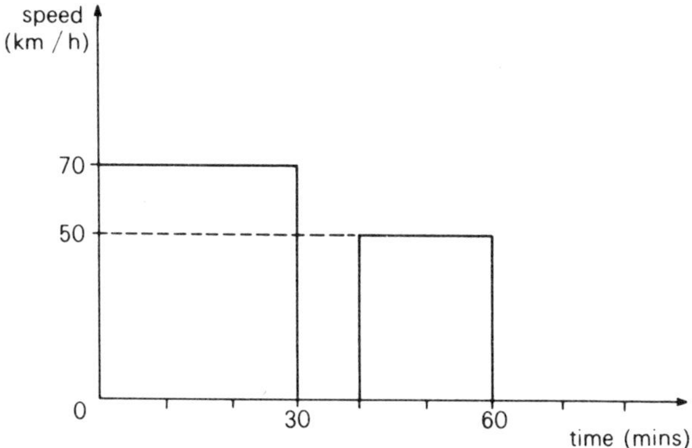

Fig. 3.3

3 A bullet travelling at 150 m/s is fired from a rifle aimed at a target 500 m away. How long does the bullet take to hit the target?

4 The rateable value of a house is £200 per annum. If the local rate is 48 p in the pound, find the rate payable per month.

5 The cost of running an electric fire is 0.9 p per unit (kilowatt-hour). If the fire is $1\frac{1}{2}$ kW, how much would it cost to run the fire continuously for 1 week?

6 A motor car uses one gallon of petrol in travelling a distance of 40 miles. If 1 km $= \frac{5}{8}$ mile and 1 gallon $= 4\frac{1}{2}$ litres, calculate the smallest number of litres of petrol needed to complete a journey of 130 kilometres.

7 A man earns £5000 per annum. Before income tax is payable, he is allowed to claim the following allowances against tax: £865 for his wife, and £240 for each child. The remainder of his income is taxed at 35 %. Calculate, to the nearest penny, the amount of tax he pays each month, if he has two children.

8 Mr Ball earns £6500 a year. He is buying his house on a mortgage and the interest he pays in a year amounts to £900. He is allowed to claim the full amount of this against tax. How much tax does he pay each month, assuming the information given in question 7? Mr Ball has a wife and 3 children.

9 A $\frac{1}{2}$p coin is 16 mm in diameter. These coins are stamped out of a strip of metal 16 mm wide and 300 cm long. What is the maximum number of coins that can be stamped from this strip? The waste metal is then melted and made into a strip 16 mm wide again, and of the same thickness. How many $\frac{1}{2}$p coins can be cut from this strip? (Area of a circle $= \pi r^2$, $\pi = 3.14$.)

10 The rateable value of Mr Roberts' house was £200 and rates were charged at 55p in the pound. Calculate the yearly rate that he has to pay. After extensive alterations, the rateable value is increased to £280. After an increase in rates, the new rate payable for the year is £66.40 more. What is the new rate?

HARDER EXAMPLES 3

1 Find the cost of running a 3 kW electric fire for eight hours each night over a period of 10 weeks, at a cost of 0.95p per unit. [1 unit = 1 kilowatt-hour.]

2 A car travelling at 20 m/s experiences a uniform deceleration which brings it to rest in 8 seconds. Sketch the speed–time graph and use it to find the distance travelled while the car is braking.

3 A main road ABCD is crossed by minor roads at A, B, C and D. At each junction there are traffic lights. The lights all change together, remaining red for half a minute and then immediately green for half a minute, and so on. The distances AB, BC and CD are each 1 km.

(i) A motorist P passes A just after the lights go green. By drawing a distance–time graph, show that if he travels at 60 km/h he will not have to stop. Take 2 cm to represent both 1 km and 1 minute. State how long it takes before he passes D.

(ii) On the same axes as your answer to (i) draw a distance–time graph showing the progress of a second motorist Q who leaves A at the same time as P and travels at 90 km/h when he can. State how long it takes before Q passes D.

(iii) In order to pass D before P does, it is necessary for Q to exceed a certain speed V km/h on at least one stage of the journey. Find the value of V.

[O & C (SMP)]

4 A tenant of a flat paid a monthly rent of £26.50 twelve times a year, and a half-yearly rate of £24.70. Find, to the nearest penny, how much he paid for rent and rates together per week. (Take one year to be 52 weeks.)

[O]

5 On a train journey of 117 kilometres the average speed for the first 27 kilometres is 45 km/h, and for the rest of the journey the average speed is 37.5 km/h. Calculate the uniform speed at which the train would have to travel in order to cover the whole distance in the same time.

[L]

6 If $x = a + 3b$ and $y = 4a - 3b$, find a and b, given that when $x = 1.2 \times 10^5$, $y = 3.6 \times 10^6$.

7 A town council have decided to increase the local rate by 6p in the pound. The total rateable value is £1 200 000. What will be the increase in income from the rates?

A householder finds out that his yearly rates will increase from £120 to £134.40. What is the rateable value of the house, and what was the original rate before the 6p increase?

8 Since a city typist moved to the coast, both the railway and the bus fares for her journeys have increased. Each year for her rail travel she buys three quarterly season tickets, one monthly season ticket and two weekly season tickets.

The quarterly tickets have increased from £45.50 each to £53.00 the monthly tickets from £16.50 each to £19.70, and the weekly tickets from £4.60 each to £5.40.

During each of the 45 weeks she travels, she takes ten bus rides, each of which has increased from 9p to 12p. Calculate

(i) the original annual cost of her travel,

(ii) the increase in annual cost,

(iii) the increase in annual cost expressed as a percentage of the original annual cost, correct to 3 significant figures.

[L]

9 The speed of light is 3×10^8 m/s. Find the distance that light travels in one year (365 days). Give your answer in the form $A \times 10^n$, where A is correct to 3 significant figures. It is found that two stars are 25 light years away. What is the distance between these two stars? Express your answer in the form $A \times 10^n$.

10 The amounts that a man pays in mortgage repayments and taxes are in the ratio 11 : 4, and total £1500 per annum. His mortgage repayments increase by 1 % and his tax decreases by 5 %. Calculate the percentage change in the total amount he has to pay.

11 An article manufactured in America cost $13.95. The official rate of exchange before devaluation in 1967 was $2.79 for £1 and after devaluation it was $2.40 for £1. Calculate the increase in the cost of the article in British currency as the result of devaluation.

[JMB]

12 A man left an inn X and walked at a steady speed of 4 km/h along a road to a station Y. A cyclist left the inn half an hour after the man, and travelling at 12 km/h reached the station 40 minutes before the man.

Calculate the distance along the road from X to Y.

If the cyclist overtook the man at a point Z on the road at 10 a.m., calculate

(i) the time at which the man left the inn,
(ii) the distance XZ.

[L]

4 Indices and logarithms

4.1 Indices

It is often necessary (and usually more economical) to abbreviate certain arithmetical expressions such as $2 \times 2 \times 2 \times 2 \times 2 \times 2 \times 2 \times 3 \times 3$. To attempt such abbreviations, we may write $2 \times 2 \times 2 \times 2 \times 2 \times 2 \times 2$ as 2^7 and 3×3 as 3^2. The small figures 7 and 2 are called *index numbers* and are respectively the number of 2's and the number of 3's that are multiplied together.

$$\therefore\ 2 \times 2 \times 2 \times 2 \times 2 \times 2 \times 2 \times 3 \times 3 = 2^7 \times 3^2.$$

We say that 2^7 is the seventh power of 2.

4.2 Rules of manipulation

1. *Multiplication*

$$a^4 \times a^3 = (a \times a \times a \times a) \times (a \times a \times a)$$
$$= a^7$$

i.e. $a^4 \times a^3 = a^{4+3}$ (the indices are *added*)

2. *Division*

$$a^4 \div a^2 = \frac{a \times a \times a \times a}{a \times a}$$
$$= a^2$$

i.e. $a^4 \div a^2 = a^{4-2}$ (the indices are *subtracted*)

3. $(a^3)^2 = (a \times a \times a)^2$

$$= (a \times a \times a) \times (a \times a \times a)$$

$= a^6$ (the indices are *multiplied*)

With these rules in mind, let us look at an example.

EXAMPLE 1

$$7^3 \div 7^5 = \frac{\not{7}^1 \times \not{7}^1 \times \not{7}^1}{\not{7}_1 \times \not{7}_1 \times \not{7}_1 \times 7 \times 7} = \frac{1}{7 \times 7} = \frac{1}{7^2}.$$

Rule 2 above implies that this should be written as 7^{-2}. Let us see if this is so.

4.3 Negative indices

$$a^4 \div a^4 = \frac{a \times a \times a \times a}{a \times a \times a \times a}$$

$$= 1$$

but

$$a^{4-4} = a^0 \qquad \text{(rule 2)}$$
$$\therefore\ a^0 = 1.$$

Now

$$a^2 \times a^{-2} = a^0$$
$$\Rightarrow a^{-2} = \frac{1}{a^2}.$$

So we are correct in assuming that $1/7^2 = 7^{-2}$ in example 1.

The general results which follow from the above are

$$a^m \times a^n = a^{m+n} \qquad \text{(I)}$$

$$a^m \div a^n = a^{m-n} \qquad \text{(II)}$$

$$(a^m)^n = a^{mn} \qquad \text{(III)}$$

These are true when m and n are positive or negative integers.

4.4 Fractional indices

We have assumed so far that the indices are integers. Is it possible to assign any meaning to an expression such as $4^{1/3}$?

If we assume our three rules still hold, then rule 3 says $(4^{1/3})^3 = 4^{(1/3)\times 3} = 4^1 = 4$.

Therefore $4^{1/3}$ must have been the cube root of 4. In general, $a^{1/q}$ is the q th root of a (i.e. the number which multiplied together q times produces a).

$$(a^{1/q})^p = a^{p/q} \qquad \text{(rule 3)}$$

Therefore $a^{p/q}$ is the qth root of a raised to the power p.
(*Note:* the qth root can be written $\sqrt[q]{a}$.)

EXAMPLE 2

$$81^{3/4} = (\sqrt[4]{81})^3 = 3^3 = 27.$$

EXAMPLE 3

$$8^{-2/3} = \frac{1}{8^{2/3}} = \frac{1}{(\sqrt[3]{8})^2} = \frac{1}{2^2} = \tfrac{1}{4}.$$

EXAMPLE 4

$$\left(\frac{11}{12}\right)^{-2} = \frac{1}{(11/12)^2} = \frac{1}{121/144} = \tfrac{144}{121}.$$

EXERCISE 4a

Simplify:

1 $a^4 \times a^7$
2 $b^3 \div b^4$
3 $3c^3 \times 4c^2$
4 $4d^5 \div 2d^2$
5 $(3e^2)^3$
6 $(3f)^2$
7 $(g^2 \times g^5) \div g^6$
8 $16h^3 \div 12h^4$
9 $5j^{-3} \div 4j^5$
10 $\sqrt{k^4}$
11 $3l^{-2} \times 4l^{-4}$
12 $(-12m)^3$
13 $-4n^2 \div 3n^{-3}$
14 $7p/21p^3$
15 $\sqrt[3]{27p^6}$
16 $\sqrt{2r^8}$
17 $(8s^2)^{1/3}$
18 $(64t^3)^{2/3}$
19 $\left(\frac{1}{u}\right)^{-3}$
20 $\left(\frac{-1}{2v^2}\right)^2$

Find the value of:

21 $\sqrt{10^8}$
22 $\sqrt{10^{-4}}$
23 $\sqrt[3]{8^2}$
24 $(0.16)^{1/2}$
25 $(2.25)^{3/2}$
26 $125^{-2/3}$
27 $27^{-1/3}$
28 $8^{-1/3}$
29 $(9/16)^{-1/2}$
30 $(1/25)^{3/2}$
31 $2^7 \times 4$
32 3^{-4}
33 $(3^{1/2})^0$
34 $32^{3/5}$
35 $8^{11} \div 8^9$
36 $(3^{1/5})^{10}$
37 $(49^{1/3})^{3/2}$
38 5^3
39 $49^{1/2}$
40 $625^{-1/2}$.

4.5 Standard form

Any number can be written in the form $A \times 10^K$ where $1 \le A < 10$ and K is an integer. Such a form is known as the standard form (see chapter 3).

EXERCISE 4b (Revision)

Write the following in standard form:

1 36.75
2 435
3 3695
4 10^3
5 10.01
6 0.298
7 0.0375
8 0.000257
9 10^{-4}
10 $1/10^{-3}$.

4.6 Logarithms

We have already seen that any number can be expressed in standard form. Now, we shall try to express any number y as a power of 10.

i.e. $y = 10^x$.

x is known as the logarithm of y to the base 10, or

$x = \log_{10} y$.

(i) Since $1 = 10^0$, then the logarithm of 1 is 0.
(ii) $10 = 10^1$, therefore the logarithm of 10 is 1.
(iii) $100 = 10^2$, therefore the logarithm of 100 is 2.

We can continue like this and plot a graph of the logarithm against the number, as in Fig. 4.1.

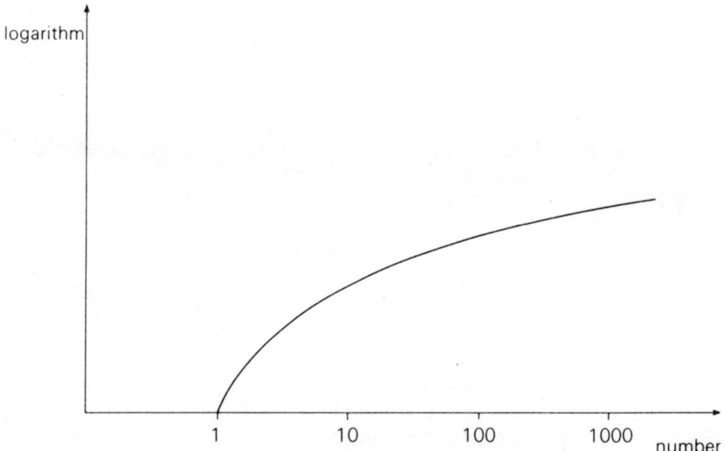

Fig. 4.1

We have taken liberties with the scales in order to get all of the diagram in. We have assumed that we can join up the points with a continuous line so that we can find the logarithm of any number. A graph for numbers between 1 and 10 is shown in Fig. 4.2. (The methods for obtaining these values accurately are beyond the scope of this book.)

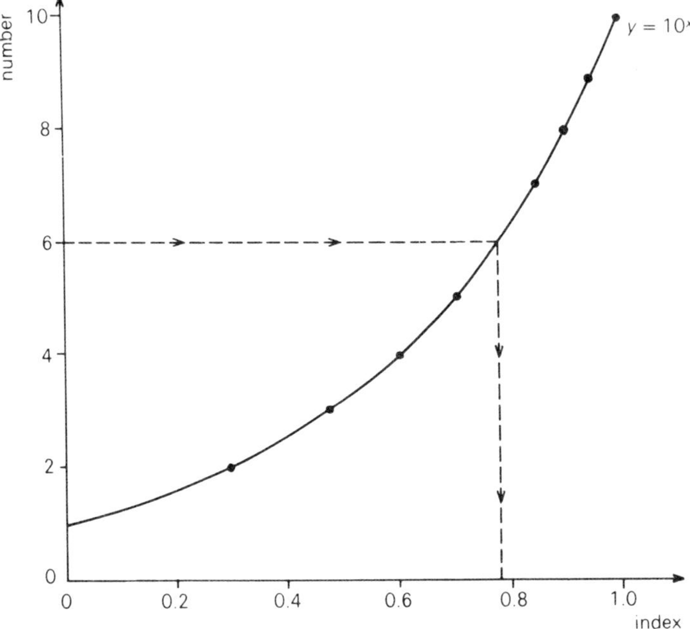

Fig. 4.2

To find the logarithm of any value y, let us take an example.

EXAMPLE 5

Find the logarithm of 6 to the base 10. From the vertical axis, we read horizontally from the value $y = 6$ until we reach the graph of the function, then we read in a vertical direction to the horizontal x-axis which will give the required answer.

The logarithm of 6 (written $\log_{10} 6$) to the base 10

≈ 0.78 (approximately equal).

EXERCISE 4c

1 From the graph, find the approximate value of $\log_{10} y$ for the following values of y:
(i) 4 (ii) 5 (iii) 2.5 (iv) 7.5 (v) 1.3 (vi) 9.8.

2 From the graph, find the number whose logarithm is:
(i) 0.2 (ii) 0.8 (iii) 0.28 (iv) 0.56 (v) 0.98 (vi) 0.615.

4.7 Logarithm tables

It is important to note that other logarithmic bases can be used, i.e.

$$y = 2^x \Rightarrow x = \log_2 y$$

or

$$y = 7^x \Rightarrow x = \log_7 y.$$

We shall in this book, however, only discuss logarithms to the base 10, and we shall discard the subscript 10 in the expression $\log_{10} y$.

It can be seen from Fig. 4.2 that only limited accuracy can be obtained from the graph, and to get better results, we shall refer to a table of logarithms. (The method of calculating these tables is beyond the scope of this book.)

EXAMPLE 6

Find the logarithm of 5.885. Figure 4.3 is an extract from the logarithmic tables:

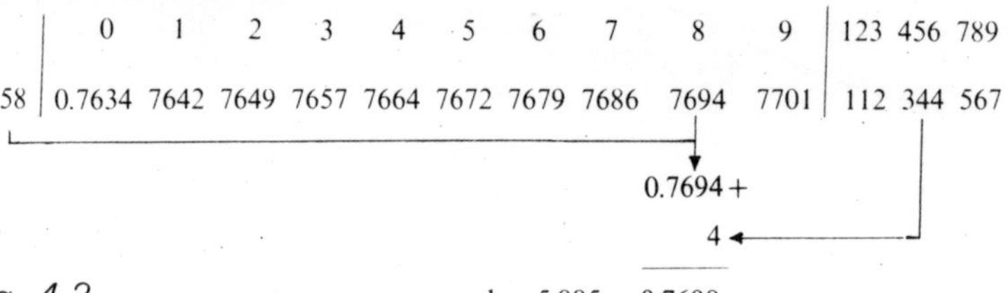

	0	1	2	3	4	5	6	7	8	9	123 456 789
58	0.7634	7642	7649	7657	7664	7672	7679	7686	7694	7701	112 344 567

Fig. 4.3 log 5.885 = 0.7698

To find the logarithm of a number greater than 10 we have to make use of the standard form and the rules for indices.

EXAMPLE 7

Find the logarithm of 23.56.

$$\begin{aligned} 23.56 &= 2.356 \times 10^1 \\ &= 10^{0.3722} \times 10^1 \\ &= 10^{0.3722+1} \\ &= 10^{1.3722} \end{aligned}$$

$$\therefore \log 23.56 = 1.3722$$

EXAMPLE 8

Find the logarithm of 897.

$$\begin{aligned} 897 &= 8.97 \times 10^2 \\ &= 10^{0.9528} \times 10^2 \\ &= 10^{2+0.9528} \\ &= 10^{2.9528} \end{aligned}$$

$$\therefore \log 897 = 2.9528.$$

EXAMPLE 9

Find the logarithm of 0.2758.

$$\begin{aligned} 0.2758 &= 2.758 \times 10^{-1} \\ &= 10^{0.4406} \times 10^{-1} \\ &= 10^{-1+0.4406} \\ \therefore \log 0.2758 &= -1+0.4406. \end{aligned}$$

It is customary to keep the part of the logarithm after the decimal point positive and in this instance the minus sign is placed above the 1.

$$\log 0.2758 = \bar{1}.4406.$$

↓ ↓

characteristic mantissa

EXAMPLE 10

Find the logarithm of 0.00587.

$$\begin{aligned} 0.00587 &= 5.87 \times 10^{-3} \\ &= 10^{0.7686} \times 10^{-3} \\ &= 10^{-3+0.7686} \\ &= 10^{\bar{3}.7686} \\ \therefore \log 0.00587 &= \bar{3}.7686. \end{aligned}$$

↓ ↓

characteristic mantissa

EXERCISE 4d

Using tables, find the logarithms of the following numbers:

1	3.275	**6**	59.5	**11**	21328	**16**	0.003202
2	7.963	**7**	10^3	**12**	493000	**17**	0.000587
3	4.28	**8**	318	**13**	0.5728	**18**	10^{-4}
4	8.6549	**9**	4531	**14**	0.7651	**19**	0.0001
5	21.57	**10**	67590	**15**	0.1212	**20**	0.004343.

4.8 Antilogarithms

If we are given a logarithm and need to find the number corresponding to this logarithm, we can use the antilogarithmic tables. We look up the mantissa in the same manner as in the previous table and use the characteristic to place the decimal point.

EXAMPLE 11

Find the number whose logarithm is $\bar{2}.7921$.

From the tables, the antilogarithm of $\bar{2}.7921$

$= 6.195 \times 10^{-2}$
$= 0.06195.$

EXERCISE 4e

Write down the number whose logarithm is:

1 0.5721 **6** 2.3758 **11** $\bar{1}.3214$ **16** $\bar{3}.2679$
2 0.3258 **7** 4.3261 **12** $\bar{1}.0762$ **17** $\bar{2}.1431$
3 0.9215 **8** 2.9871 **13** $\bar{2}.0098$ **18** $\bar{1}.9835$
4 0.3541 **9** 3.6721 **14** $\bar{2}.7894$ **19** $\bar{5}.1000$
5 1.5728 **10** 1.5875 **15** $\bar{3}.5410$ **20** $\bar{4}.3210$

4.9 Multiplication using logarithms

EXAMPLE 12

Find the value of 14.25×13.78.

$$\begin{aligned} 14.25 \times 13.78 &= 10^{1.1538} \times 10^{1.1393} \\ &= 10^{2.2931} \\ \text{antilog } 2.2931 &= 196.3 \\ \Rightarrow 14.25 \times 13.78 &= 196 \text{ (to 3 significant figures).} \end{aligned}$$

Logarithms only give three-figure accuracy, so answers should be given to 3 significant figures.

It is helpful to set out problems like this in the following way:

14.25×13.78
$= 196$ (to 3 sig. figs).

no.	log
14.25	1.1538
13.78	1.1393
196.3	←2.2931

Note: in order to multiply, we *add* the logarithms.

4.10 Division using logarithms

EXAMPLE 13

Find the value of $16.28 \div 8.31$.

$$\begin{aligned} 16.28 \div 8.31 &= 10^{1.2116} \div 10^{0.9196} \\ &= 10^{0.2920} \\ \text{antilog } 0.2920 &= 1.959 \\ \Rightarrow 16.28 \div 8.31 &= 1.96 \text{ (to 3 sig. figs.)} \end{aligned}$$

or

$16.28 \div 8.31$
$= 1.96$ (to 3 sig. figs).

no.		log
16.28		1.2116
8.31		0.9196
1.959	←	0.2920

Note: in order to divide, we *subtract* the logarithms.

4.11 Negative characteristics

Two examples will help to show how negative characteristics are handled.

EXAMPLE 14

0.3215×0.06751
$= 0.0217$ (to 3 sig. figs).

no.		log
0.3215		$\bar{1}.5072$
0.06751		$\bar{2}.8294$
0.02171	←	$\bar{2}.3366$

↓

$1+2+1$ (carried) $= (-1)+(-2)+1 = -2$

EXAMPLE 15

$0.2148 \div 0.7628$
$= 0.282$ (to 3 sig. figs).

no.		log
0.2148		$\bar{1}.3320$
0.7628		$\bar{1}.8825$
0.2815	←	$\bar{1}.4495$

$(-1)-(-1+1) = -1$ ←

EXERCISE 4f

Evaluate:

1 3.75×4.285
2 13.5×17.52
3 14.98×0.2351
4 391×1.586
5 0.0258×4.765
6 0.0718×0.0326
7 13000×0.00721
8 0.0058×25.4
9 0.3715×0.02839

10 323×0.0753
11 $4.285 \div 3.75$
12 $17.52 \div 13.5$
13 $14.98 \div 0.2351$
14 $391 \div 1.586$
15 $0.0258 \div 4.765$
16 $0.0718 \div 0.0326$
17 $13000 \div 0.00721$
18 $0.0058 \div 25.4$
19 $0.3715 \div 0.02839$
20 $323 \div 0.0753$.

4.12 Powers and roots

EXAMPLE 16

$$\begin{aligned} (4.78)^3 &= (10^{0.6794})^3 \\ &= 10^{2.0382} \\ \text{antilog } 2.0382 &= 109.1 \\ \Rightarrow \quad (4.78)^3 &= 109 \text{ (to 3 sig. figs)} \end{aligned}$$

or

no.	log
4.78	0.6794
	×3
109.1 ←	2.0382

$(4.78)^3 = 109$ (to 3 sig. figs).

Note: the logarithm is *multiplied* by the *power*.

EXAMPLE 17

$$\begin{aligned} (0.6728)^4 &= (10^{\bar{1}.8279})^4 \\ &= 10^{(-1+0.8279)\times 4} \\ &= 10^{-4+3.3116} \\ &= 10^{-1+0.3116} \\ &= 10^{\bar{1}.3116} \\ \text{antilog } \bar{1}.3116 &= 0.2049 \\ \Rightarrow \quad (0.6728)^4 &= 0.205 \text{ (to 3 sig. figs)} \end{aligned}$$

or

$(0.6728)^4$
$= 0.205$ (to 3 sig. figs).

no.	log
0.6728	$\bar{1}.8279$
	×4
0.2049 ←	$\bar{1}.3116$

EXAMPLE 18

$$\begin{aligned} \sqrt[3]{(8.576)} &= (8.576)^{1/3} \\ &= 10^{0.9333\times 1/3} \\ &= 10^{0.3111} \\ \text{antilog } 0.3111 &= 2.046 \\ \Rightarrow (8.576)^{1/3} &= 2.05 \text{ (to 3 sig. figs)} \end{aligned}$$

or

$(8.576)^{1/3}$

$= 2.05$ (to 3 sig. figs).

no.	log.
8.576	0.9333
	$\div 3$
2.046 ←	0.3111

Note: the logarithm is *divided* by the required root.

EXAMPLE 19

$$\begin{aligned}\sqrt[4]{(0.2579)} &= (0.2579)^{1/4}\\ &= 10^{\bar{1}.4114 \times 1/4}\\ &= 10^{(-1+0.4114)/4}\\ &= 10^{(-4+3.4114)/4}\\ &= 10^{-1+0.8528}\\ &= 10^{\bar{1}.8528}\end{aligned}$$

antilog $\bar{1}.8528 = 0.7125$

$\Rightarrow \sqrt[4]{(0.2579)} = 0.713$ (to 3 sig. figs)

or

$\sqrt[4]{(0.2579)}$

$= 0.713$ (to 3 sig. figs).

no.	log
0.2579	$\bar{1}.4114$
	$\div 4$
0.7125 ←	$\bar{1}.8528$

Care must be taken over the division of a logarithm with a negative characteristic by an integer. If the characteristic is not exactly divisible by the integer, the number under the bar should be increased to the next integer which is divisible by the divisor. The mantissa must then be balanced accordingly. For example,

$$\begin{aligned}\frac{\bar{2}.5873}{3} &= \frac{\bar{3}+1.5873}{3}\\ &= \bar{1}+0.5291\\ &= \bar{1}.5291.\end{aligned}$$

EXERCISE 4g

Evaluate the following:

1 $(3.798)^3$
2 $(15.8)^4$
3 $(198)^2$
4 $(0.0586)^3$
5 $(0.257)^4$
6 $\sqrt[3]{89.7}$
7 $\sqrt[5]{(368)}$
8 $\sqrt[4]{7.2}$
9 $\sqrt[2]{0.5721}$
10 $\sqrt[3]{(0.07283)}$
11 $\sqrt[6]{(0.479)}$
12 $(0.589)^{1/3}$
13 $(82.5)^{1/4}$
14 $(0.0321)^{1/2}$
15 $(71.3)^{3/4}$
16 $(2.986)^{2/3}$
17 $(17.58)^{3/5}$
18 $(0.037)^{2/3}$
19 $(0.25)^{3/2}$
20 $(0.649)^{4/3}$.

4.13 Other tables

Other tables exist to help in the solution of arithmetical problems and they are used in much the same way as the logarithmic and antilogarithmic tables.

(i) *Squares and square roots.* It is necessary to find a rough approximation to our problem as in the case of squares we have to locate the decimal point and in the case of square roots we have to decide on one of two tables to use and then locate the decimal point.

EXAMPLE 20

Find the square of 42.5

$$\begin{aligned} \text{Rough approximation} &= 40 \times 40 \\ &= 1600. \\ \text{From the tables } (42.5)^2 &= 1806 \\ \Rightarrow (42.5)^2 &= 1810 \text{ (to 3 SF).} \end{aligned}$$

EXAMPLE 21

Find the square root of 37.5

$$\begin{aligned} \text{Rough approximation} &= 6 \qquad (6 \times 6 = 36). \\ \text{From the tables } \sqrt{37.5} &= 1936 \text{ or } 6124. \end{aligned}$$

We can discount the former.

$$\therefore \sqrt{37.5} = 6.12 \text{ (to 3 SF).}$$

EXAMPLE 22.

Find the square root of 0.05372. To find the approximate square root it may be useful to employ the following method. From the decimal point divide the number in the following way:

0. ¦ 05 ¦ 37 ¦ 2

We then find the approximate square root of the first compartment which does not contain two zeros, working from the decimal point left to right. The rest of the compartments are filled with zeros. In the first compartment, we have 05. The approximate square root of 5 is 2. Hence the example above can be tackled in the following way:

Rough approximation = 0. ¦ 2 ¦ 0 ¦
0. ¦ 05 ¦ 37 ¦ 2

From the tables $\sqrt{(0.05372)} = 0.232$ (to 3 SF).

Example 21 can be done in a similar way. The approximation is found by dividing the digits into pairs working from the decimal point to the left. A rough approximation to the first compartment reading left to right is then found and the remaining compartments filled with zeros, i.e.

¦ 6. ¦ 0
¦ 37. ¦ 5

(ii) *Reciprocals.* Suppose we wish to find the value of

$$\frac{1}{13.7}+\frac{1}{7.2}.$$

We have a similar problem with placing the decimal point to the one we have in square and square root tables.

From the reciprocal tables the figures for 1/13.7 are 7299. To place the decimal point, we use the fact that

$$\frac{1}{100}<\frac{1}{13.7}<\frac{1}{10}$$

$$\text{i.e. } 0.01<\frac{1}{13.7}<0.1$$

$$\Rightarrow \frac{1}{13.7}=0.07299.$$

Similarly

$$\frac{1}{10}<\frac{1}{7.2}<1$$

$$\text{i.e. } 0.1<\frac{1}{7.2}<1$$

$$\Rightarrow \frac{1}{7.2}=0.1389$$

$$\therefore \frac{1}{13.7}+\frac{1}{7.2} = 0.07299+0.1389$$
$$= 0.21189$$
$$= 0.212 \text{ (to 3 SF).}$$

EXAMPLE 23

Find $\frac{5}{21.37}$

$$\frac{5}{21.37} = 5 \times \frac{1}{21.37}$$

$$\frac{1}{100} < \frac{1}{21.37} < \frac{1}{10}$$

$$\text{i.e. } 0.01 < \frac{1}{21.37} < 0.1$$

$$\Rightarrow \frac{5}{21.37} = 5 \times 0.04680 \qquad \text{(not 0.04710)}$$
$$= 0.23400$$
$$= 0.234 \text{ (to 3 SF)}$$

EXERCISE 4h

Evaluate the following, using reciprocal tables:

1 $\frac{1}{51.58}$

2 $\frac{1}{4.7}$

3 $\frac{1}{0.498}$

4 $\frac{5}{14.58}$

5 $\frac{12}{721.5}$

6 $\frac{3}{91.2}$

7 $\frac{1}{5.3}+\frac{1}{6.5}$

8 $\frac{1}{29.5}+\frac{1}{63.5}$

9 $\frac{3}{41.5}+\frac{2}{9.8}$

10 $\frac{15}{37.51}-\frac{2}{11.9}$.

EXAMPLE 24

Evaluate $\sqrt[3]{\frac{0.275 \times 1.583}{21.2 \div 13.75}}$ and express the answer in standard form.

$$\sqrt[3]{\frac{0.275 \times 1.583}{21.2 \div 13.75}} = 0.6560$$
$$= 6.56 \times 10^{-1}.$$

no.	log
0.275	$\bar{1}.4393$
1.583	0.1995
	1.6388
	0.1880 ←
	$\bar{1}.4508$
	÷ 3
0.6560 ←	$\bar{1}.8169$
21.2	1.3263
13.75	1.1383
	0.1880

HARDER EXAMPLES 4

1 Evaluate:
(i) $64^{-2/3}$ (ii) $125^{2/3}$ (iii) $1/(\frac{4}{9})^{-3/2}$

2 Write $(2 \times 10^3) \times (21 \times 10^{-4})$ as a single number expressed in standard form.

3 Write $(16 \times 10^{-2}) \div (8 \times 10^3)$ as a single number expressed in standard form.

4 Write $(4 \times 10^{-4}) + (3 \times 10^{-3})$ as a single number expressed in standard form.

5 (i) Express 0.0016 in standard form.

(ii) Express $\sqrt[2]{0.0016}$ in the same form.

6 Calculate $\dfrac{14.27 \times 13.69}{15.82^2 - 2.13^2}$.

7 Calculate $\dfrac{23.52 - 17.64}{100 \times 11.76}$.

8 Calculate $92.98^2 - 7.02^2$.

9 Calculate $\dfrac{(0.875)^2}{\sqrt{(0.325)}}$.

10 Calculate $\dfrac{0.72 + \sqrt{0.72}}{0.72}$.

11 Which is larger, $\sqrt[3]{0.0075}$ or $(0.75)^2$?

12 Calculate $\dfrac{7}{51.56}-(0.198)^2$.

13 Calculate $\dfrac{56.58 \times (0.158)^3}{\sqrt{2.429}}$.

14 Find the value of $\sqrt{a^2+b^2}$ if $a = 21.5$ and $b = 32$.
15 Find the value of $V = \frac{4}{3}\pi r^3$ where $\pi = 3.14$ and $r = 4.98$.
16 Find the value of S if $S = 4\pi r^2$ where $\pi = 3.14$ and $r = 0.125$.
17 $T = 2\pi\sqrt{l/g}$. Find T if $\pi = 3.14$, $l = 17.5$ and $g = 980$.
18 Find x if $\dfrac{1}{x} = \dfrac{1}{2.85} - \dfrac{1}{9.76}$

5 Algebra

5.1 Like and unlike terms

We can simplify $3x+2x$ because, apart from the number in front (known as the coefficient) of x, the two terms are the same (like terms).

$$3x+2x = 5x$$

However, $2x+x^2$ cannot be simplified, although x appears in each. They are said to be unlike terms.

EXAMPLE 1

Simplify $3x+ax+x^2+2x-5ax$.

We join the like terms: $3x+2x$, $ax-5ax$.
Therefore the answer is $5x+x^2-4ax$.

EXERCISE 5a

Simplify, where possible, the following:

1 $5a-6b-3a$
2 $2x^2-x+3x^2$
3 $5a-2ab+b$
4 $3a^2-2a+6a^2$
5 $x^3-5ax+3x^2$
6 $4a-3b-7a+5b^2$
7 $6a^3-a^2+a^3$
8 $x^2-2x-5x^2+3x$
9 $7x^2-xy-x^2+3xy$
10 $8at-t^2+at$
11 $t^2-3tm+mt$
12 $4m^2-mn-6n$
13 $12pq+p^2q^2+pq^2$
14 $3pq-8qp-9qp^2$
15 $15xy-7xy+11x^2y^2$
16 $12fg+f^2g-fg^2+2gf$
17 $ab+ba+ab^2+ba^2$
18 $4xt-tx^2+4x^2t-3tx$.

5.2 Removal of brackets (the distributive rule)

$a(b+c)$ means $a\times(b+c)$
i.e. add b and c and multiply your answer by a.

Consider a numerical example.

$$4 \times (3+7) = 4 \times 10 = 40.$$

However,

$$4 \times 3 + 4 \times 7 = 12 + 28 = 40$$
$$\therefore\ 4 \times (3+7) = 4 \times 3 + 4 \times 7.$$

Each number inside the bracket is multiplied by the number outside:

$$a(b+c) = ab + ac. \qquad \text{(A)}$$

Equation (A) is the *distributive law.*

EXAMPLE 2

Simplify $3x(a+x) - a(a-x)$.

$$3x(a+x) = 3xa + 3x^2$$
$$a(a-x) = a^2 - ax$$

subtract everything from the
↓ bracket which follows

$$\therefore\ 3x(a+x) - a(a-x) = 3xa + 3x^2 - (a^2 - ax)$$

↓ subtraction of a negative quantity is the same as addition

$$= 3xa + 3x^2 - a^2 + ax$$
$$= 3x^2 + 4ax - a^2.$$

EXERCISE 5b

Simplify the following:

1 $a(b+c) + c(b+a)$
2 $2a(a-b) + 5ab$
3 $m(2n+m) - 3m(m+n)$
4 $2x(y+z) - 4x(y+t)$
5 $a(a^2+b) - b(a^2+b)$
6 $a^2(a-b) - a^3(1+b)$
7 $a(2-a) - a^2(1+a)$
8 $3a(1-b) + ab$
9 $x(x+1) - 3(2-x)$
10 $x^2(x-1) + x^2(x+1)$
11 $a(ab-3a) - b(ab-b)$
12 $2xy(x+y) - 4x(y^2-x)$
13 $5m(n+p) - pm(m-n)$
14 $ax^2(x-y) + ay(x^2-2)$
15 $3xy(x-y) + 2xy(y-x)$.

5.3 The product of two brackets

$(a+b)(c+d)$ means $(a+b) \times c + (a+b) \times d$
$= c \times (a+b) + d \times (a+b)$.

Note: we have used here the fact that for any two numbers x and y

$$xy = yx \qquad \text{(B)}$$

This is the *commutative law* of multiplication.

$$(a+b)(c+d) = ca+cb+da+db.$$

Each term in each bracket multiplies each term in the other bracket.

EXAMPLE 3

Expand and simplify $(3x-y)(x+y)$.

$$\begin{aligned}(3x-y)(x+y) &= 3x \times x + 3x \times y - y \times x - y \times y \\ &= 3x^2+3xy-yx-y^2 \\ &= 3x^2+2xy-y^2.\end{aligned}$$

EXERCISE 5c

Expand and simplify:

1 $(x+1)(x+2)$

2 $(a+d)(2a+b)$

3 $(a+b)(3a+2b)$

4 $(x-y)(x+2y)$

5 $(x-3y)(2x+5y)$

6 $(t+3m)(t-2m)$

7 $(a+x)(2a+x)$

8 $(x-3a)(a-2x)$

9 $(a+4y)(y-a)$

10 $(b+c)(3b-c)$

11 $(b+4c)(a+y)$

12 $(x^2-y^2)(x+3y)$.

5.4 $(a+b)^2$ and (a^2-b^2)

$$(a+b)^2 = (a+b)(a+b) = a^2+ab+ba+b^2$$

$$= a^2+2ab+b^2 \qquad \text{(C)}$$

first term squared ↑ (for a^2); twice the product ↑ (for $2ab$); second term squared ↑ (for b^2)

$$\begin{aligned}(a+b)(a-b) &= a^2-ab+ba-b^2 \\ &= a^2-b^2. \qquad \text{(D)}\end{aligned}$$

This is known as the *difference of two squares.*

EXAMPLE 4

Expand and simplify (i) $(3x-2y)^2$; (ii) $(5x-2y)(5x+2y)$.

(i) Equation (C) gives:

$$\begin{aligned}(3x-2y)^2 &= (3x)^2+2(3x)\times(-2y)+(-2y)^2\\ &= 9x^2-12xy+4y^2.\end{aligned}$$

(ii) Equation (D) gives:

$$\begin{aligned}(5x-2y)(5x+2y) &= (5x)^2-(2y)^2\\ &= 25x^2-4y^2.\end{aligned}$$

EXERCISE 5d

Expand and simplify:

1 $(3a+b)^2$
2 $(a-4b)^2$
3 $(2x-3y)^2$
4 $(5x-y)^2$
5 $(\frac{1}{4}+x)^2$
6 $(5a-3b)(5a+3b)$
7 $(2x-y)(2x+y)$
8 $(m-3n)(m+3n)$
9 $(a-\frac{1}{2})(a+\frac{1}{2})$
10 $(b-5)(b+5)$.

5.5 Factorisation

The reverse process of expanding brackets in algebraic expressions is known as factorisation.

(i) *Common factors.*

EXAMPLE 5

Factorise $3x+9y$. The common factor of these two terms is 3.

$$\therefore\ 3x+9y = 3(x+3y).$$

EXAMPLE 6

Factorise $4x^2-12xy$. The common factor of these two terms is $4x$.

$$\therefore\ 4x^2-12xy = 4x(x-3y).$$

EXAMPLE 7

Factorise $3a+12+ab+4b$. To tackle this problem we group the expression into pairs and look for a common factor for each pair.

$$\begin{aligned}\therefore\ 3a+12+ab+4b &= (3a+12)+(ab+4b)\\ &= 3(a+4)+b(a+4) \qquad \text{(E)}\\ &= (a+4)(3+b)\end{aligned}$$

since $(a+4)$ is a common factor of the two terms in (E).

EXAMPLE 8

Factorise $2mn-2m+p-np$.

$$2mn - 2m + p - np = 2m(n-1) + p(1-n).$$

If we remember that

$$(1-n) = -(n-1)$$

then

$$2m(n-1) + p(1-n) = 2m(n-1) - p(n-1)$$
$$= (n-1)(2m-p).$$

EXERCISE 5e

Factorise:

1 $ax + 4x^2$
2 $3bz^2 - 9b^2z$
3 $gx - kx^2$
4 $3x^3 + 2x^2 + 5x$
5 $6x^4 - 3x^2$
6 $5x^2 + 10x - 25$
7 $a^2b^3 - b^2a^3$
8 $7(x-2) - 3(x-2)$
9 $a(c-d) + b(c-d)$
10 $x(y-z) - w(z-y)$
11 $a(4x+1) + 4x + 1$
12 $a(x-y) + b(x-y)$
13 $qr + rs - ps - pq$
14 $6xy - 2y + 12mx - 4m$
15 $cx - dx + dq - cq$
16 $qr + st - qt - rs$
17 $a^2 - 3a - ab + 3b$
18 $x^2 - 3x + 3y - yx$
19 $ab + 3a + bc + 3c$
20 $ax + bx + cx - ay - by - cy.$

(ii) *Difference of two squares.* We have already seen that

$$(x+y)(x-y) = x^2 - y^2.$$

Can we recognise the right-hand side of this identity when it occurs and so be able to factorise it?

EXAMPLE 9

Factorise $x^2 - 9$.

$$x^2 - 9 = x^2 - 3^2$$
$$= (x-3)(x+3).$$

EXAMPLE 10

Factorise $16p^2 - 25q^2$.

$$16p^2 - 25q^2 = (4p)^2 - (5q)^2$$
$$= (4p - 5q)(4p + 5q).$$

EXAMPLE 11

Evaluate $100^2 - 99^2$.

$$100^2 - 99^2 = (100 - 99)(100 + 99)$$
$$= 1 \times 199$$
$$= 199.$$

EXERCISE 5f

Resolve, where possible, into factors:

1 x^2-16
2 $4x^2-25$
3 $2x^2-2$
4 $81x^2-100$
5 $1-x^2$
6 $27-3x^2$
7 $1-16x^2$
8 R^2-r^2
9 kx^2-y^2
10 $18x^2-98$
11 $25-36(x-y)^2$
12 $(x-3)^2-9$
13 $36x^2-16y^2$
14 x^2-4y^2
15 $(y+1)^2-9$
16 $(y+7)^2-4$
17 $121x^2y^2-4$
18 $72x^2-8$
19 $8x^2-16$
20 $x^2-(a+b)^2$.

Evaluate the following:

21 $(18.3)^2-(8.3)^2$
22 314^2-305^2
23 $(4.7)^2-(3.7)^2$
24 100^2-99^2
25 66^2-36^2
26 575^2-425^2
27 369^2-331^2
28 $(4.9)^2-(3.1)^2$
29 52^2-48^2
30 1000^2-999^2.

5.6 Factorisation of a quadratic expression

We have already met the problem of the expansion of two binomials into a quadratic expression, i.e.

$$\begin{aligned}(x+3)(x+2) &= x^2+2x+3x+6\\ &= x^2+5x+6.\end{aligned}$$

Each term on the right-hand side is produced in the following way:

x^2: $(x \quad)(x \quad)$

$\downarrow \quad \downarrow$

$x \times x = x^2$

constant: $+6$: $(\quad +3)(\quad +2)$

$\downarrow \quad \downarrow$

$+3 \times +2 = +6$

$+5x$: $(\quad 3)(x \quad)$ and $(x \quad)(\quad +2)$

$\downarrow \qquad \downarrow$

$+3x \qquad +2x \qquad = +5x$

The product of the numerical quantities gives the constant term and the sum of the numerical quantities gives the coefficient of x.

EXAMPLE 12

Factorise $x^2+11x+18$. We need two numbers whose product is

+18 and whose sum is +11. By inspection the numbers are +9 and +2. So

$$x^2+11x+18=(x+9)(x+2).$$

Check:

$$\begin{aligned}(x+9)(x+2)&=x^2+2x+9x+18\\&=x^2+11x+18.\quad\checkmark\end{aligned}$$

EXAMPLE 13

Factorise $x^2+5x-14$. We need two numbers whose product is −14 and whose sum is +5. By inspection the numbers are +7 and −2. So

$$x^2+5x-14=(x+7)(x-2).$$

Check:

$$\begin{aligned}(x+7)(x-2)&=x^2-2x+7x-14\\&=x^2-5x-14.\quad\checkmark\end{aligned}$$

EXAMPLE 14

Factorise x^2-6x+9. We need two numbers whose product is +9 and whose sum is −6. By inspection the numbers are −3 and −3. So

$$x^2-6x+9=(x-3)(x-3).$$

Check:

$$\begin{aligned}(x-3)(x-3)&=x^2-3x-3x+9\\&=x^2-6x+9.\quad\checkmark\end{aligned}$$

EXERCISE 5g

Factorise the following where possible:

1	x^2+2x+1	**12**	x^2-3x+4	**23**	$x^2+70x-144$
2	x^2+6x+8	**13**	$x^2-7x+10$	**24**	$x^2+19x-120$
3	$x^2+7x+10$	**14**	$x^2-12x+35$	**25**	x^2-x-2
4	x^2+5x+4	**15**	$x^2-30x+200$	**26**	x^2-4x-5
5	x^2+8x+6	**16**	$x^2-11x+24$	**27**	x^2-2x-8
6	$x^2+10x+21$	**17**	x^2+3x-4	**28**	x^2-7x-8
7	$x^2+15x+36$	**18**	x^2+2x-3	**29**	$x^2-15x-16$
8	$x^2+32x+60$	**19**	x^2+4x-5	**30**	$x^2-21x-72$
9	x^2-2x+1	**20**	x^2+5x-6	**31**	$x^2-4x-12$
10	x^2-4x+3	**21**	$x^2+11x-26$	**32**	$x^2-5x-24$.
11	x^2-5x+6	**22**	$x^2+9x-22$		

When the coefficient of x^2 is greater than 1, the factorisation process is approached in much the same manner. Care must be taken, however, over the first term in each bracket.

EXAMPLE 15

Factorise $6x^2-11x+3$. The coefficients of x in the two factors are either 3 and 2 or 6 and 1. The numerical quantities must have the same sign and, since the middle term of the expression is negative, the sign of the numerical quantities must be negative.

The only numerical quantities are -3 and -1. We have to choose between the following pairs of factors:

	'x' term
$(3x-3)(2x-1)$	$-9x$
$(3x-1)(2x-3)$	$-11x$
$(6x-3)(x-1)$	$-9x$
$(6x-1)(x-3)$	$-19x$

It can be seen that the only result which agrees with the middle term is the second one of these. Check:

$$(3x-1)(2x-3) = 6x^2-9x-2x+3$$
$$= 6x^2-11x+3. \quad \checkmark$$

EXERCISE 5h

Factorise the following:

1	$2x^2+13x+15$	**8**	$4x^2+4x-15$	**15**	$6x^2-5x+1$
2	$3x^2+10x+8$	**9**	$6x^2+x-5$	**16**	$3x^2-4x-4$
3	$4x^2+8x+3$	**10**	$9x^2+x-8$	**17**	$9x^2-7x-2$
4	$6x^2+7x+2$	**11**	$4x^2-4x+1$	**18**	$2x^2-3x-5$
5	$6x^2+17x+10$	**12**	$2x^2-7x+3$	**19**	$4x^2-5x-6$
6	$8x^2+2x-3$	**13**	$5x^2-9x+4$	**20**	$4x^4-3x^2-1.$
7	$2x^2+3x-9$	**14**	$4x^2-9x+2$		

5.7 Solution of linear equations

These are equations of the form $ax+b=0$ where x is the unknown value and a and b are known. In such equations, no powers of x higher than the first exist, that is, there are no x^2, x^3 or higher terms involving x.

In solving such equations, it may be helpful to keep at the back of our minds the comparison between the two sides of an equation and two scales of a weighing machine. It is important to keep both sides balanced at all times.

EXAMPLE 16

Solve $x+3=7$. Subtract 3 from both sides

$$\Rightarrow x+3-3 = 7-3$$
$$\Rightarrow x = 4.$$

EXAMPLE 17

Solve $2x+11 = 29$. Subtract 11 from both sides

$$\Rightarrow 2x+11-11 = 29-11$$
$$\Rightarrow 2x = 18.$$

Divide both sides by 2

$$\Rightarrow x = 9.$$

EXAMPLE 18

Solve $4x-7 = 19$. Add 7 to both sides

$$\Rightarrow 4x-7+7 = 19+7$$
$$\Rightarrow 4x = 26.$$

Divide both sides by 4

$$\Rightarrow x = \frac{26}{4}$$
$$\Rightarrow x = 6\tfrac{1}{2}.$$

EXAMPLE 19

Solve $\frac{3}{4}(x+5) = 7$. Multiply both sides by 4

$$\Rightarrow 4 \times \tfrac{3}{4}(x+5) = 4 \times 7$$
$$\Rightarrow 3(x+5) = 28.$$

Remove the brackets

$$\Rightarrow 3x+15 = 28.$$

Subtract 15 from both sides

$$\Rightarrow 3x+15-15 = 28-15$$
$$\Rightarrow 3x = 13$$
$$\Rightarrow x = 4\tfrac{1}{3}.$$

EXERCISE 5i

Solve the following equations:

1 $x+7 = 14$ **4** $3x-13 = 7$ **7** $4-3x = \frac{1}{7}$

2 $x-15 = -3$ **5** $5-2x = 11$ **8** $5+2x = 7-5x$

3 $2x+5 = 12$ **6** $7-\frac{2}{3}x = 6$ **9** $3-2x = 7-4x$

10 $\frac{x}{5} = 4$

11 $\frac{x}{6} = x-5$

12 $(x-5)+13 = 21$

13 $(x+8)-15 = 11$

14 $3(x-3) = 4$

15 $5(x-2) = 3(4-x)$

16 $4(x-3) = 7$

17 $\frac{1}{3}(x+4) = 16$

18 $\frac{5}{7}(x-5) = 20$

19 $\frac{1}{2}(x-2) = -5$

20 $\frac{4}{5}(x+4) = 8$

21 $\frac{x-1}{2} - \frac{x-2}{3} = 1$

22 $\frac{2x}{3} - \frac{3x}{4} = \frac{1}{5}$

23 $\frac{x+2}{6} - \frac{x+3}{3} = x$

24 $\frac{2}{3}(x+3) = \frac{4}{5}(x+7)$

25 $\frac{x}{3}+\frac{x}{4}+\frac{x}{5} = 21$

26 $\frac{1}{2}x - \frac{1}{4}x = 3.$

5.8 Solution of quadratic equations

(i) *By factors.* If $P \times Q = 0$, then either $P = 0$ or $Q = 0$, e.g.

$$3 \times 0 = 0$$
$$0 \times 5 = 0 \qquad \text{etc.}$$

Let us see how we can use this to solve quadratic equations by factors.

EXAMPLE 20

Solve the equation $x^2-3x+2 = 0$.

Factorising, we have $\underbrace{(x-2)}_{P}\underbrace{(x-1)}_{Q} = 0$.

This is of the form $P \times Q = 0$. Therefore, either

$$x-2 = 0$$
$$\Rightarrow x = 2$$

or

$$x-1 = 0$$
$$\Rightarrow x = 1.$$

EXAMPLE 21

Solve the equation $\frac{1}{x+1}+\frac{5}{2x+1} = \frac{4}{3}$.

The common denominator is $3(x+1)(2x+1)$.

$$\therefore \frac{1}{\cancel{(x+1)}} \times 3\cancel{(x+1)}(2x+1) + \frac{5}{\cancel{(2x+1)}} \times 3(x+1)\cancel{(2x+1)}$$

$$= \frac{4}{\cancel{3}} \times \cancel{3}(x+1)(2x+1)$$

$$\Rightarrow 3(2x+1)+15(x+1) = 4(x+1)(2x+1)$$
$$\Rightarrow \quad 6x+3+15x+15 = 4(2x^2+3x+1) = 8x^2+12x+4$$
$$\Rightarrow 8x^2-9x-14 = 0$$
$$\Rightarrow (8x+7)(x-2) = 0$$
$$\therefore \text{ either } \quad x-2 = 0$$
$$\Rightarrow x = 2$$
$$\text{or} \quad 8x+7 = 0$$
$$\Rightarrow x = -\tfrac{7}{8}.$$

EXERCISE 5j

Solve the following equations:

1 $x^2-4x+3 = 0$
2 $x^2+5x+6 = 0$
3 $t^2-3t-10 = 0$
4 $p^2-7p+12 = 0$
5 $u^2-10u+9 = 0$
6 $x^2-x-20 = 0$
7 $2y^2-3y+1 = 0$
8 $3x^2-5x-2 = 0$
9 $6m^2-17m+12 = 0$
10 $2b^2 = 5b+3$
11 $\dfrac{1}{x+2}+\dfrac{1}{x+3} = \dfrac{7}{12}$
12 $\dfrac{6}{x+2}-\dfrac{1}{x} = 1$
13 $\dfrac{1}{x+2}-\dfrac{2}{x-3} = \dfrac{3}{2}$
14 $\dfrac{1}{x}+x = 2$
15 $\dfrac{1}{x-1}-\dfrac{1}{x+1} = 6.$

(ii) *By formula.* If the quadratic equation cannot be factorised, then we can use the formula

$$x = \frac{-b \pm \sqrt{b^2-4ac}}{2a}$$

where $ax^2+bx+c = 0$ is the equation to be solved.

Proof:

$$ax^2+bx+c = 0$$

Take c from both sides:

$$ax^2+bx = -c.$$

Divide by a:

$$x^2+\frac{bx}{a}=\frac{-c}{a}.$$

Add $\left(\frac{b}{2a}\right)^2$ to both sides:

$$x^2+\frac{bx}{a}+\left(\frac{b}{2a}\right)^2=-\frac{c}{a}+\left(\frac{b}{2a}\right)^2$$

$$\Rightarrow\left(x+\frac{b}{2a}\right)^2=\frac{b^2}{4a^2}-\frac{c}{a}$$

$$\Rightarrow\left(x+\frac{b}{2a}\right)^2=\frac{b^2-4ac}{4a^2}.$$

Take the square root of both sides:

$$x+\frac{b}{2a}=\frac{\pm\sqrt{b^2-4ac}}{2a}.$$

Subtract $b/2a$ from both sides:

$$x=\frac{-b}{2a}\pm\frac{\sqrt{b^2-4ac}}{2a}$$

$$\text{i.e. } x=\frac{-b\pm\sqrt{b^2-4ac}}{2a}.$$

EXAMPLE 22

$5x^2-8x+2=0$

$a=5,\ b=-8,\ c=2$

Substituting in formula:

$$x=\frac{-b\pm\sqrt{b^2-4ac}}{2a}$$

$$=\frac{-(-8)\pm\sqrt{(-8)^2-4\times5\times2}}{2\times5}$$

$$=\frac{8\pm\sqrt{64-40}}{10}$$

$$=\frac{8\pm4.899}{10}$$

$$= \frac{12.899}{10} \quad \text{or} \quad \frac{3.101}{10}$$

$= 1.29$ or 0.31 (to 2 decimal places).

EXERCISE 5k

Solve the following equations, giving your answers to 2 decimal places:

1 $x^2-2x-1=0$

2 $x^2+5x+3=0$

3 $x^2+18x+70=0$

4 $x^2-5x+2=0$

5 $p^2+7p+8=0$

6 $u^2-2u-7=0$

7 $7m^2-2m-3=0$

8 $5y^2+2y-1=0$

9 $4k^2+12k+5=0$

10 $3x^2-5x-6=0$

11 $4q^2+12q-5=0$

12 $8x^2-6x-3=0$

13 $2x^2=x+19$

14 $x^2+8x=3$

15 $4x^2=9x+1$

16 $7=8x-x^2$

17 $x+\dfrac{1}{3x}=2$

18 $\dfrac{1}{x+2}+\dfrac{1}{x+3}=\dfrac{2}{3}$

19 $u+1+\dfrac{3}{u+1}=8$

20 $\dfrac{y+2}{y+5}=\dfrac{2y+3}{y-1}$.

5.9 Simultaneous equations

$$x+y=5$$

Possible solutions for this equation are infinite, but what values solve this equation and the equation $x-2y=2$?

If we draw a graph of both of these equations (Fig. 5.1), we see that they cross at a unique point (4, 1).

If we substitute the values $x=4$ and $y=1$ into the two equations, we will find that they are both satisfied. The equations are said to be simultaneously true for these values of x and y.

To solve such equations, we can use the following method. We shall eliminate one of the unknowns by a process of adding or subtracting the two equations in the following manner.

EXAMPLE 23

$$3x+4y=7 \qquad (1)$$
$$3x-4y=-1 \qquad (2)$$

Adding (1) and (2)

$$6x=6$$
$$\Rightarrow x=1.$$

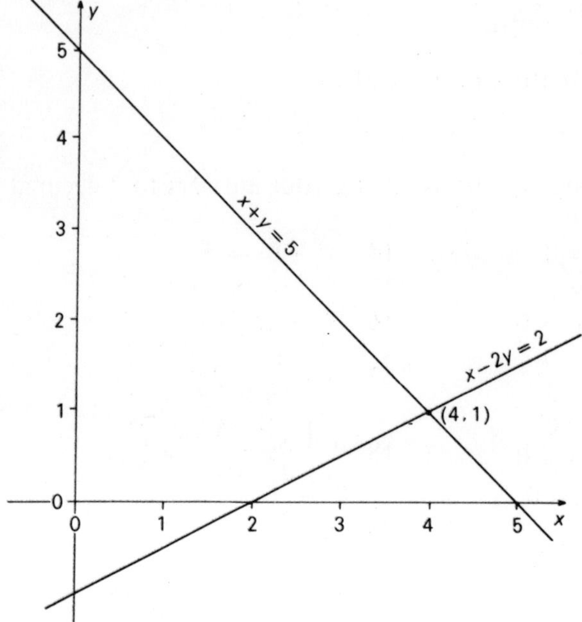

Fig. 5.1

Substituting in equation (1)

$$3 \times 1 + 4y = 7$$
$$\Rightarrow y = 1.$$

Checking in equation (2)

$$3 \times 1 - 4 \times 1 = -1. \quad \checkmark$$

EXAMPLE 24

$$3p - 2q = 4 \qquad (1)$$
$$2p + 3q = 7 \qquad (2)$$

We cannot eliminate either p or q by simple addition or subtraction this time, so we have to change the number which multiplies p or q. i.e. multiply equation (1) by 2

$$6p - 4q = 8 \qquad (3)$$

multiply equation (2) by 3

$$6p + 9q = 21 \qquad (4).$$

Subtract (3) from (4)

$$13q = 13$$
$$\Rightarrow q = 1.$$

Substituting in equation (3)

$$6p - 4 \times 1 = 8$$
$$\Rightarrow p = 2.$$

Checking in equation (4)

$$6 \times 2 + 9 \times 1 = 21. \quad \checkmark$$

EXERCISE 5I

Solve the following equations:

1 $x+y = 7$, $2x-y = 8$

2 $3x-2y = 4$, $5x-2y = 0$

3 $4x-3y = 5$, $7x-5y = 9$

4 $2a+b = 8$, $5a-b = 6$

5 $k+3l = 2$, $2k-4l = 1$

6 $3p+2q = 13$, $2p+3q = 12$

7 $4x-5y = 21$, $6x+7y = -12$

8 $3x+2y = 17$, $4y-x = -8$

9 $2a = 3b+5$, $5b = 3a-4$

10 $3m = 2n-7$, $3n = 13+6m$.

5.10 Change of subject of formulae

Consider the equation $y = 2x+3$. What is the value of x when $y = 5$? To answer this question we are solving the equation

$$5 = 2x+3.$$

Taking 3 from both sides:

$$5-3 = 2x$$
$$\Rightarrow +2 = 2x.$$

Dividing both sides by 2:

$$1 = x.$$

In the original equation, y was expressed in terms of x, but it may be useful to express x in terms of y. To do this, we follow the same procedure as in the solution to the above problem using the letter y instead of a numerical quantity, i.e.

$$y = 2x+3.$$

Taking 3 from both sides:

$$y-3 = 2x.$$

Dividing both sides by 2:

$$\frac{y-3}{2} = x \quad \left(\text{or } x = \frac{y-3}{2}\right).$$

x is now expressed in terms of y. We have changed the subject of the formula from y to x.

EXAMPLE 25

Make l the subject of the formula $A = \pi rl + \pi r^2$.

Subtract from both sides πr^2:

$$\therefore\ A - \pi r^2 = \pi rl.$$

Divide both sides by πr:

$$\frac{A - \pi r^2}{\pi r} = \frac{\cancel{\pi r} l}{\cancel{\pi r}}$$

or

$$l = \frac{A - \pi r^2}{\pi r}.$$

EXAMPLE 26

Make y the subject of the formula $x = \dfrac{y}{y-1}$.

Remove the denominator by multiplying both sides of the equation by $y-1$:

$$x(y-1) = \frac{\cancel{(y-1)}}{\cancel{(y-1)}} y$$

$$\Rightarrow x(y-1) = y.$$

Remove brackets:

$$xy - x = y.$$

Arrange all terms involving y on one side of the equation by subtracting y from both sides:

$$xy - x - y = 0.$$

Add x to both sides:

$$xy - y = x.$$

Factorising:

$$y(x-1) = x.$$

Dividing both sides by $(x-1)$:

$$y = \frac{x}{x-1}.$$

EXAMPLE 27

Make g the subject of the formula $T = 2\pi\sqrt{\frac{l}{g}}$.

Remove the square root by squaring both sides of the equation:

$$T^2 = 4\pi^2\frac{l}{g}$$

Remove the denominator by multiplying both sides of the equation by g:

$$gT^2 = 4\pi^2\frac{l}{\cancel{g}} \times \cancel{g}$$

$$gT^2 = 4\pi^2 l.$$

Divide by T^2:

$$\frac{a\cancel{T^2}}{\cancel{T^2}} = \frac{4\pi^2 l}{T^2}$$

$$\therefore\ g = \frac{4\pi^2 l}{T^2}.$$

EXERCISE 5m

Make x the subject of the formula in the following questions:

1 $a = x^2 - 5$

2 $ax + b = cx - d$

3 $a(x+1) = b(x+3)$

4 $\frac{1}{x} = \frac{a}{x-1}$

5 $b = \frac{x-c}{a}$

6 $\frac{x+a}{a} = b - a$

7 $\frac{1}{x} + \frac{bx}{x+1} = b$

8 $\frac{x}{x+a} = \frac{b}{c}$

9 $\frac{1}{x} = \frac{1}{a} - \frac{1}{b}$

10 $\dfrac{x-a}{x-b} = 5$

11 $\dfrac{1}{x} + b = c$

12 $y = mx + c$

13 $(x-a) = \dfrac{b}{x+a}$

14 $x^2 + y^2 = a^2$

15 $y = a\sqrt{\dfrac{x}{bc}}$

16 $b = \dfrac{4}{x}$

17 $b = \sqrt{\dfrac{x}{a+x}}$

18 $\sqrt{(x-a)} + b = \dfrac{1}{c}$

19 $x^2 = (x-a)(x-b)$

20 $y = \sqrt{\dfrac{x-a}{x-b}}.$

21 Make T the subject of the formula

$$I = \frac{PRT}{100}.$$

22 (i) Make f the subject of the formula

$$\frac{1}{f} = \frac{1}{u} + \frac{1}{v}.$$

(ii) Find f when $u = 0.5$ and $v = 0.75$.

23 (i) Make r the subject of the formula

$$v = \pi r^2 h.$$

(ii) Find r when $v = 20$ and $h = 2.5 (\pi = 3.14)$.

24 Make s the subject of the formula

$$v^2 = u^2 + 2gs.$$

25 Make u the subject of the formula

$$s = ut + \tfrac{1}{2}gt^2.$$

26 Make L the subject of the formula

$$T = \sqrt{\frac{L}{L+D}}.$$

27 Make d the subject of the formula

$$S = \frac{n}{2}\left\{2a + (n-1)d\right\}.$$

28 Make g the subject of the formula

$$t = 2\pi\sqrt{\frac{h^2+g^2}{h^2}}.$$

29 (i) Make R the subject of the formula

$$S = \pi(R+r)h.$$

(ii) Find R when $S = 235$, $r = 7.5$, $h = 9$, $\pi = 3.14$.

30 Make t the subject of the formula

$$m = \frac{1+at}{1-at}.$$

5.11 Abstract operations

In everyday life we are used to the symbols $+, -, \times$ and $\div$ and their uses, e.g. $a \div b$ means divide the quantity b into the quantity a. But what is meant by the symbol $*$ or the symbol $\circ$? What do we mean by $2 * 3$ or $(7 \circ 5) \circ 3$? Clearly there is no meaning, unless the operation is defined in terms of known operations. Usually these operations are the basic arithmetical operations above.

EXAMPLE 28

If $a * b$ denotes $a+b-ab$, find $2 * 3$.

$$\begin{aligned} 2 * 3 &= 2+3-2\times 3 \\ &= 5-6 \\ &= -1. \end{aligned}$$

EXAMPLE 29

If $a \circ b$ means $(a-b)/b$, find $(7 \circ 5) \circ 3$ and $7 \circ (5 \circ 3)$.

$$\begin{aligned} (7 \circ 5) \circ 3 &= \left(\frac{7-5}{5}\right) \circ 3 \\ &= \tfrac{2}{5} \circ 3 \\ &= \frac{\frac{2}{5} - 3}{3} \\ &= \frac{-2\frac{3}{5}}{3} \end{aligned}$$

$$= -\frac{13}{15}$$

$$\begin{aligned} 7 \circ (5 \circ 3) &= 7 \circ \left(\frac{5-3}{3}\right) \\ &= 7 \circ \tfrac{2}{3} \\ &= \frac{7-\frac{2}{3}}{\frac{2}{3}} \\ &= \frac{\frac{19}{3}}{\frac{2}{3}} = \frac{19}{2}. \end{aligned}$$

5.12 Properties of algebraic systems

Closure. A set S is said to be closed under the operation $*$ if for all $a, b \in \mathrm{S}$

$$a * b \in \mathrm{S}.$$

For example, the set of integers under addition.
Is the set of positive integers closed under subtraction?

Associative. The operation $*$ is said to be *associative* if for all $a, b, c \in \mathrm{S}$

$$(a * b) * c = a * (b * c).$$

For the set of integers,

$$a + (b + c) = (a + b) + c$$

e.g.

$$5 + (3 + 9) = (5 + 3) + 9 = 17.$$

EXAMPLE 30

For the set of integers, is $(a - b) - c = a - (b - c)$ always true?

Let us try some numbers: $a = 1$, $b = 2$, $c = 3$.

$$\begin{aligned} (a-b)-c &= (1-2)-3 \\ &= -1-3 = -4 \\ a-(b-c) &= 1-(2-3) \\ &= 1-(-1) \\ &= 1+1 = 2. \end{aligned}$$

Therefore, subtraction is not associative.

Look back at example 29. Clearly this is not associative. What can you say about multiplication and division?

Commutative. The operation $*$ is said to be *commutative* if for all $a, b \in S$

$$a * b = b * a.$$

For the set of integers,

$$a + b = b + a \qquad \text{(e.g. } 3 + 5 = 5 + 3\text{)}$$

but

$$a \div b \neq b \div a \qquad (3 \div 5 \neq 5 \div 3).$$

EXERCISE 5n

1 State whether the following sets are closed under the given operations:
- (i) {even numbers} under addition
- (ii) {odd numbers} under addition
- (iii) {odd numbers} under multiplication
- (iv) {prime numbers} under addition
- (v) {real numbers} under multiplication
- (vi) {1, 2, 3, 4} under subtraction.

2 Which of the following operations are associative?
- (i) division
- (ii) $*$, where $a * b = a^2 + b^2$
- (iii) matrix addition
- (iv) matrix multiplication
- (v) †, where $a \dagger b = \dfrac{a+b}{2}$
- (vi) set intersection
- (vii) $\#$, where $a \# b = a + b + ab$
- (viii) $\barwedge$, where $a \barwedge b = \dfrac{a+b}{a-b}$.

3 Which of the following operations are commutative?
- (i) multiplication
- (ii) $*$, where $a * b = a^2 + b^2$
- (iii) $\circ$, where $a \circ b = a^2 - b^2$
- (iv) multiplication of all 2×2 matrices
- (v) $\oplus$, where $a \oplus b$ denotes the average of a and b
- (vi) \$, where $a \,\$\, b = \dfrac{a^2 + b^2}{a + b}$

4 If $a * b$ means $a+3b$, find the values of $2 * 3$ and $3 * 2$. What can you say about the operation?

5 Given that $a * b$ means $a^2 - b^2$, calculate $4 * 5$.

6 If $a \oplus b$ denotes the lowest common multiple of a and b, find the value of $(2 \oplus 6) \oplus 3$ and $2 \oplus (6 \oplus 3)$.

7 If $a \dagger b$ denotes $\dfrac{a+b}{b}$, evaluate $3 \dagger (4 \dagger 5)$ and $(3 \dagger 4) \dagger 5$.

(i) Find a value y such that $y \dagger 2 = 7$.
(ii) Find a value x such that $3 \dagger x = 4$.

8 If the operation $*$ is defined such that $x * y = x^y$, calculate
(i) $3 * 2$ (ii) $2 * (3 * 2)$.

[W]

9 If the operation $\oplus$ is defined such that $x \oplus y = 2x + 3y$, calculate
(i) $2 \oplus 3$ (ii) $3 \oplus 2$ (iii) $(3 \oplus 2) \oplus 3$.
What do the results of (i) and (ii) show?

[W]

10 The operation $*$ is defined on the set of real numbers by the relation

$$a * b = a^2 + b^2 + 2ab.$$

Calculate (i) $2.2 * 1.8$ (ii) $1.8 * 2.2$.
Give an example to show that this operation is not associative.

Identity. A set S is said to have an *identity* $e \in \mathrm{S}$ under the operation $*$ if for all $a \in \mathrm{S}$

$$a * e = a = e * a.$$

If S is the set of integers, the identity under addition is 0, i.e.

$$3 + 0 = 3 = 0 + 3.$$

If S is the set of real numbers, the identity under multiplication is 1, i.e.

$$15.67 \times 1 = 15.67 = 1 \times 15.67.$$

Inverses. A set S is said to have the property of inverses under the operation $*$ if for all $a \in \mathrm{S}$ there is an $a^{-1} \in \mathrm{S}$ such that

$$a * a^{-1} = e = a^{-1} * a.$$

If S is the set of integers,

$$3 + (-3) = 0 = (-3) + 3.$$

-3 is the inverse of 3 under addition (and *vice versa*). Does the set S have the property of inverses under multiplication? If $3 \times a = 1$, then $a = \frac{1}{3}$. But $\frac{1}{3} \notin S$, so the set under the given operation does not possess the property of inverses.

5.13 Operation tables

We can also define algebraic operations in tabular form. The operation $*$ on the set $S = \{1, 2, 3, 4\}$ is defined as shown in the table:

$*$	1	2	**3**	4
1	2	3	4	1
2	3	4	**1**	2
3	4	1	2	3
4	1	2	3	4

To interpret $a * b$, we read across row a and down column b to the point of intersection, e.g. $2 * 3$ is the intersection of row '2' with the column '3'. From the table, (see entries in bold type), $2 * 3 = 1$.

Is the operation commutative? Can you see a quick way of deciding if the operation is commutative by looking at the table? Clearly the identity element is 4.

What are the inverses of the four elements?

EXAMPLE 31

The operation $\circ$ on the set $S = \{e, a, b, c\}$ is defined by the operation table

$\circ$	e	a	b	c
e	e	a	b	c
a	a	e	c	b
b	b	c	e	a
c	c	b	a	e

(i) Is the operation commutative?
(ii) Find the identity.
(iii) State the inverses.
(iv) Find y if $a \circ y = b$.

(i) The table is symmetrical about the diagonal from top left to bottom right, thus implying that the operation is commutative.
(ii) The identity element is e.
(iii) Inverses:

element	inverse	
e	e	$(e \circ e = e)$
a	a	$(a \circ a = e)$
b	b	$(b \circ b = e)$
c	c	$(c \circ c = e)$

Each element is a self-inverse.
(iv) $a \circ y = b$. Look along the row labelled a to the element b in the body of the table. The appropriate column is the value of y. Therefore $y = c$.

EXERCISE 5o

1 A combination is defined by the following table:

$\circ$	1	2	3	4
1	1	2	3	4
2	2	4	1	3
3	3	1	4	**2**
4	4	3	2	1

where the entry in bold type is $3 \circ 4$.
Solve the equations (i) $4 \circ x = 2$ (ii) $y \circ 2 = 4$.

2 The operation † is defined on S = {0, 1, 2, 3} by

$a \dagger b$ is the remainder of $\dfrac{a+b}{4}$.

Construct the operation table.
(i) Is S closed with respect to †?
(ii) Is † commutative on S?
(iii) What is the identity?

3 The diagram shows the composition table for an operation ? defined on {0, 1, 2}:

?	0	1	2
0	1	2	0
1	2	0	1
2	0	1	2

If $(2 ? 1) ? x = 0$, find x.

4 A combination $*$ is defined by the following table:

*	2	4	6	8
2	8	2	4	6
4	2	4	6	8
6	4	6	8	2
8	6	8	2	4

(i) State the identity element of the set {2, 4, 6, 8}.
(ii) Solve the equation $(8 * 2) * y = 4 * 6$.

5 Given that $a * b$ denotes $a^2 - b^2$
(i) evaluate $2\frac{1}{2} * 1\frac{1}{2}$ (ii) find a when $a * 7 = 1 * a$.
[C]

6 Given that $a * b$ denotes $\dfrac{a+b}{a}$, evaluate
(i) $2 * 1$ (ii) $1 * 2$ (iii) $(2 * 1) * (1 * 2)$.
[C]

7 If the operation $*$ is defined as $a * b$ meaning $a + 2b$, evaluate $5 * 3$ and $3 * 5$.
Find values of x and y such that $x * y = y * x$.

8 Given that $x\,!\,y$ means the remainder when x is divided by y, evaluate $7!4$ and $13!3$.
Find two possible values of y such that $12!\,y = 2$.
Find a possible value of x such that $x!7 = 3$.
Why is there no solution to the equation $z!5 = 21$?

9 p and q are integers and $p * q$ denotes the remainder when the product pq is divided by 10. Find a value of p, greater than 4, such that

$$p * p = (8 * p) + 3.$$

[C]

10 In this question $a \oplus b$ means $a + b + 2$ and $a * b$ means $\dfrac{a - 2b - 2}{b + 2}$.
(i) Show algebraically that $\oplus$ is associative and commutative.
(ii) (a) Evaluate $(25 * 7) * 1$ and $25 * (7 * 1)$.
(b) Evaluate $(-1 * 0)$ and $(0 * -1)$.
What do these results show?
(iii) Find an identity element for $\oplus$ (i.e. a number x such that $a \oplus x = a$).

6 Inequalities

6.1 Inequations

What is the difference between the solution of

$$3x+2 = 8 \qquad (1)$$

and

$$3x+2 < 8? \qquad (2)$$

Clearly, the work on solution of equations from the last chapter will help us to solve (1), but how do we solve (2)? An expression of the form $3x+2<8$ is known as an *inequation* and to solve it we need to look at the properties of the inequality sign.

6.2 Properties of '<' and '>'

$a<b$ means a is less than b
$c>d$ means c is greater than d.
For example: $5<7, -3>-6$.

(i) *Addition and subtraction of a constant*

$$5+4<7+4, -3+2>-6+2$$
$$5-3<7-3, -3-4>-6-4$$

Are these statements correct? Try some other values yourself. You should find that the addition or subtraction of the same quantity to both sides does not alter the identity. That is, if $a<b$, then $a+c<b+c$ or $a-c<b-c$.

EXAMPLE 1

Find the solution of

$$x+3<7.$$

Subtracting 3 from both sides,

$$x+3-3<7-3$$
$$\Rightarrow x<4.$$

(ii) *Multiplication and division by a constant*

$$3\times 5<3\times 7,\ 3\times -3>3\times -6$$
$$\tfrac{1}{3}\times 5<\tfrac{1}{3}\times 7,\ \tfrac{1}{3}\times -3>\tfrac{1}{3}\times -6$$

Are these statements correct? What happens when you multiply or divide by a negative quantity? For example,

$$5<7, \text{ but is } -3\times 5<-3\times 7?$$
$$-3>-6, \text{ but is } (-\tfrac{1}{3})\times -3>(-\tfrac{1}{3})\times(-6)?$$

By making up some examples of your own you should find that multiplying or dividing by a positive quantity, the sign remains the same, but if the quantity is negative the inequality sign alters. For example, $2<5$, but $-3\times 2>-3\times 5$.

EXAMPLE 2

Solve the inequation $3x+2<8$.

Subtracting 2 from both sides:

$$3x+2-2<8-2$$
$$\Rightarrow 3x<6.$$

Dividing both sides by 3:

$$x<2.$$

EXAMPLE 3

Solve the inequation $4-2x>5$.

Subtracting 4 from both sides:

$$-4+4-2x>5-4$$
$$\Rightarrow -2x>1.$$

Dividing both sides by -2:

$x<-\frac{1}{2}$ (the inequality changes round).

EXAMPLE 4

$$3x+7<2x-5.$$

Subtracting $2x$ from both sides:

$$3x-2x+7<-5+2x-2x$$
$$\Rightarrow x+7<-5.$$

Subtracting 7 from both sides:

$$x+7-7<-5-7$$
$$\Rightarrow x<-12.$$

EXERCISE 6a

Solve the following inequations for x:

1 $x+3<17$
2 $2-x>15$
3 $3>5x-2$
4 $2+2x<3$
5 $7<4-3x$
6 $x-7>2x+3$
7 $5-2x>3x-7$
8 $8-3x<1-6x$
9 $10+6x<4x-1$
10 $-7x<-3+2x$
11 $3(x+1)>15$
12 $2(x-3)<x+5$
13 $5(2-x)<16+x$
14 $4(x+2)>3(x-3)$
15 $2(x-1)<5-x$
16 $10(x-5)>7(6-x)$
17 $5(4-x)>3(6-x)$
18 $4(2x+5)<3(2+3x)$
19 $3(2-x)+5(2x+3)<4(x-3)$
20 $2(2-5x)<7(3-4x)$.

6.3 Solution sets

The possible values which satisfy an inequation can be considered as a *solution set*. For example, $3x+2<8$ has the solution set $\{x:x<2\}$.

The solution set is dependent on the possible values of the variable (the universal set). For example, if in the above problem x is a positive integer, the solution set is $\{1\}$.

EXAMPLE 5

Find the solution set of $2x-3<3x+4$ if x is a negative integer.

Subtracting $2x$ from both sides:

$$2x-2x-3<3x-2x+4$$
$$\Rightarrow -3<x+4.$$

Subtracting 4 from both sides:

$$-3-4<x+4-4$$
$$\Rightarrow -7<x.$$

The solution set is $\{-6,-5,-4,-3,-2,-1\}$.

EXAMPLE 6

Find the solution set of $\frac{1}{x}<\frac{1}{3}$ $(x\neq 0)$.

Multiply both sides by 3:

$\frac{3}{x} < 1.$

If $x > 0$, $3 < x$: the solution set is $\{x:x > 3\}$.
If $x < 0$, $3 > x$: the solution set is $\{x:x < 0\}$ (note < and > change round).

Therefore, the complete solution set is $\{x:x < 0 \text{ and } x > 3\}$.

EXAMPLE 7

Find the solution set of $-15 < 2x - 3 < 7$.

Consider this as two problems:

$-15 < 2x - 3$ (A)	and	$2x - 3 < 7$ (B).
Adding 3 to both sides:		
$-12 < 2x$		$2x < 10$.
Dividing both sides by 2:		
$-6 < x$		$x < 5$.
Solution set is $\{x: -6 < x\}$		Solution set is $\{x:x < 5\}$.

For both inequations to be satisfied the solution set is $\{-6 < x < 5\}$.
Rather than separating into two inequalities we can save a little work as follows:

$$-15 < 2x - 3 < 7.$$

Adding 3 to all parts:

$$-12 < 2x < 10.$$

Dividing all parts by 2:

$$-6 < x < 5.$$

The solution set is $\{-6 < x < 5\}$.

EXAMPLE 8

Find the solution set of $2x - 3 \leq 3x + 4$ if x is a negative integer.

This inequation reduces to $-7 \leq x$.
The solution set is $\{-7, -6, -5, -4, -3, -2, -1\}$ (-7 is included because of the equality in the expression).

EXERCISE 6b

Find the solution set of the following inequations:

1 $17 - x > 2x + 5$

2 $3-2x<5-4x$
3 $3x-3 \leq x+15$ (x is a positive integer)
4 $2-x<15+4x$ (x is a negative integer)
5 $2<x+5<15$
6 $3<2(x-5)<7$ (x is an integer)
7 $-14<3x-5<7$
8 $-3<5(x+4)<9$
9 $-2 \leq 5-x \leq 3$ (x is an integer)
10 $-3<5(x+4)<9$ (x is a positive integer)
11 $\frac{1}{x}>2$
12 $\frac{2}{5}<\frac{1}{x}$
13 $\frac{2}{3}<\frac{3}{x-1}$ $(x \neq 1)$
14 $\frac{1}{2-x}<\frac{1}{4}$ $(x \neq 2)$.

6.4 Quadratic inequalities

EXAMPLE 9

Find the range of possible values of x for which $x^2-5x-6<0$.
Factorising the quadratic:

$(x-6)(x+1)<0.$

This can only be true if one of the factors is positive and the other negative. The possibilities are:

(i) $x-6>0$ and $x+1<0$.
(ii) $x-6<0$ and $x+1>0$.

(i) would necessitate $x>6$ and $x<-1$, which is impossible.
(ii) would necessitate $x<6$ and $x>-1$. It is possible to satisfy both these conditions. The range of possible values is $-1<x<6$. Therefore, the solution set is $\{-1<x<6\}$.

EXAMPLE 10

Find the range of possible values of x for which

$x^2-8x+16<36.$

Subtracting 36 from both sides:

$$x^2 - 8x - 20 < 0.$$

Factorising the quadratic:

$$(x-10)(x+2) < 0.$$

The possibilities are:

(i) $x - 10 > 0$ and $x + 2 < 0$.
(ii) $x - 10 < 0$ and $x + 2 > 0$.

(i) would necessitate $x > 10$ and $x < -2$, which is impossible.
(ii) would necessitate $x < 10$ and $x > -2$. Both these conditions can be satisfied. The range of possible values is $-2 < x < 10$. Therefore, the solution set is $\{-2 < x < 10\}$.

EXERCISE 6c

Solve the following inequations:

1 $x^2 - 3x - 4 < 0$
2 $x^2 + 3x - 10 < 0$
3 $x^2 - 7x + 6 < 0$
4 $x^2 - 11x + 18 > 0$
5 $6 - x - x^2 > 0$
6 $x^2 - 12x + 28 < 8$
7 $x^2 - 5x + 3 < -3$
8 $x^2 + 6x - 12 > -5$
9 $x^2 - 12 < 3$
10 $2x^2 - 5x + 11 < 9$.

6.5 Further examples

EXAMPLE 11

If $3 \leq P \leq 4.5$ and $1.5 \leq Q \leq 9$, find the maximum value of (i) PQ, (ii) P/Q.

(i) maximum value of PQ = maximum value of P $\times$ maximum value of Q

$$= 4.5 \times 9$$
$$= 40.5.$$

(ii) maximum value of $\dfrac{P}{Q} = \dfrac{\text{maximum value of } P}{\text{minimum value of } Q}$

$$= \frac{4.5}{1.5}$$
$$= 3.$$

EXAMPLE 12

Solve the inequation $3x - 1 \leq 2x + 17 \leq 3x$.

Solving as two separate inequalities:

(i) $3x-1 \leq 2x+17$ $\Rightarrow x \leq 18$ (ii) $2x+17 \leq 3x$ $\Rightarrow 17 \leq x$

(i) and (ii) $\Rightarrow$ $17 \leq x \leq 18$.

HARDER EXAMPLES 6

1 If $3 \leq b \leq 7$ and $2 \leq c \leq 5$, what is the minimum value of $\dfrac{2b}{c+2}$?

2 If $-2 \leq x \leq 5$ and $3 \leq y \leq 7$, find the maximum value of (i) $x+y$ (ii) $x-y$.

3 If $a = 0.32$ and $b = 5.71$ are correct to two decimal places, find the maximum values of (i) $a+b$ (ii) ab.

4 If $x = 4.8$ and $y = 3.7$ are correct to one decimal place, find the limits a and b such that $a < x+y < b$.

5 Solve the inequation $4x + \dfrac{(3x+1)}{3} < \frac{1}{2}$.

6 Solve the inequation $\frac{1}{2}x + \frac{2}{3}(x-5) < 3$.

7 Find the solution set of the inequation $n^2 - 3n - 4 \leq 0$ where n is integral.

[JMB]

8 Find the solution set of the inequation $6 - 5x - x^2 > 0$.

9 If S = $\{n : n$ is a positive integer and $\dfrac{n-7}{n} > \frac{3}{4}\}$ what is the least element of S?

10 Find a fraction p/q which satisfies the inequality $\dfrac{1}{\sqrt{3}} < \dfrac{p}{q} < \dfrac{1}{\sqrt{2}}$.

11 Find the integral values of x which satisfy $-3 \leq 5x - 7 \leq 35$.

12 If r is an integer such that $800 < 4r^3 < 1000$, find r.

[C]

13 If $2x - 1 \leq 7 + x \leq 3x + 1$, find a and b such that $a \leq x \leq b$.

14 If $5x - 2 < \frac{1}{3}(x+2) < 3x + 7$ and x is an integer, find the smallest and largest value of x.

15 A = {prime numbers}
B = $\{x : x$ is an integer and $-3 \leq 2x - 5 \leq 27\}$.
Find $A \cap B$.

16 Find an integer n such that $n^2 \leq \sqrt{300} \leq (n+1)^2$.

17 Write down any pair of positive integers a and b satisfying the following inequalities: $a + b < 5$, $a - b > 1$.

18 If A $= \{(x, y): x$ and y are positive integers and $x+y \leq 5\}$, find $n(\text{A})$.

19 The measurements of a given rectangle are correct to one decimal place. If the length is 3.5 cm and the breadth is 4.5 cm, express the smallest possible area as a percentage of the largest possible area.

20 If $2(x+1) \leq x+7 \leq 5x+2$, express the smallest possible value of x as a percentage of the largest possible value.

21 F is the set of all fractions p/q which lie between 0 and 1, where p and q are positive integers having no common factor and $2 \leq q \leq 5$. List all the members of F and indicate the smallest and the largest members.

[C]

22 The universal set is the set of positive integers. Given that A $= \{x: 14 < 3x < 27\}$ and B $= \{x: x+7 < 17 < x+11\}$, list all the members of $\text{A} \cup \text{B}$ and $\text{A} \cap \text{B}$, stating clearly which is which.

[C]

7 Area and volume

It is assumed here that the reader has a simple idea of the concept of area and volume.

7.1 Areas of simple plane figures

The formulae for finding the areas of some simple plane figures are given in Fig. 7.1.

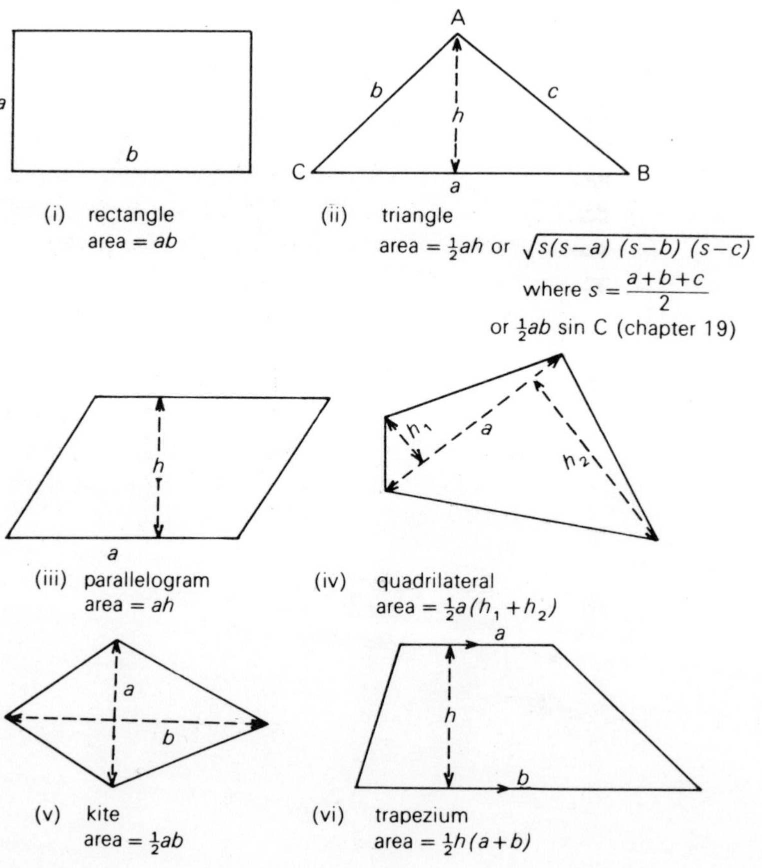

Fig. 7.1

7.2 Compound areas

To work out the area of a more complicated shape, you must divide it up into simple shapes whose areas you can easily find. In Fig. 7.2 the shape has been divided into trapezia by 3 parallel lines.

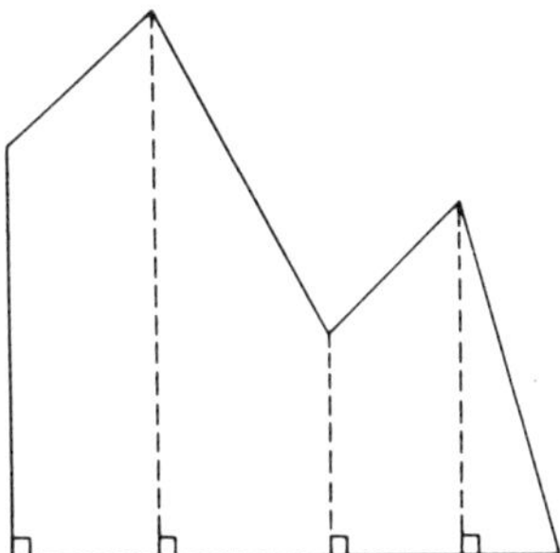

Fig. 7.2

7.3 The circle

Figure 7.3 shows a circle of radius 3 cm which has 8 isosceles triangles inscribed inside the circle, OA_1A_2, OA_2A_3, etc., and 8 similar triangles with bases just touching the outside of the circle, OB_1B_2, OB_2B_3, etc. By accurate drawing, or by more advanced trigonometry, we find that the sum of the lengths A_1A_2, A_2A_3, etc., is 18.4 cm, and the sum of the lengths B_1B_2, B_2B_3, etc., is 19.9 cm.

Therefore the circumference of the circle lies between 18.4 and 19.9 cm. If we increase the number of triangles we get closer and closer to the circumference of the circle. The table (Fig. 7.4) gives a list of circumferences obtained using this method for various radii.

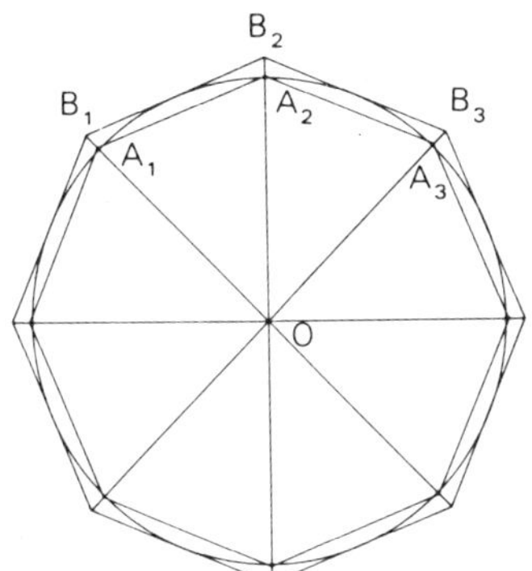

Fig. 7.3

radius r (cm)	circumference C (cm)	$C \div 2r$
2	12.4	3.1
3	19.2	3.2
4	24.8	3.1
5	32.0	3.2
6	38.4	3.2

Fig. 7.4

The final column in this table has been obtained by dividing the circumference of the circle by its diameter ($2r$, i.e. twice the radius). In each case, you will notice that the answer is approximately the same, the average being 3.16. This suggests that circumference $= 3.16 \times$ diameter. The value of 3.16 is an approximation to one of mathematics' most important numbers, π. The method given here is a fairly crude method, and π has been calculated to many hundreds of decimal places. It is a non-terminating decimal. For our purpose $\pi = 3.142$ (sometimes taken as $\frac{22}{7}$).

$$\text{Circumference } C = \pi d \quad (d = \text{diameter})$$
$$= 2\pi r.$$

7.4 The area of a circle

Proceeding as in section 7.3, we find that

total area of larger triangles = 29.8 cm^2
total area of smaller triangles = 25.5 cm^2.

It follows that the area of the circle lies somewhere between these two values.

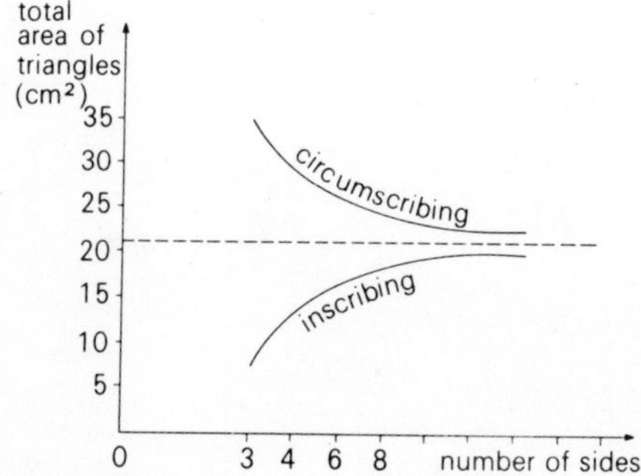

Fig. 7.5

The graph in Fig. 7.5 shows how the areas approach a limit when the method of section 7.3 is applied. It is found that this limiting value is 28.26 cm^2, which is 3.14 × the square of the radius. Our friend π appears again.

Area of a circle $= \pi \times r^2$.

EXERCISE 7a

In questions 1–10, take $\pi = \frac{22}{7}$, and find the circumference and area of the circle.

1 radius = 7 cm
2 radius = 3.5 cm
3 radius = 21 cm
4 radius = 2 m
5 diameter = $1\frac{3}{4}$ cm
6 diameter = $2\frac{1}{3}$ cm
7 radius = 1.4 cm
8 radius = 77 m
9 diameter = 84 cm
10 radius = 0.07 m.

In questions 11–20, take $\pi = 3.142$, and find the circumference and area of the circle, giving your answer correct to 2 decimal places.

11 radius = 1.3 cm
12 diameter = 6.3 cm
13 radius = 1.72 m
14 diameter = 5.83 cm
15 radius = 4.7 cm
16 radius = 5.85 m
17 diameter = 9.05 cm
18 radius = 19.86 cm
19 radius = 8.11 m
20 diameter = 16.3 cm

7.5 Compound shapes

Shapes which have circular parts are divided up in the same way as before. The shape in Fig. 7.6 can easily be divided into a trapezium, two semi-circles, and a rectangle.

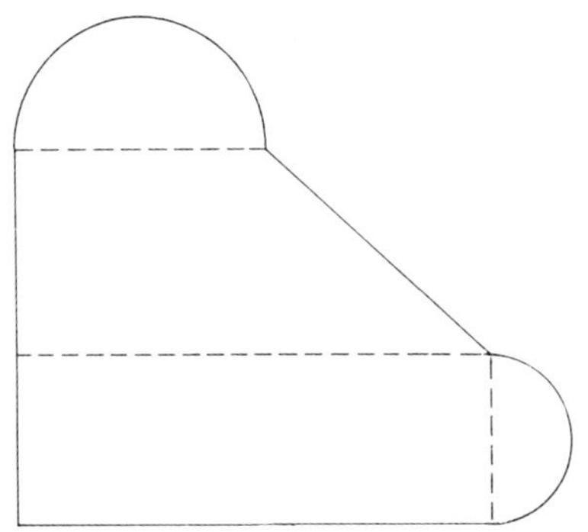

Fig. 7.6

7.6 Surface areas of simple shapes

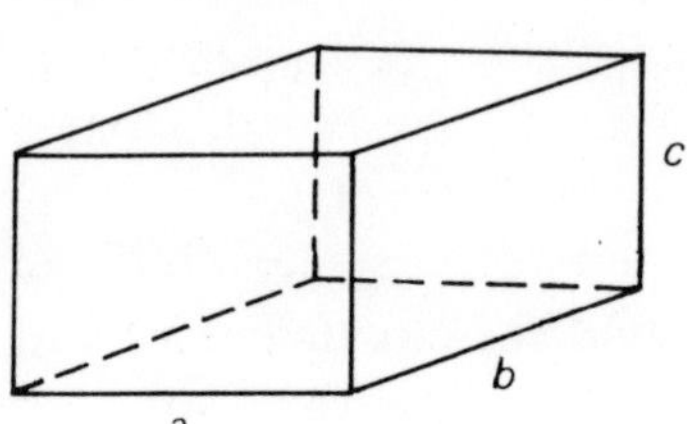

The cuboid has 2 faces which measure $b \times c$, 2 which measure $a \times c$, and 2 which measure $a \times b$.

$\therefore$ surface area $= 2(ab + ac + bc)$.

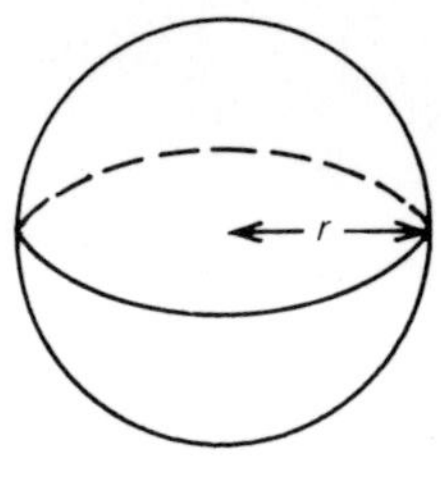

Sphere: surface area $= 4\pi r^2$ [too difficult to prove here].

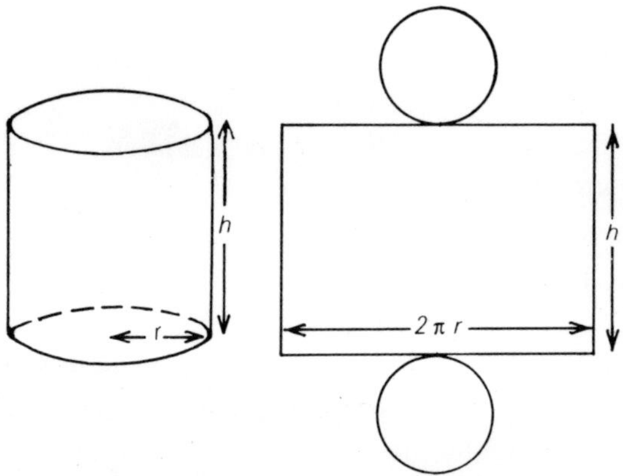

Cylinder: curved surface area $= 2\pi rh$.
Total surface area including the ends $= 2\pi r(r + h)$.

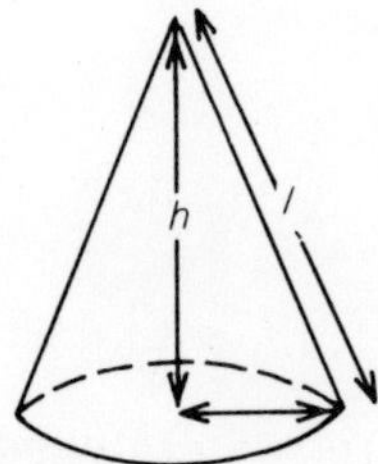

Curved surface area of a cone $= \pi rl$.

Fig. 7.7

EXERCISE 7b

1 Find the surface area of a cube of side 8 cm.

2 A rectangular box without a lid has a base which measures 4 cm by 11 cm, and the box is 8 cm high. Find its surface area.

3 10 identical spheres of radius 7 cm are to be painted. What is the total area of surface to be painted? $[\pi = \frac{22}{7}]$

4 Find the surface area of a sphere of diameter 8.72 cm $[\pi = 3.14]$. Give your answer correct to two decimal places.

5 A tin can, closed at both ends, has height 11 cm and base radius 4 cm. The metal is cut from a sheet which measures 12 cm by 50 cm. What area of metal remains, assuming no overlap is necessary to solder the tin together? $[\pi = 3.14]$

6 The curved surface area of a cone is 100 cm^2. If the radius of the base is 4 cm, find the slant height $[\pi = 3.14]$.

7.7 Volumes of regular shapes

Using the information given in section 7.6,

(i) volume of cuboid $= abc$

(ii) volume of sphere $= \frac{4}{3}\pi r^3$

(iii) volume of cylinder $= \pi r^2 h$

(iv) volume of cone $= \frac{1}{3}\pi r^2 h$.

7.8 Shapes with constant cross-section

Any shape which has two parallel ends which are perpendicular to the sides and which has the same cross-section throughout has a volume given by Volume $= Ah$ where A is the area of the cross-section.

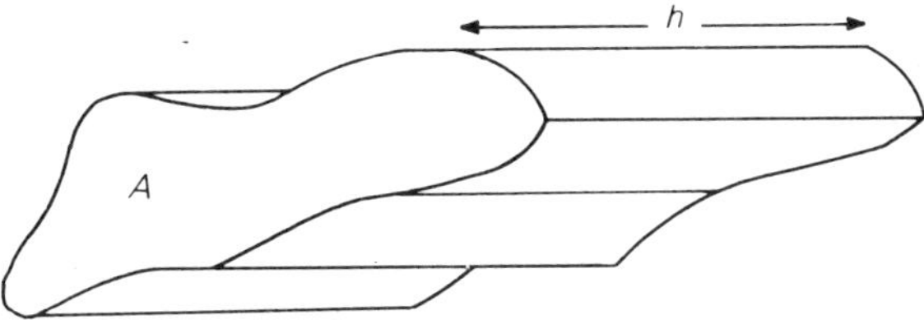

Fig. 7.8

In the same way that more involved areas were treated, the volume of a more complicated shape must be broken down into simpler parts.

7.9 Pyramids

Figure 7.9 shows a pyramid with an irregular-shaped base of area A. Its volume is given by Volume $= \frac{1}{3}A \times h$, where h is the perpendicular height.

If the base is triangular in shape, the pyramid is called a tetrahedron.

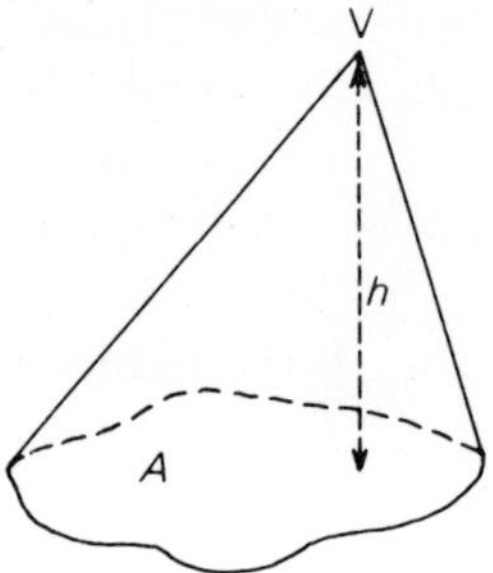

Fig. 7.9

EXERCISE 7c

Take $\pi = 3.14$ unless stated otherwise.

1 Find the volume of a cylinder, base radius 14 cm and height 3 cm ($\pi = \frac{22}{7}$).
2 A block of metal is in the form of a cuboid with measurements 5 cm × 6 cm × 2 cm. A hole of radius 1 cm is drilled to a depth of $\frac{1}{2}$ cm in one of the faces. Find the volume of metal which remains.
3 A sphere of radius 21 cm is melted down and recast into a cube. What is the length of the side of the cube? ($\pi = \frac{22}{7}$)
4 Find the volume of a cone of height 10 cm, base radius 5 cm.
5 A pipe of thickness 0.1 cm has an internal radius of 1 cm. Find the volume of a pipe 1 m in length.
6 A sphere has a volume of 100 cm^3. Find its radius.
7 A pyramid has a square base of side 4 cm. Find its volume, given that the height is numerically equal to the area of the base.
8 The sides of the base of a tetrahedron are 3 cm, 3 cm and 4 cm. If the height of the tetrahedron is 4 cm, find its volume.
9 A cylinder of height 12 cm has a volume of 60 cm^3. Find the radius of the base.
10 Find the volume of a sphere whose surface area is 1 m^2.

EXERCISE 7d

1 Calculate the areas of the shapes given in Fig. 7.10. All measurements are in cm. Take $\pi = 3.14$.

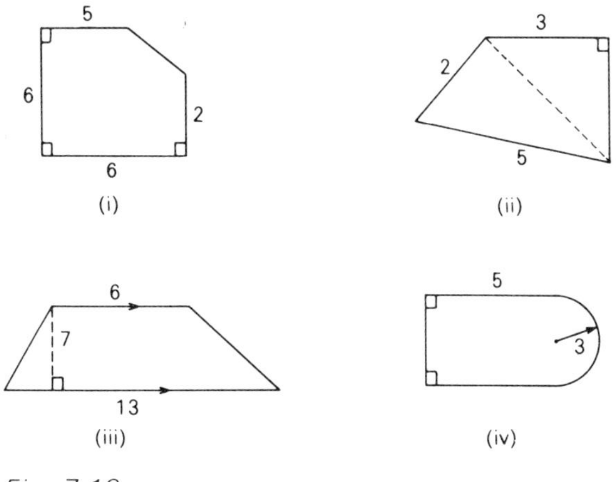

Fig. 7.10

2 Taking $\pi = \frac{22}{7}$, find the areas of the circles with the following radii:
(i) 7 cm; (ii) $3\frac{1}{2}$ m; (iii) 0.07 mm; (iv) 4.9 m.

3 Taking $\pi = 3.14$, calculate the areas of the circles with the following radii, giving your answers correct to 3 significant figures:
(i) 5.7 cm; (ii) 0.013 m; (iii) 58 mm; (iv) 6.15 cm.

4 Find formulae for the areas of the shapes given in Fig. 7.11.

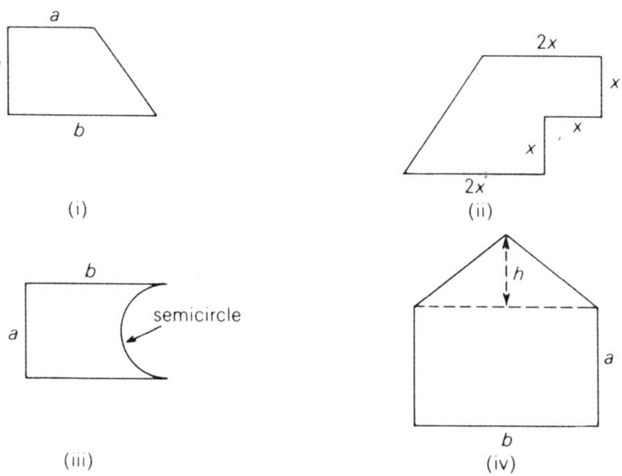

Fig. 7.11

5 A floor which measures 15 m by 8 m is to be laid with tiles measuring 50 cm × 25 cm. How many are required? A carpet is laid in the room so that a space of 1 m exists between its edges and the edges of the room. What fraction of the floor is uncovered?

6 

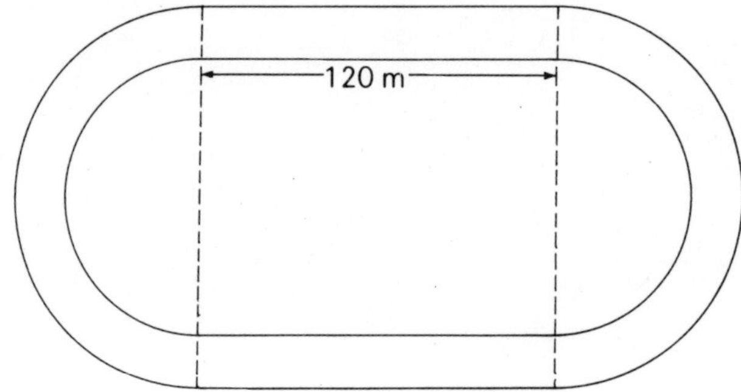

Fig. 7.12

Figure 7.12 shows a diagram of a running track. The distance round the inside of the track is 400 m. If the ends are circular, and the width of the track is 4 m, find the area of the running surface ($\pi = 3.14$).

7 A ball bearing of radius 1 cm, is dropped into a measuring cylinder of radius 4 cm containing water. If the original reading on the cylinder is 60 cm^3, what will be the reading after the ball bearing is fully immersed?

8

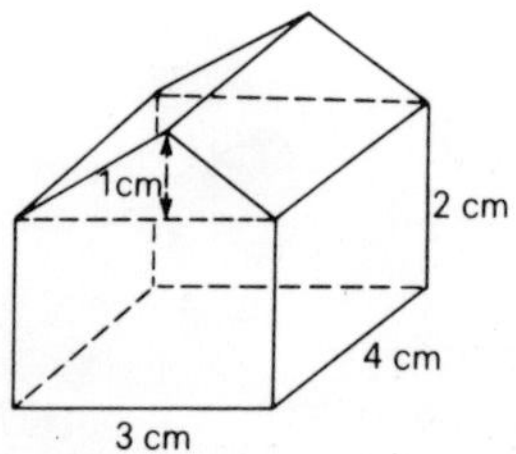

Fig. 7.13

Figure 7.13 shows the dimensions of a model house. What is its volume?

9 Water flows along a pipe of radius 0.6 cm, at 8 cm/s. This pipe is draining the water from a tank which holds 1000 litres when full. How long would it take to completely empty the tank?

10 A cube of volume V_1 is cut from a sphere of volume V_2. What is the largest value of V_1/V_2?

HARDER EXAMPLES 7

1 One diagonal of a rhombus is of length 12 cm. If the length of the sides of the rhombus is 6.5 cm, find the area of the rhombus.

2

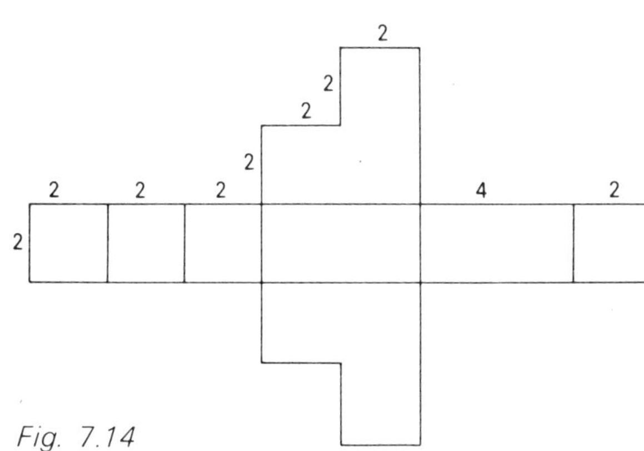

Fig. 7.14

Figure 7.14 shows the net for constructing a certain shape; all measurements are in cm. What is the volume of this shape?

3 The rectangle R_1 has length l_1 and breadth b_1; the rectangle R_2 has length l_2 and breadth b_2. The length and breadth of the rectangle R are respectively L and B. If $3l_1 = l_2$, $2l_2 = 5L$, $3b_1 = 2b_2$, and $4B = b_1$, express the areas of the three rectangles as a ratio in its simplest form.

4 A rectangle PQRS has PQ = 10 cm and QR = 24 cm. Calculate the area of the equilateral triangle PRT.

5 PQRS is a quadrilateral in which the diagonal PR = 6.0 cm, angles PRQ and RPS are right angles, QR = 1.2 cm, PS = 4.0 cm.

(i) Calculate the area of the quadrilateral.

(ii) Tiles of the size and shape of PQRS are fitted together to cover the plane. Draw, on squared paper, a diagram showing four neighbouring tiles of this tessellation. The diagram should be full size.

(iii) About how many tiles would be required to cover a square of side 10 metres?

[O&C]

6 What is the volume of the largest ball that can be placed into a square box whose inside measurements are 14 cm × 14 cm

× 14 cm? What is the volume of the space remaining in the box? ($\pi = \frac{22}{7}$)

7 Two right circular cones X and Y are made, X having three times the radius of Y, and Y having half the volume of X. Calculate the ratio of the heights of X and Y.

8 Two cooking pans are cylindrical in shape and have the same height. The smaller pan has diameter 15 cm and holds 1 litre. Calculate the diameter of the larger pan which holds 2 litres.

[SU]

9 A roll of copper wire is made of wire 200 metres long with a circular cross-section of diameter 1.4 millimetres. Using $\pi = \frac{22}{7}$, calculate the volume of the coil.

If 1 cubic centimetre of copper weighs 8.8 grams, calculate the weight of copper in the roll of wire in kilograms correct to three significant figures.

[W]

10 The triangle ABC, in which AB = 9 cm, BC = 15 cm and CA = 16 cm, lies in a horizontal plane. AV is a vertical line of length 12 cm. Calculate

(i) the area of triangle ABC;
(ii) the volume of the pyramid VABC;
(iii) the area of the triangle VBC;
(iv) the length of the perpendicular from A to the plane VBC.

[O]

11 A sky-rocket consists of a cone of height a and base radius r which is attached to a cylindrical part of height h and radius r. Write down an expression for the volume V in terms of r, a and h.

Show that $a = \dfrac{3(V - \pi r^2 h)}{\pi r^2}$.

If $a = 3h$, find the ratio of V to the volume of the cylinder.

12 A wooden cone has a circular base of radius 5 cm, and its vertical height is 12 cm. Calculate the slant height of the cone.

The cone stands with its axis vertical on a table. A thin hoop of radius 3 cm is placed on the cone so that it lies horizontally touching the curved surface of the cone all round. Calculate

(i) how far the centre of the hoop is below the vertex;
(ii) what fraction of the volume of the cone lies above the level of the hoop;
(iii) what fraction of the curved surface of the cone lies above the level of the hoop.

[O]

8 Ratio and percentage

Suppose we draw a line of length 50 cm. What fraction of the line is 12 cm? Clearly this can be written as $\frac{12}{50}$.

We can also write this as 'the ratio of the 12 cm line to the 50 cm line is 12 to 50'. The alternative notation for a ratio uses a colon (:) and $\frac{12}{50}$ is written as 12:50.

As $\frac{12}{50}$ is equivalent to $\frac{6}{25}$, the ratio can also be written as 6:25 which has now reduced the ratio to its simplest form.

The ratio of the longer line to the shorter line is 50 cm to 12 cm which in the alternative notation is 25:6.

8.1 Comparison of ratios

EXAMPLE 1

Express the following ratio in its simplest form: 30 cm^3 to 1 litre.

$$\text{The ratio is } 3\not{0}:100\not{0} \quad (\text{since 1 litre} = 1000\ \text{cm}^3)$$
$$= 3:100.$$

EXAMPLE 2

In one year Alan saves £450, Bill saves £300 and Charles saves £100.

(i) What is the ratio of Alan's savings to Bill's?
(ii) What is the ratio of Bill's savings to Charles's?

(i) Ratio is £450:£300 = 3:2.
(ii) Ratio is £300:£100 = 3:1.

As the ratio 3:2 is equivalent to 9:6 and the ratio 3:1 is equivalent to 6:2, we can compare all three savings in the ratio 9:6:2. Check that the ratio of Alan's savings to Charles's is 9:2.

EXAMPLE 3

Decrease £3.00 in the ratio 4:5.

The new amount is in the ratio of 4 to 5 to £3.00.

New amount $= \frac{4}{5} \times$ £3.00 = £2.40.

EXERCISE 8a

Express the following ratios in their simplest forms.

1 40 cm : 3 m
2 60 g : 5 kg
3 550 ml : 1.20 litre
4 $3^2 : 6^2$
5 13.65 km : 22.1 km
6 24 kg : 280 kg
7 4p : £3
8 £4.12½ : £7.62½
9 96 mm : 123 mm
10 12 minutes : 1½ hours
11 30° : 45° : 135°
12 log 2 : log 4 : log 8.

Express the following ratios in the form (a) $1 : k$, (b) $k : 1$.

13 22.2°C : 37.5°C
14 £7.40 : £12.80
15 $2^4 : 4^3$
16 18 minutes : 3 hours
17 1.54 g : 1.12 g
18 37 ml : 3.5 litres

19 Increase £3.50 in the ratio 9:7.
20 Increase 4.40 kg in the ratio 15 : 11.
21 Decrease 56 in the ratio 5 : 8.
22 Decrease 51.3 m in the ratio 2 : 3.
23 If $2 : x = 5 : 8$, find x.
24 If $3 : (y-2) = 2 : 5$, find y.
25 If $a : b = 2 : 3$ and $b : c = 4 : 5$, find the ratio $a : c$.
26 If $p : q = 7 : 9$ and $q : r = 4 : 3$, find the ratio $p : q : r$.
27 The ratio of the three highest peaks in the Mathemanian Alps to each other is 2 : 5 : 8. The smallest mountain is 1200 metres in height. Find the height of the other two mountains.
28 The circumferences of two circles are in the ratio 1 : 5. The smaller circle has a radius of 7 cm. What is the radius of the larger circle?
29 The radius of a circle was doubled. Express the area of the original circle as a ratio of the new one.
30 A block of wood 8 cm by 5 cm by 2.5 cm weighs 350 g. Find the weight of a block of the same material with dimensions 6 cm by 3.5 cm by 1 cm.
31 At the beginning of the school year, Alan and Bill had their heights measured. The former was 145 cm and the latter 155 cm. In the year Alan grew 7 cm and Bill grew 8 cm. Which ratio of growth to height is the greater?
32 It was decided to construct a scale model of a cube by reducing all the lengths to a quarter of the original lengths. Find the ratio of the volume of the scale model to that of the actual cube.

8.2 Proportional parts

If two people A and B divide a cake into 5 equal parts and A takes 3 pieces to B's 2 pieces, then the ratio of A's quantity to B's is 3 : 2. If, on the other hand, we are given the information that the ratio of the quantities of cake for A and B are 3 : 2 then we can consider the whole cake having been divided into 5, 10, 15, 20, . . . pieces, with A receiving 3, 6, 9, 12, . . . pieces and B receiving 2, 4, 6, 8, . . . pieces. In other words A receives three-fifths and B receives two-fifths.

Similarly if three people share a cake in the ratio 2 : 5 : 7 then the proportions of the cake are $\frac{2}{14}$, $\frac{5}{14}$, $\frac{7}{14}$ $(2+5+7 = 14)$.

EXAMPLE 4

Divide 156 in the ratio 3 : 4 : 5.

$$\frac{3}{\not{1}\not{2}_1} \times \not{1}\not{5}\not{6}^{13} = 39 \qquad (\textit{note:}\quad 3+4+5 = 12)$$

$$\frac{4}{\not{1}\not{2}_1} \times \not{1}\not{5}\not{6}^{13} = 52$$

$$\frac{5}{\not{1}\not{2}_1} \times \not{1}\not{5}\not{6}^{13} = 65.$$

EXAMPLE 5

The prize money for the first three prizes in a raffle was divided in the ratio 3 : 5 : 7. The smallest prize was £4.50. How much was the largest prize?

Let the total prize money $= \pounds x$.

$$\text{The smallest prize, } \pounds 4.50 = \tfrac{3}{15} \times \pounds x$$
$$= \tfrac{1}{5} \times \pounds x.$$

Therefore

$$\pounds x = 5 \times \pounds 4.50$$
$$= \pounds 22.50.$$
$$\text{The largest prize} = \tfrac{7}{15} \times \pounds 22.50$$
$$= 7 \times \pounds 1.50$$
$$= \pounds 10.50.$$

EXERCISE 8b

1 Divide 56 in the ratio 5 : 9.

2 Divide £24 in the ratio 1 : 2 : 3.
3 Divide 18 in the ratio 5 : 3 : 1.
4 Divide £128 in the ratio 2 : 3 : 7.
5 Divide 324 in the ratio $1\frac{1}{3} : 1 : \frac{2}{3}$.
6 Divide 13 in the ratio $\frac{1}{2} : \frac{1}{3} : \frac{1}{4}$.
7 The sides of a triangle are in the ratio 3 : 4 : 5 and the perimeter is 105 cm. Find the lengths of the sides.
8 Divide £7 into two parts so that one is three-quarters of the other.
9 A, B and C share £180 in the ratio $1\frac{1}{2} : \frac{5}{6} : \frac{2}{3}$. What does each receive?
10 Divide 360° into 4 angles in the ratio 1 : 2 : 5 : 7.
11 Two angles of a pentagon are 58° and 83°. The other three are in the ratio 5 : 6 : 8. Find these angles.
12 A journey of 15 km is covered in three stages. The distances covered in each stage are in the ratio 2 : 3 : 5. Find the distances covered in each stage.
13 Alan, Bill and Charles received £144 between them. Alan had four times as much as Charles and Bill had three times as much as Charles. Find how much each person received.
14 A number of sweets were divided in the ratio 3 : 5 : 7. John's share of 25 was the smallest portion. How many sweets were there altogether?
15 A sum of money is divided between three men, A, B and C, in the ratio 7 : 5 : 1. If B has £2.40 more than C, how much does A receive?
16 Four equal angles of a pentagon are twice the size of the fifth angle. Find all five angles.
17 Danny offered to pay his debts from his existing capital of £1200. He owed Alan £600, Bill £750 and Charles £650. How much did each receive?
18 Mark invests one-eighth of his salary in three companies, A, B and C, in the ratio $3 : 2\frac{1}{4} : 1\frac{1}{2}$. His smallest investment per month is £3.00. Find his monthly salary.

8.3 Percentages

A *percentage* is a fraction written in terms of hundredths, e.g. 63 % in an examination means 63 out of 100 or simply the candidate obtaining 63/100 of the total marks.

$$\therefore\ 63\,\% = \frac{63}{100}.$$

EXAMPLE 6

(i) Fractions and decimals to a percentage.

(a) Express $\frac{5}{8}$ as a percentage.

$$\frac{5}{8} = \frac{5 \times 12\frac{1}{2}}{8 \times 12\frac{1}{2}} = \frac{62\frac{1}{2}}{100} = 62\tfrac{1}{2}\,\%.$$

$$\downarrow$$

$$(8 \times 12\tfrac{1}{2} = 100)$$

(b) Express 0.579 as a percentage.

$$0.579 = \frac{579}{1000} = \frac{57.9}{100} = 57.9\,\%.$$

(ii) Percentages to fractions and decimals.

(a) Change 65 % to a fraction.

$$65\,\% = \frac{{}^{13}\not{65}}{\not{100}_{20}} = \frac{13}{20}.$$

(b) Change 221 % to a decimal

$$221\,\% = \frac{221}{100} = 2.21.$$

EXERCISE 8c

1 Convert the following to percentages:

(i) $\frac{2}{3}$ (ii) 2.79 (iii) $\frac{1}{6}$ (iv) 0.0375
(v) $\frac{1}{200}$ (vi) $\frac{1}{1000}$ (vii) 3.28 (viii) $\frac{11}{20}$
(ix) $\frac{3}{2}$ (x) 2 (xi) $\frac{49}{500}$ (xii) $\frac{3}{7}$.

2 Convert the following percentages to fractions and decimals:

(i) $33\frac{1}{3}$ % (ii) 72 % (iii) $12\frac{1}{2}$ % (iv) $8\frac{1}{3}$ %
(v) 0.6 % (vi) 225 % (vii) 0.25 % (viii) 0.5 %
(ix) $\frac{2}{3}$ % (x) 12 % (xi) $27\frac{1}{2}$ % (xii) 36.5 %.

Express the following as percentages:

3 A distance of 30 cm to a distance of 1 km

4 £3.75 of £42.50

5 420 school children in a town of 12000

6 2.8 g of 7 g

7 2p of 25 p

8 15 g of 1.5 kg

9 7 cm of 2 m

10 Find $12\frac{1}{2}$ % of £12.

11 Find 15% of 340 g.
12 Find 35% of 18 km.
13 Find 47% of £3.50.
14 Find 0.2% of 1 litre.
15 Find 1.2% of 450°.
16 Find 72% of 19.6 m.
17 Find 17% of £2.35.

8.4 Percentage changes

EXAMPLE 7

Increase £120 by 7%.

$$\begin{aligned} 7\% \text{ of } £120 &= \tfrac{7}{100} \times £120 \\ &= 7 \times £1.20 \\ &= £8.40 \\ \therefore \text{ new value } &= £128.40. \end{aligned}$$

EXAMPLE 8

Decrease 35 metres by 33%.

$$\begin{aligned} 33\% \text{ of } 35 \text{ m} &= \tfrac{33}{100} \times 35 \text{ m} \\ &= \tfrac{33}{100} \times 3500 \text{ cm} \\ &= 1155 \text{ cm} \\ &= 11.55 \text{ m} \\ \therefore \text{ New measurement} &= 23.45 \text{ m}. \end{aligned}$$

8.5 Profit and loss

To make £1 profit on selling an article which cost £5 is obviously worth more than making the same profit on an article costing £50. To give some measure of profit (and loss) we make the following definitions:

$$\text{percentage profit} = \frac{\text{profit}}{\text{cost price}} \times 100$$

$$\text{percentage loss} = \frac{\text{loss}}{\text{cost price}} \times 100.$$

When considering a percentage difference it is sometimes advisable to consider the ratio between the two figures.

EXAMPLE 9

An article purchased for £12 was sold at a 55% profit. Find the selling price.

Ratio of cost price to selling price is 100 : 155

$$\Rightarrow \text{cost price} : \text{selling price} = 100 : 155$$

$$\frac{\text{cost price}}{\text{selling price}} = \frac{100}{155}$$

$$\Rightarrow \text{cost price} = \frac{100}{155} \times \text{selling price}$$

$$\therefore \text{selling price} = \frac{^{31}\cancel{155}}{\cancel{100}_{\cancel{20}_5}} \times £\cancel{12}^3$$

$$= £\frac{93}{5}$$

$$= £18.60.$$

EXAMPLE 10

100 shares were sold for £175, making a 40 % profit. Find the buying price.

$$\text{buying price} : \text{selling price} = 100 : 140$$

i.e.

$$\frac{\text{buying price}}{\text{selling price}} = \frac{100}{140}$$

$$\therefore \text{buying price} = \frac{100}{140} \times £175$$

$$= \frac{5}{\cancel{7}_1} \times £\cancel{175}^{25}$$

$$= £125$$

EXERCISE 8d

1 Increase 360 by 15 %.
2 Increase 3.8 kg by $12\frac{1}{2}$ %.
3 Increase 8.5 cm by 13 %.
4 Decrease £1.35 by 20 %.
5 Decrease 80 litres by 63 %.
6 Decrease 7.28 m by 34 %. (to nearest cm).

7 Increase the radius of a circle of radius 15 cm by 20 %. What is the % increase in area?

8 A shoe manufacturer decides to streamline its shoeboxes by reducing the width by 8 % and the height by 5 %. The original measurements were:

length 35 cm
width 15 cm
height 10 cm

Find the new volume and express this as a percentage of the old volume.

9 Mr Jones' yearly salary of £2500 was increased by 13 %. Find the new monthly salary.

10 An article costing £17.50 is sold at a $22\frac{1}{2}$ % profit. Find the selling price.

11 A retailer makes 35 % profit on the refrigerators he sells. The selling price is £81. What is the cost price?

12 An article costing £28 is sold at a 12 % loss. Find the selling price.

13 An article was sold for a profit of 39 %. Find the cost price if the profit was £7.80.

14 Mr White sold 12 dozen eggs for £7.92, making 10 % profit. How much did he pay for them?

15 The rail fare between Stenworth and Shoreton was increased by 12 %, making a 54p increase in the price. What is the new fare?

16 Cobar mining company shares were down 8 % from the previous week's price of £12.75 each. Find this week's price.

17 Mr Buckwell sold his car for £270, making a loss on the buying price of 25 %. Find the original cost of the car.

18 John made a 36 % loss on an investment and was £30.24 out of pocket. How much was his investment?

8.6 Simple interest

If an investor places an amount of money in a deposit account at the bank, then, after a period of time, the bank will credit the account with profit made from the investment of this money. This profit is called the *interest*. If the interest is returned to the investor rather than credited to the account, it is called *simple interest*.

The rate of interest is the proportion of the amount invested which is returned after a given time, e.g. 5 % per annum gives a return of $\frac{5}{100}$ of the capital invested after one year.

The capital itself will henceforth be referred to as the *principal*.

EXAMPLE 11

An investor draws the simple interest on an amount of £500 invested at a rate of 7 % per annum. What was the total interest after 3 years; after 5 years?

time	interest	total interest
1st year	$\frac{7}{100} \times £500 = £35$	£35
2nd year	$\frac{7}{100} \times £500 = £35$	$2 \times £35$
3rd year	$\frac{7}{100} \times £500 = £35$	$3 \times £35$

Total interest after 3 years is £105.

Following the pattern of the table, after 5 years the interest is £175.

We could replace all the working above by considering that

simple interest = number of years × (rate per annum × principal).

If we let

I = interest
T = number of years
P = principal
R = rate per cent per annum

then

$$I = \frac{P \times R \times T}{100}.$$

The total of the principal and interest is known as the *amount* and is given by

$$\text{Amount} = P + I = P + \frac{PRT}{100}.$$

EXAMPLE 12

£150 is invested for a period of 8 months at a rate of 8% per annum. What was (i) the interest received and (ii) the total amount, after 8 months?

$$I = \frac{P \times R \times T}{100}$$

$P = £150$, $R = 8$, $T = \frac{2}{3}$

(As the rate per cent is on a yearly basis, the time has to be converted to years.)

$$I = \frac{£150 \times 8 \times \frac{2}{3}}{100}$$

$$I = \frac{£800}{100}$$
$$= £8.$$

$\therefore$ total amount after 8 months $= £150 + £8$
$= £158.$

Other ways of looking at the simple interest formula:

$$\text{from } I = \frac{PRT}{100}: P = \frac{100I}{RT}, R = \frac{100I}{PT}, T = \frac{100I}{PR}.$$

EXAMPLE 13

If the interest after 4 years on a principal of £450 was £63, find the rate of interest.

From the formula $R = \dfrac{100 \times I}{P \times T}$,

$$R = \frac{100 \times 63}{450 \times 4} = \tfrac{7}{2} = 3\tfrac{1}{2}\,\%.$$

EXERCISE 8e

1 Find the simple interest on £360 invested for $2\frac{1}{2}$ years at $7\frac{1}{2}$ % per annum.
2 Find the simple interest on £270 invested for 15 months at 12 % per annum.
3 Find the rate of interest on a principal of £25 invested for a period of 2 years with a return of £7 interest.
4 Find the rate of interest on a principal of £310 invested for a period of $4\frac{1}{2}$ years with a return of £46.50 interest.
5 £420 was invested at a rate of $7\frac{1}{2}$ % per annum and gave a return of £189 interest. How long was the money invested?
6 £336 was invested at a rate of 5 % per annum and gave a return of £28. How long was the money invested?
7 A sum of money invested for a period of 3 years at a rate of $5\frac{1}{2}$ % per annum produces a return of £82.50 interest. Find this sum.
8 A sum of money invested for a period of 21 months at a rate of 64 % per annum produces a return of £70 interest. Find this sum.
9 Find the principal if the final amount after 5 years at $3\frac{1}{2}$ % per annum is £282.
10 Find the principal if the final amount after $4\frac{1}{4}$ years at $3\frac{1}{2}$ % per annum is £68.50.

8.7 Compound interest

If instead of the interest being returned to the investor, it is reinvested, the form of interest becomes known as *compound interest*.

EXAMPLE 14

Find the compound interest on £750 at 5 % per annum for 3 years.

To calculate the amount after 3 years we must calculate the amount each year which in turn becomes the principal for the following year to the nearest 1p.

	principal	interest	amount
after 1st year	£750	$\frac{5}{100}\times$ £750 = £37.50	£787.50
after 2nd year	£787.50 ←	$\frac{5}{100}\times$ £787.50 = £39.37$\frac{1}{2}$	£826.87$\frac{1}{2}$
after 3rd year	£826.87$\frac{1}{2}$ ←	$\frac{5}{100}\times$ £826.87$\frac{1}{2}$ = £41.34$\frac{1}{2}$	£868.22

Total interest = £118.22

Final amount = £868.22.

EXAMPLE 15

Find the compound interest of £18 at 8 % per annum for $1\frac{3}{4}$ years to the nearest 1p.

	principal	interest	amount
after 1st year	£18	$\frac{8}{100}\times$ £18.00 = £1.44	£19.44
after $1\frac{3}{4}$ years	£19.44 ←	$\frac{8}{100}\times\frac{3}{4}\times$ £19.44 = £1.16	£20.60

$$\left(\text{using}\ \frac{PRT}{100}\right)$$

Total interest = £2.60

Final amount = £20.60

EXERCISE 8f

Find the compound interest to the nearest 1p and the amount for the following:

1 £320 invested for 4 years at 5% per annum
2 £75 invested for 15 months at 10% per annum
3 £164 invested for 2 years at $3\frac{1}{2}$% per annum
4 £325.50 invested for 6 years at 8% per annum
5 £100 invested for $2\frac{1}{2}$ years at $3\frac{3}{4}$% per annum.

6 An ink blot of area 10 cm^2 increases at the rate of $4\frac{1}{2}$% per second. How large is it after 3 seconds?
7 A car travelling at an average speed of 50 km/h increases its speed at the rate of 5% per hour. How fast is it travelling after 3 hours?
8 £1000 is invested for 4 years at 6% per annum compound interest. The interest is subject to an income tax deduction of $37\frac{1}{2}$p in the £1. Find the amount after 4 years.

HARDER EXAMPLES 8

1 A house was bought for £11 000 and sold for £12 750. Find the percentage profit.
2 Mr Smith sold his car which he bought for £600 at a 35% loss to Mr Brown. Mr Brown in turn made a loss of 25% when he sold the car to Mr Jones. How much did Mr Jones pay for the car?
3 A shopkeeper buys eggs at £5.00 per gross. Find his profit per cent if he sells them at 60p a dozen.
4 A shopsoiled cassette player worth £46 new is sold for a loss of 35%. Find the selling price.
5 Spiro toffee is sold at 5p a bar or 3 bars for 14p. Find the percentage saving if 3 bars are bought.
6 Mr Jones calls in a contractor to give a quote on a swimming pool. The quote was as follows:

£200	for labour
£ 18	for cement
£200	for bricks
£ 10	for sand.

It is known that the contractor buys his materials 30% cheaper than the retail price. What is Mr Jones' percentage saving if he does the job himself?
7 A refrigerator is sold for £38.50 at a loss of $12\frac{1}{2}$%. Find the original buying price.
8 Davey Jones' car insurance premium increased this year by £7.50. This represented an increase of $17\frac{1}{2}$%. Find his premium for the previous year.
9 Alan sells Bill a tape recorder at a 25% profit. Bill in turn sells

it to Charles at the price Alan paid for it. What was Bill's percentage loss?

10 Two numbers x and y are such that $x = y^2$. If y is now increased by 10%, find the corresponding percentage increase in x.

11 By selling an article for £18 a shopkeeper makes a profit of 44% on the cost price. At what price must it be sold in order to make a profit of 40%?

[L]

12 A firm makes articles which are sold in retail shops at a net profit to the firm of 50p per article. If an article is unsold, it is returned to the factory and the firm disposes of it at a price which gives a net profit to the firm of 25p. From past records, the manufacturer estimates that the number of articles produced by the factory in a certain period will be 1000 ($\pm$100), and of the articles sent to the shops, 10 per cent ($\pm$2 per cent) will be unsold. Calculate the extreme values between which the manufacturer's net profit for the period will lie.

[JMB]

13 The production of a car factory was 26 600 vehicles in 1968 and 30 058 vehicles in 1969. Express the increase in production as a percentage of the 1968 production.

[O]

14 A shopkeeper marks his goods at prices which would give him profits ranging from 10% to 20% on their cost prices to him. Find the range of possible marked prices of an article whose cost price was £2.50.

A sale is held: the sale prices of all goods are reduced below the marked prices by amounts ranging between 10% and 20% of the marked prices. Find the range of possible sale prices of the article whose cost price is £2.50.

[L]

15 £250 is invested at 6% compound interest per annum. Calculate:

(i) the amount after one year;
(ii) the amount after two years;
(iii) the interest that will be added on at the end of the third year. (Give your answer to the nearest penny.)

[NI]

9 Matrices

9.1 Definition

The test results in Physics, Chemistry and Biology for Alan, Bill and Charles were as follows:

	Physics	Chemistry	Biology
Alan	8	5	6
Bill	7	4	9
Charles	3	4	7

This information can be written in the form:

$$\begin{pmatrix} 8 & 5 & 6 \\ 7 & 4 & 9 \\ 3 & 4 & 7 \end{pmatrix}$$

which is called a *matrix*. Each number in a matrix is called an element. The following week, the same pupils had a repeat test in all three subjects and the results were:

	Physics	Chemistry	Biology
Alan	6	6	4
Bill	8	3	5
Charles	5	7	6

In matrix form this information is represented

$$\begin{pmatrix} 6 & 6 & 4 \\ 8 & 3 & 5 \\ 5 & 7 & 6 \end{pmatrix}$$

We can find the combined totals for any pupil in any subject by adding the corresponding elements of the matrices together:

$$\begin{pmatrix} 8 & 5 & 6 \\ 7 & 4 & 9 \\ 3 & 4 & 7 \end{pmatrix} + \begin{pmatrix} 6 & 6 & 4 \\ 8 & 3 & 5 \\ 5 & 7 & 6 \end{pmatrix} =$$

	Physics	Chemistry	Biology
Alan	14	11	10
Bill	15	7	14
Charles	8	11	13

This is known as matrix addition. Clearly, if we wish to subtract two matrices, we subtract the corresponding elements.

9.2 The order of a matrix

The matrix

$$\begin{pmatrix} 3 & 5 & 2 \\ 1 & 4 & 5 \end{pmatrix}$$

has 2 rows and 3 columns, and we define the *order* of this matrix as 2 by 3 (2×3).

EXAMPLE 1

If $A = \begin{pmatrix} 2 & 1 \\ 3 & 2 \end{pmatrix}$, $B = \begin{pmatrix} -1 & 2 \\ 4 & -3 \end{pmatrix}$ and $C = \begin{pmatrix} 2 & -1 \\ 1 & 2 \\ 5 & 4 \end{pmatrix}$,

find (i) $A+B$ (ii) $A-B$ (iii) $A+A+A$ (iv) $A+C$.

(i) $$\begin{pmatrix} 2 & 1 \\ 3 & 2 \end{pmatrix} + \begin{pmatrix} -1 & 2 \\ 4 & -3 \end{pmatrix} = \begin{pmatrix} 2+(-1) & 1+2 \\ 3+4 & 2+(-3) \end{pmatrix} = \begin{pmatrix} 1 & 3 \\ 7 & -1 \end{pmatrix}$$

(ii) $$\begin{pmatrix} 2 & 1 \\ 3 & 2 \end{pmatrix} - \begin{pmatrix} -1 & 2 \\ 4 & -3 \end{pmatrix} = \begin{pmatrix} 2-(-1) & 1-2 \\ 3-4 & 2-(-3) \end{pmatrix} = \begin{pmatrix} 3 & -1 \\ -1 & 5 \end{pmatrix}$$

(iii) $$\begin{pmatrix} 2 & 1 \\ 3 & 2 \end{pmatrix} + \begin{pmatrix} 2 & 1 \\ 3 & 2 \end{pmatrix} + \begin{pmatrix} 2 & 1 \\ 3 & 2 \end{pmatrix} = \begin{pmatrix} 6 & 3 \\ 9 & 6 \end{pmatrix}$$

or
$$3\begin{pmatrix} 2 & 1 \\ 3 & 2 \end{pmatrix} = \begin{pmatrix} 6 & 3 \\ 9 & 6 \end{pmatrix}$$

Similarly

$$\begin{aligned} B+B+B+B &= 4B \\ &= 4\begin{pmatrix} -1 & 2 \\ 4 & -3 \end{pmatrix} \\ &= \begin{pmatrix} -4 & 8 \\ 16 & -12 \end{pmatrix} \end{aligned}$$

(iv) $$\begin{pmatrix} 2 & 1 \\ 3 & 2 \end{pmatrix} + \begin{pmatrix} 2 & -1 \\ 1 & 2 \\ 5 & 4 \end{pmatrix} = \begin{pmatrix} 4 & 0 \\ 4 & 4 \\ ? & ? \end{pmatrix}$$

As matrix A has no third row, we have no elements to add to the third row of C. So matrices with different orders cannot be added together.

9.3 Equality of matrices

Two matrices are said to be equal only if the corresponding elements in each matrix are the same.

EXAMPLE 2

Find a and b when:

$$\begin{pmatrix} a \\ 1 \end{pmatrix} + \begin{pmatrix} 2 \\ -b \end{pmatrix} = \begin{pmatrix} 5 \\ 2 \end{pmatrix}$$

Now

$$\begin{pmatrix} a \\ 1 \end{pmatrix} + \begin{pmatrix} 2 \\ -b \end{pmatrix} = \begin{pmatrix} a+2 \\ 1-b \end{pmatrix}$$

$$\therefore \quad \begin{pmatrix} a+2 \\ 1-b \end{pmatrix} = \begin{pmatrix} 5 \\ 2 \end{pmatrix}$$

$$\therefore \quad a+2 = 5, \Rightarrow a = 3$$

and

$$1-b = 2, \Rightarrow b = -1.$$

EXERCISE 9a

1 Let $A = \begin{pmatrix} 2 & -2 \\ -2 & 2 \end{pmatrix}$, $B = \begin{pmatrix} 3 & -1 \\ 3 & -1 \end{pmatrix}$,

$C = \begin{pmatrix} 3 & 2 & -1 \\ 4 & 0 & 1 \end{pmatrix}$, $D = \begin{pmatrix} 5 & 4 & -1 \\ 2 & 3 & 4 \end{pmatrix}$,

$E = \begin{pmatrix} 5 & 2 \\ -2 & -4 \\ -1 & -5 \end{pmatrix}$, $F = \begin{pmatrix} 4 & 1 \\ -3 & 2 \\ 4 & 5 \end{pmatrix}$,

$G = \begin{pmatrix} 1 & 2 & 3 \\ 0 & 1 & -2 \\ 2 & -1 & 3 \end{pmatrix}$

Evaluate, if possible:

(i) $C+D$ (ii) $E+2F$ (iii) $A-2B$
(iv) $D-E$ (v) $3G$ (vi) $A+B+C$

2 $A = \begin{pmatrix} -1 & 2 \\ 1 & -3 \end{pmatrix}$, $B = \begin{pmatrix} 4 & -5 \\ 2 & 1 \end{pmatrix}$, $C = \begin{pmatrix} -1 & 2 \\ 4 & 1 \end{pmatrix}$

Verify that $A+(B+C) = (A+B)+C$.

3 Find the matrix M that satisfies the equation:

$$2M - \begin{pmatrix} 3 & -2 \\ 4 & 5 \end{pmatrix} = \begin{pmatrix} -5 & 8 \\ -2 & 5 \end{pmatrix}.$$

4 Find the matrix Y if:

$$\begin{pmatrix} 2 & -1 \\ 1 & 3 \end{pmatrix} + 2Y = \begin{pmatrix} 5 & 4 \\ 1 & 2 \end{pmatrix}$$

5 Find the matrix A if:

$$\begin{pmatrix} 1 & 0 \\ -3 & -2 \end{pmatrix} - 2A = A + \begin{pmatrix} -2 & -3 \\ 0 & 4 \end{pmatrix}$$

6 Find the matrix Q if:

$$\begin{pmatrix} 4 & -2 & -1 \\ 0 & 2 & -4 \end{pmatrix} = \begin{pmatrix} 4 & 1 & -5 \\ 2 & -3 & 0 \end{pmatrix} - 3Q$$

9.4 Multiplication of matrices

In the Mathemanian football league, the points system was 5 points for a win, 3 for a draw and 1 for a match lost. Stenworth's record for one year was matches won 10, drawn 6, lost 8. How many points did Stenworth gain? We can represent the two sets of information by matrices:

$$(5 \quad 3 \quad 1) \quad \text{and} \quad (10 \quad 6 \quad 8)$$

Clearly the number of points obtained can be evaluated by multiplication and addition:

$$5 \times 10 + 3 \times 6 + 1 \times 8 = 76 \text{ points.}$$

This process is known as matrix multiplication and is usually written in the form:

$$(5 \quad 3 \quad 1)\begin{pmatrix} 10 \\ 6 \\ 8 \end{pmatrix} = (5 \times 10 + 3 \times 6 + 1 \times 8) = (76)$$

In the same season, Stenworth's chief rival, Shoreton Town, won 8, drew 9 and lost 7. The number of points obtained by Shoreton can be evaluated in a similar manner

$$(5 \quad 3 \quad 1)\begin{pmatrix} 8 \\ 9 \\ 7 \end{pmatrix} = (5 \times 8 + 3 \times 9 + 1 \times 7) = (74)$$

The two sets of results can be combined in the following manner:

$$(5 \quad 3 \quad 1)\begin{pmatrix} 10 & 8 \\ 6 & 9 \\ 8 & 7 \end{pmatrix} = (76 \quad 74) \ldots (1)$$

The following year, the points system was altered so that there were only 4 points for a win, and by a coincidence both teams again produced the same results. The matrix representation is therefore:

$$(4 \quad 3 \quad 1)\begin{pmatrix} 10 & 8 \\ 6 & 9 \\ 8 & 7 \end{pmatrix} = (66 \quad 66) \ldots (2)$$

We can combine (1) and (2) as follows:

$$\begin{array}{ccc} (5 \quad 3 \quad 1) & \begin{pmatrix} 10 & 8 \\ 6 & 9 \\ 8 & 7 \end{pmatrix} & = (76 \quad 74) \\ \uparrow & & \uparrow \\ (4 \quad 3 \quad 1) & & (66 \quad 66) \end{array}$$

so that we have:

$$\begin{pmatrix} 5 & 3 & 1 \\ 4 & 3 & 1 \end{pmatrix} \begin{pmatrix} 10 & 8 \\ 6 & 9 \\ 8 & 7 \end{pmatrix} = \begin{pmatrix} 76 & 74 \\ 66 & 66 \end{pmatrix}$$

To obtain the elements of the resultant matrix in a multiplication problem, we must combine rows of the first matrix with columns of the second matrix.

EXAMPLE 3

If $C = \begin{pmatrix} 2 & -1 \\ 2 & 1 \end{pmatrix}$ and $D = \begin{pmatrix} -1 & -3 \\ 4 & 2 \end{pmatrix}$, calculate the matrix product CD.

$$\begin{pmatrix} \mathbf{2} & \mathbf{-1} \\ 2 & 1 \end{pmatrix} \begin{pmatrix} \mathbf{-1} & -3 \\ \mathbf{4} & 2 \end{pmatrix} = \begin{pmatrix} \mathbf{-6} & \\ & \end{pmatrix}$$

Here, only part of the problem is done to show the positioning of the resultant element. If we superimpose one matrix on to the other, the intersection of the row and column in bold type places the resultant element of -6. The complete solution is:

$$\begin{pmatrix} 2 & -1 \\ 2 & 1 \end{pmatrix} \begin{pmatrix} -1 & -3 \\ 4 & 2 \end{pmatrix} = \begin{pmatrix} -6 & -8 \\ 2 & -4 \end{pmatrix}$$

EXAMPLE 4

If

$$S = \begin{pmatrix} 2 & 1 & 3 \\ -1 & 4 & 1 \end{pmatrix} \text{ and } T = \begin{pmatrix} 2 & -1 \\ 1 & 3 \end{pmatrix}$$

calculate the matrix products (i) TS, and (ii) ST.

$$TS = \begin{pmatrix} 2 & -1 \\ 1 & 3 \end{pmatrix} \begin{pmatrix} 2 & 1 & 3 \\ -1 & 4 & 1 \end{pmatrix}$$

$$= \begin{pmatrix} 5 & -2 & 5 \\ -1 & 13 & 6 \end{pmatrix}$$

$$ST = \begin{pmatrix} \mathbf{2} & \mathbf{1} & \mathbf{3} \\ -1 & 4 & 1 \end{pmatrix} \begin{pmatrix} \mathbf{2} & -1 \\ \mathbf{1} & 3 \end{pmatrix}$$

$$= \begin{pmatrix} ? & \\ & \end{pmatrix}$$

The element in the first row and column is computed using the first row of S and the first column of T, that is,

$$(2 \times 2) + (1 \times 1) + (3 \times ?)$$

As we cannot determine the value of the third bracket, no resultant element exists and consequently there is no resultant matrix.

EXAMPLE 5

If

$$P = \begin{pmatrix} 1 & 3 \\ -2 & 1 \\ 1 & 0 \end{pmatrix} \quad \text{and} \quad Q = \begin{pmatrix} 1 & -1 & 0 \\ 0 & 1 & -1 \end{pmatrix}$$

calculate the matrix product PQ.

$$PQ = \begin{pmatrix} 1 & 3 \\ -2 & 1 \\ 1 & 0 \end{pmatrix} \begin{pmatrix} 1 & -1 & 0 \\ 0 & 1 & -1 \end{pmatrix} = \begin{pmatrix} 1 & 2 & -3 \\ -2 & 3 & -1 \\ 1 & -1 & 0 \end{pmatrix}$$

This example shows that matrices can only be multiplied together if the number of columns in the first matrix is the same as the number of rows in the second matrix. Moreover, it can be seen that the order of the resultant matrix (3×3) is derived from the number of rows in the first matrix and the number of columns in the second matrix.

Check that this is also the case for the other examples.

In general, if the order of the matrix A is $p \times q$ and the order of the matrix B is $q \times r$, then the order of the resultant matrix is $p \times r$.

EXERCISE 9b

Evaluate, where possible, the following matrix products:

1 $(1 \quad 2)\begin{pmatrix} -1 \\ 1 \end{pmatrix}$

2 $(1 \quad 0)\begin{pmatrix} 0 \\ 1 \end{pmatrix}$

3 $(1 \quad -1 \quad 0)\begin{pmatrix} -2 \\ 2 \\ 1 \end{pmatrix}$

4 $(0 \quad 0 \quad 0)\begin{pmatrix} 0 \\ 0 \\ 0 \end{pmatrix}$

5 $(1 \quad 2)\begin{pmatrix} -1 \\ 0 \\ 1 \end{pmatrix}$

6 $(1 \quad 3 \quad 1)\begin{pmatrix} 2 \\ -1 \end{pmatrix}$

7 $(3 \quad -1)\begin{pmatrix} 2 & -1 \\ 1 & 2 \end{pmatrix}$

8 $(2 \quad -1)\begin{pmatrix} 1 & 0 \\ 0 & 1 \end{pmatrix}$

9 $\begin{pmatrix} 1 & -1 \\ -1 & 1 \end{pmatrix}(2 \quad -2)$

10 $\begin{pmatrix} -1 & 1 \\ 1 & -1 \end{pmatrix}\begin{pmatrix} 2 \\ 1 \end{pmatrix}$

11 $\begin{pmatrix} 4 & 1 \\ 2 & 3 \end{pmatrix}\begin{pmatrix} -1 & 1 \\ 1 & -2 \end{pmatrix}$

12 $\begin{pmatrix} 2 & 3 \\ 1 & 4 \end{pmatrix}\begin{pmatrix} 1 & -1 & 0 \\ 0 & 1 & 0 \end{pmatrix}$

13 $\begin{pmatrix} -3 & -1 \\ -1 & 2 \end{pmatrix}\begin{pmatrix} 1 & 2 \\ 0 & 1 \\ 1 & 0 \end{pmatrix}$

14 $\begin{pmatrix} 2 & 1 & 3 \\ 1 & 4 & 1 \end{pmatrix}\begin{pmatrix} 2 & 1 \\ 4 & 5 \\ 3 & 4 \end{pmatrix}$

15 $\begin{pmatrix} 1 & 2 \\ -1 & -1 \\ 1 & 0 \end{pmatrix}\begin{pmatrix} 2 & 1 & 0 \\ 1 & -1 & 1 \end{pmatrix}$

16 $\begin{pmatrix} -1 & 3 & 5 \\ 2 & -1 & -2 \end{pmatrix}\begin{pmatrix} 1 & 2 \\ 0 & 3 \\ -4 & -5 \end{pmatrix}$

17 $\begin{pmatrix} 2 & -1 & 0 \\ 1 & 0 & -3 \end{pmatrix}\begin{pmatrix} 2 & -2 \\ -1 & 1 \end{pmatrix}$

18 $\begin{pmatrix} 2 & 1 & -1 \\ 0 & 1 & 0 \end{pmatrix} \begin{pmatrix} 2 & -3 & 2 \\ 1 & -1 & 0 \\ 0 & 4 & 1 \end{pmatrix}$

19 $\begin{pmatrix} 1 & -1 & 0 \\ 0 & 0 & 1 \\ 1 & -2 & 1 \end{pmatrix} \begin{pmatrix} 2 & 1 & 0 \\ 0 & 0 & 1 \end{pmatrix}$

20 $\begin{pmatrix} 1 & 0 & 1 \\ 0 & 1 & 1 \\ 1 & 0 & 1 \end{pmatrix} \begin{pmatrix} 2 & -1 & 2 \\ 3 & 1 & -1 \\ 2 & 1 & 0 \end{pmatrix}$

21 If $A = \begin{pmatrix} 2 & -1 \\ 1 & 3 \end{pmatrix}$ and $B = \begin{pmatrix} 1 & -1 \\ 2 & 0 \end{pmatrix}$ calculate (i) AB; (ii) BA.

22 If $R = \begin{pmatrix} -1 & 0 \\ 3 & 2 \end{pmatrix}$, $S = \begin{pmatrix} 0 & 2 \\ 1 & 3 \end{pmatrix}$, $T = \begin{pmatrix} -1 & 0 \\ 2 & -1 \end{pmatrix}$ calculate (i) $R(ST)$; (ii) $(RS)T$. What do you notice about your answers?

In your answer for question 21 you will see that $AB \neq BA$ and generally this is the case for the multiplication of matrices. We say that matrix multiplication is non-commutative. Question 22 shows the associative property of the multiplication of matrices (see chapter 5).

9.5 The identity matrix

The 2×2 *identity matrix* for matrix multiplication is the matrix I such that

$$IA = A = AI.$$

This matrix is $\begin{pmatrix} 1 & 0 \\ 0 & 1 \end{pmatrix}$

Similarly, the 3×3 identity matrix is $\begin{pmatrix} 1 & 0 & 0 \\ 0 & 1 & 0 \\ 0 & 0 & 1 \end{pmatrix}$

9.6 Transpose of a matrix

If we change the rows of a matrix A to columns we form the transpose of the original matrix, denoted by A'. For example, if

$$A = \begin{pmatrix} 3 & 2 & -1 \\ 2 & 4 & 0 \end{pmatrix}$$

then

$$A' = \begin{pmatrix} 3 & 2 \\ 2 & 4 \\ -1 & 0 \end{pmatrix}.$$

9.7 The inverse of a matrix

The *inverse matrix* of A, denoted by A^{-1}, has the property $AA^{-1} = I = A^{-1}A$. Suppose we wish to find the inverse of the matrix $\begin{pmatrix} 3 & 1 \\ 2 & 1 \end{pmatrix}$.

Let us denote the required matrix by $\begin{pmatrix} a & b \\ c & d \end{pmatrix}$

$$\begin{pmatrix} a & b \\ c & d \end{pmatrix}\begin{pmatrix} 3 & 1 \\ 2 & 1 \end{pmatrix} = \begin{pmatrix} 1 & 0 \\ 0 & 1 \end{pmatrix}$$

$$\Rightarrow \begin{pmatrix} 3a+2b & a+b \\ 3c+2d & c+d \end{pmatrix} = \begin{pmatrix} 1 & 0 \\ 0 & 1 \end{pmatrix}$$

As two matrices can only be equal to each other if corresponding elements are the same,

$$a+b = 0 \Rightarrow a = -b$$

and

$$3c+2d = 0 \Rightarrow c = -\tfrac{2}{3}d$$

$$\therefore \begin{pmatrix} 3a+2b & a+b \\ 3c+2d & c+d \end{pmatrix} = \begin{pmatrix} -b & 0 \\ 0 & \frac{1}{3}d \end{pmatrix} = \begin{pmatrix} 1 & 0 \\ 0 & 1 \end{pmatrix}$$

$\therefore\ -b = 1 \Rightarrow b = -1$

and

$\frac{1}{3}d = 1 \Rightarrow d = 3.$

But

$a = -b \Rightarrow a = 1$

and

$c = -\frac{2}{3}d \Rightarrow c = -2.$

Therefore, the inverse of $\begin{pmatrix} 3 & 1 \\ 2 & 1 \end{pmatrix}$ is $\begin{pmatrix} 1 & -1 \\ -2 & 3 \end{pmatrix}$

Check: $\begin{pmatrix} 1 & -1 \\ -2 & 3 \end{pmatrix}\begin{pmatrix} 3 & 1 \\ 2 & 1 \end{pmatrix} = \begin{pmatrix} 1 & 0 \\ 0 & 1 \end{pmatrix}$ ✓

If we start with the original matrix and

(i) $\begin{pmatrix} \mathbf{3} & 1 \\ 2 & \mathbf{1} \end{pmatrix}$ interchange these two terms

(ii) $\begin{pmatrix} 3 & \mathbf{1} \\ \mathbf{2} & 1 \end{pmatrix}$ alter the signs of the other two terms.

then we find that we have the inverse matrix.
Does this work every time?

EXAMPLE 6

Find the inverse of the matrix $\begin{pmatrix} -4 & -3 \\ 5 & 3 \end{pmatrix}$

If we follow the above procedure we must try $\begin{pmatrix} 3 & 3 \\ -5 & -4 \end{pmatrix}$

$$\begin{pmatrix} 3 & 3 \\ -5 & -4 \end{pmatrix}\begin{pmatrix} -4 & -3 \\ 5 & 3 \end{pmatrix} = \begin{pmatrix} 3 & 0 \\ 0 & 3 \end{pmatrix} = 3\begin{pmatrix} 1 & 0 \\ 0 & 1 \end{pmatrix}$$

We see then, that the rule does not work.

$$\frac{1}{3}\begin{pmatrix} 3 & 3 \\ -5 & -4 \end{pmatrix}\begin{pmatrix} -4 & -3 \\ 5 & 3 \end{pmatrix} = \begin{pmatrix} 1 & 0 \\ 0 & 1 \end{pmatrix}$$

The inverse of $\begin{pmatrix} -4 & -3 \\ 5 & 3 \end{pmatrix}$ is $\frac{1}{3}\begin{pmatrix} 3 & 3 \\ -5 & -4 \end{pmatrix}$

The general form of the inverse of the matrix $\begin{pmatrix} a & b \\ c & d \end{pmatrix}$ is given by:

$$\frac{1}{ad-bc}\begin{pmatrix} d & -b \\ -c & a \end{pmatrix}$$

$ad-bc$ is known as the determinant of the matrix $\begin{pmatrix} a & b \\ c & d \end{pmatrix}$

EXAMPLE 7

Find the inverse of the matrix $\begin{pmatrix} 3 & 5 \\ -2 & 1 \end{pmatrix}$

The inverse is

$$\frac{1}{(3\times 1)-(-2\times 5)}\begin{pmatrix} 1 & -5 \\ 2 & 3 \end{pmatrix} = \frac{1}{3+10}\begin{pmatrix} 1 & -5 \\ 2 & 3 \end{pmatrix}$$

$$= \frac{1}{13}\begin{pmatrix} 1 & -5 \\ 2 & 3 \end{pmatrix}$$

Check: $\frac{1}{13}\begin{pmatrix} 1 & -5 \\ 2 & 3 \end{pmatrix}\begin{pmatrix} 3 & 5 \\ -2 & 1 \end{pmatrix} = \frac{1}{13}\begin{pmatrix} 13 & 0 \\ 0 & 13 \end{pmatrix} = \begin{pmatrix} 1 & 0 \\ 0 & 1 \end{pmatrix}$ √

EXERCISE 9c

1 If $A = \begin{pmatrix} 1 & 3 \\ 5 & 1 \end{pmatrix}$ and $B = \begin{pmatrix} 2 & -1 \\ 1 & 5 \end{pmatrix}$, show that $AB \neq BA$.

2 Given that $A = \begin{pmatrix} 3 & 1 \\ -1 & 2 \end{pmatrix}$, $B = \begin{pmatrix} 1 & 0 \\ -3 & 2 \end{pmatrix}$,

$C = \begin{pmatrix} 1 & 1 \\ 0 & -3 \end{pmatrix}$

show that $A(B+C) = AB+AC$. Deduce also that $AB \neq BA$ and, either by using this fact or by direct evaluation, show that $A^2-B^2 \neq (A+B)(A-B)$.

3 Given the three matrices in the previous problem, verify that $A(BC) = (AB)C$.

4 Solve the equation: $A+I = \begin{pmatrix} 2 & -3 \\ 1 & 0 \end{pmatrix}$

5 Solve the equation: $2I - 3B = \begin{pmatrix} 5 & 1 \\ 4 & -2 \end{pmatrix}$

6 If $A = \begin{pmatrix} 1 & -2 \\ 3 & 4 \end{pmatrix}$, evaluate $(A-I)(A+I)$.

7 If $A = \begin{pmatrix} 1 & -1 \\ 0 & 1 \end{pmatrix}$ and $B = \begin{pmatrix} a & b \\ 0 & c \end{pmatrix}$ and $AB = A+I$, find a, b and c.

8 State the transpose of the following matrices:

(i) $\begin{pmatrix} 1 \\ -1 \end{pmatrix}$ (ii) $\begin{pmatrix} 1 & 0 \\ 0 & -1 \end{pmatrix}$ (iii) $\begin{pmatrix} 2 & 1 & -3 \\ 0 & -2 & 0 \end{pmatrix}$

9 If $A = \begin{pmatrix} 2 & -1 \\ 3 & 5 \end{pmatrix}$, find (i) $A+A'$; (ii) AA'.

10 Find the inverse of the following matrices:

(i) $\begin{pmatrix} 1 & 1 \\ -1 & 1 \end{pmatrix}$ (ii) $\begin{pmatrix} 3 & 1 \\ 5 & 2 \end{pmatrix}$ (iii) $\begin{pmatrix} 3 & -2 \\ 1 & 3 \end{pmatrix}$

(iv) $\begin{pmatrix} 5 & -1 \\ -9 & 2 \end{pmatrix}$ (v) $\begin{pmatrix} -2 & -1 \\ 4 & 3 \end{pmatrix}$ (vi) $\begin{pmatrix} 1 & -3 \\ 1 & 7 \end{pmatrix}$

9.8 Solution of simultaneous equations by matrices

The equations

$$3p - 2q = 4$$
$$2p + 3q = 7$$

can be written as:

$$\begin{pmatrix} 3 & -2 \\ 2 & 3 \end{pmatrix}\begin{pmatrix} p \\ q \end{pmatrix} = \begin{pmatrix} 4 \\ 7 \end{pmatrix}$$

The inverse of $\begin{pmatrix} 3 & -2 \\ 2 & 3 \end{pmatrix}$ is $\frac{1}{13}\begin{pmatrix} 3 & 2 \\ -2 & 3 \end{pmatrix}$

Multiplying on the left by this inverse gives:

$$\frac{1}{13}\begin{pmatrix} 3 & 2 \\ -2 & 3 \end{pmatrix}\begin{pmatrix} 3 & -2 \\ 2 & 3 \end{pmatrix}\begin{pmatrix} p \\ q \end{pmatrix} = \frac{1}{13}\begin{pmatrix} 3 & 2 \\ -2 & 3 \end{pmatrix}\begin{pmatrix} 4 \\ 7 \end{pmatrix}$$

$$\Rightarrow \begin{pmatrix} 1 & 0 \\ 0 & 1 \end{pmatrix}\begin{pmatrix} p \\ q \end{pmatrix} = \frac{1}{13}\begin{pmatrix} 3 & 2 \\ -2 & 3 \end{pmatrix}\begin{pmatrix} 4 \\ 7 \end{pmatrix}$$

$$\Rightarrow \begin{pmatrix} p \\ q \end{pmatrix} = \frac{1}{13}\begin{pmatrix} 26 \\ 13 \end{pmatrix}$$

$$= \begin{pmatrix} 2 \\ 1 \end{pmatrix}$$

From this $p = 2$, $q = 1$.

EXERCISE 9d

Solve the following equations by using matrices:

1 $2x+y = 7$, $x-y = 5$
2 $5x-y = 7$, $2x-y = 2\frac{1}{2}$
3 $3a+2b = 17$, $a-4b = -6$
4 $5p-2q = 7$, $4p+q = -23$
5 $17a-3b = 49$, $4a-9b = 6$
6 $7x-3y = 49$, $2x+4y = -20$
7 $6a-b = -12$, $13a+5b = -4\frac{1}{2}$
8 $4x+9y = 8$, $9x-y = 1$
9 $3m-2n = 14$, $n-5m = 0$
10 $2x = 5-3y$, $y = 7+10x$

11 $\begin{pmatrix} 2 & 0 \\ 4 & -3 \end{pmatrix}\begin{pmatrix} x \\ y \end{pmatrix} = \begin{pmatrix} 5 \\ 4 \end{pmatrix}$

12 $\begin{pmatrix} 1 & -2 \\ 0 & -3 \end{pmatrix}\begin{pmatrix} a \\ b \end{pmatrix} = \begin{pmatrix} 3 \\ 6 \end{pmatrix}$

13 $\begin{pmatrix} 2 & -1 \\ 1 & 2 \end{pmatrix}\begin{pmatrix} x \\ y \end{pmatrix} = \begin{pmatrix} 7 \\ 1 \end{pmatrix}$

14 $\begin{pmatrix} 3 & -2 \\ 7 & 4 \end{pmatrix}\begin{pmatrix} k \\ l \end{pmatrix} = \begin{pmatrix} 19 \\ 1 \end{pmatrix}$

15 $\begin{pmatrix} 10 & -7 \\ 25 & -3 \end{pmatrix}\begin{pmatrix} s \\ t \end{pmatrix} = \begin{pmatrix} 27 \\ 53 \end{pmatrix}$

16 $\begin{pmatrix} 2 & -5 \\ -4 & 15 \end{pmatrix}\begin{pmatrix} p \\ q \end{pmatrix} = \begin{pmatrix} -1 \\ 4 \end{pmatrix}$

17 $\begin{pmatrix} 4 & 2 \\ 5 & 7 \end{pmatrix}\begin{pmatrix} x \\ y \end{pmatrix} = \begin{pmatrix} 7 \\ 15\frac{1}{2} \end{pmatrix}$

18 $\begin{pmatrix} \frac{1}{2} & \frac{1}{3} \\ 2 & 3 \end{pmatrix}\begin{pmatrix} c \\ d \end{pmatrix}=\begin{pmatrix} 2 \\ 13 \end{pmatrix}$

19 $\begin{pmatrix} 5 & -6 \\ 10 & 6 \end{pmatrix}\begin{pmatrix} b \\ c \end{pmatrix}=\begin{pmatrix} 10 \\ -4 \end{pmatrix}$

20 $\begin{pmatrix} 7 & -2 \\ 19 & 5 \end{pmatrix}\begin{pmatrix} x \\ y \end{pmatrix}=\begin{pmatrix} 1 \\ 7 \end{pmatrix}$

HARDER EXAMPLES 9

1 If $A=\begin{pmatrix} 1 \\ 2 \\ -3 \end{pmatrix}$ and $B=\begin{pmatrix} 4 \\ 1 \\ -5 \end{pmatrix}$, find (i) $A+B$; (ii) $A-B$; (iii) $3A-2B$.

2 Solve the equation: $3A+\begin{pmatrix} 2 & -1 \\ 1 & -5 \end{pmatrix}=\begin{pmatrix} 5 & 8 \\ -7 & -5 \end{pmatrix}$

3 If $3\begin{pmatrix} x \\ y \end{pmatrix}=\begin{pmatrix} -6 \\ 9 \end{pmatrix}$, find x and y.

4 Calculate $\begin{pmatrix} 3 & -2 \\ 1 & 0 \end{pmatrix}\begin{pmatrix} 5 & -3 \\ 2 & 1 \end{pmatrix}$

5 If $A=\begin{pmatrix} 1 & 2 \\ -1 & 1 \end{pmatrix}$, find A^2. Show that $A^2=2A-3I$.

6 $P=\begin{pmatrix} 1 & 2 \\ 1 & 0 \end{pmatrix}$, $Q=\begin{pmatrix} 0 & 2 \\ 1 & 1 \end{pmatrix}$

Evaluate (i) $(P+Q)^2$; (ii) P^2+Q^2

7 If $A=\begin{pmatrix} 2 & -3 \\ 1 & 1 \end{pmatrix}$, $B=\begin{pmatrix} 1 & -1 \\ 0 & 1 \end{pmatrix}$, $C=\begin{pmatrix} a & b \\ c & 0 \end{pmatrix}$, and $AB=BC$, find a, b and c.

8 Calculate x and y if $\begin{pmatrix} x \\ y \end{pmatrix}=\begin{pmatrix} 2 & 2 \\ 3 & -1 \end{pmatrix}\begin{pmatrix} 2 \\ 3 \end{pmatrix}$

9 Solve the equation: $\begin{pmatrix} -1 & 3 \\ 0 & 2 \end{pmatrix}\begin{pmatrix} x \\ y \end{pmatrix}=\begin{pmatrix} 4 \\ 6 \end{pmatrix}$

10 Solve the equation: $\begin{pmatrix} 0 & 4 \\ -2 & 1 \end{pmatrix}\begin{pmatrix} x \\ y \end{pmatrix}=\begin{pmatrix} -8 \\ 7 \end{pmatrix}$

11 Find x and y if: $\begin{pmatrix} 2 & -1 \\ 0 & 3 \end{pmatrix}\begin{pmatrix} 3 \\ y \end{pmatrix} = \begin{pmatrix} x \\ 1 \end{pmatrix}$

12 If $\begin{pmatrix} 3 & 2 \\ 2 & -1 \end{pmatrix}\begin{pmatrix} x \\ y \end{pmatrix} = \begin{pmatrix} x \\ -5 \end{pmatrix}$, find x and y.

13 Find a and b if $\begin{pmatrix} a & 2 \\ -1 & 5 \end{pmatrix}\begin{pmatrix} 1 \\ b \end{pmatrix} = \begin{pmatrix} -1 \\ 19 \end{pmatrix}$

14 Given that: $\begin{pmatrix} a & 3 \\ 1 & 2 \end{pmatrix}\begin{pmatrix} -2 & 1 \\ 4 & b \end{pmatrix} = \begin{pmatrix} 2 & c \\ 6 & 3 \end{pmatrix}$, find a, b and c.

15 Given that: $\begin{pmatrix} 3 & 1 \\ 1 & 0 \end{pmatrix}\begin{pmatrix} 4 & a \\ a & 2 \end{pmatrix} = \begin{pmatrix} 15 & b \\ c & 3 \end{pmatrix}$, find the value of each of a, b and c.

16 $a = \begin{pmatrix} 2 \\ 1 \end{pmatrix}$, $b = \begin{pmatrix} 3 \\ 4 \end{pmatrix}$, $c = \begin{pmatrix} x \\ y \end{pmatrix}$

If $3c = 2a - b$, find x and y. Find a 2 by 2 matrix A such that $Ac = a + b$.

17 Expand: $\begin{pmatrix} 1 & 2 \\ -1 & 3 \end{pmatrix}\begin{pmatrix} x \\ y \end{pmatrix}$

Evaluate $\begin{pmatrix} 3 & -2 \\ 1 & 1 \end{pmatrix}\begin{pmatrix} 1 & 2 \\ -1 & 3 \end{pmatrix}$

and hence solve the equation: $\begin{pmatrix} 1 & 2 \\ -1 & 3 \end{pmatrix}\begin{pmatrix} x \\ y \end{pmatrix} = \begin{pmatrix} 0 \\ -5 \end{pmatrix}$

18 State the inverse of the matrix $\begin{pmatrix} 2 & -1 \\ -5 & -1 \end{pmatrix}$

Hence or otherwise solve the equations: $2x - y = 7$, $-5x - y = -16$.

19 Calculate the determinant of the matrix $\begin{pmatrix} 3 & 2 \\ 4 & 3 \end{pmatrix}$

Write down the inverse of this matrix. Use your result to find the solution of the equations: $3x + 2y = 8$, $4x + 3y = 10\frac{1}{2}$. [L]

20 If A is a 2×2 matrix, find four distinct solutions of the equation $A^2 = A$. [AEB]

10 Mappings and functions

10.1 Definition

Look at these two sets of numbers: A = $\{-2, -1, 0, 1, 2\}$ and B = $\{2, 3, 4, 5, 6\}$. Can you see any relationship between them? The answer of course is that the difference between any two corresponding numbers is 4. Now look at the same numbers written in a different way:

$$\begin{array}{ccc} x & & y \\ -2 & \longrightarrow & 2 \\ -1 & \longrightarrow & 3 \\ 0 & \longrightarrow & 4 \\ 1 & \longrightarrow & 5 \\ 2 & \longrightarrow & 6 \end{array}$$

We use the arrows to show that there is a simple rule for changing the value of x into the value of y. We call this rule a *mapping*, or *function*. Our rule this time is that

$$y = x+4.$$

Another way of writing this is $x \rightarrow x+4$, which is read as 'x maps on to $x+4$'.

In mathematics, we go on to write that the function which maps x on to $x+2$ is written as

$$\mathrm{f}: x \rightarrow x+2.$$

Although it appears that the letter f is used to denote function, we can use any letter, so that $\mathrm{g}: x \rightarrow x+3$ denotes a different function which maps x on to $x+3$.

If we want to ask the question 'what does f map the number 1 on to?', we write this f(1). Looking at the diagram, or remembering what the rule is, we can see that $\mathrm{f}(1) = 5$.

EXERCISE 10a

Evaluate the following for the different functions given in each question:

1	$a:x \to x+3$	(i) a(0)	(ii) a(3)	(iii) a(−3)
2	$b:x \to x-2$	(i) b(5)	(ii) b(10)	(iii) b(−100)
3	$c:x \to 2x+1$	(i) c(3)	(ii) c(−2)	(iii) c(8)
4	$h:x \to 3x-2$	(i) h(4)	(ii) h(−1)	(iii) h(−3)
5	$s:x \to x^2$	(i) s(−2)	(ii) s(2)	(iii) s(0)
6	$t:x \to x^2-4$	(i) t(3)	(ii) t(−2)	(iii) t(6)
7	$m:x \to 2x^2+1$	(i) m(4)	(ii) m(−2)	(iii) m(−3)
8	$n:x \to x+1/x$	(i) n(7)	(ii) n(2)	(iii) n(−2)
9	$k:x \to x^2-1$	(i) k(3)	(ii) k(−2)	(iii) k(2)
10	$y:x \to x^2+x+1$	(i) y(2)	(ii) y(−3)	(iii) y(−1)

10.2 Alternative notation

Another way of writing the function $f:x \to x^2+1$ would be $f(x) = x^2+1$. The letter x is called a dummy variable and does not affect the answer. We could write the above function as $f(t) = t^2 + 1$. The rule here tells us that whatever the letter inside the brackets following the letter f, it will square it and add 1 to the result.

This particular notation is preferable, in the authors' opinion, in that it enables more difficult ideas to be represented.

EXAMPLE 1

$F(t)$ denotes the sum of the positive integers which are less than t. Find (i) F(1), (ii) F(8), (iii) F(−4).

(i) Since 1 is the first positive integer, there are no integers less than t, which in this case is equal to 1.

$\therefore$ F(1) = 0.

(ii) Here, $t = 8$, so the integers less than t are $\{1, 2, 3, 4, 5, 6, 7\}$.

$\therefore$ F(8) = 1+2+3+4+5+6+7 = 28.

(iii) Here, $t = -4$. Although this is a question about positive integers, F(−4) still has a meaning, because there are no positive integers less than −4.

$\therefore$ F(−4) = 0.

EXAMPLE 2

R(x) denotes the remainder when x is divided by 10 [x is an integer]. Find (i) R(73), (ii) R($10x$). (iii) Under what conditions is $R(x+y) = R(x)+R(y)$?

(i) If you divide 73 by 10, there is a remainder of 3.

$\therefore$ R(73) = 3.

(ii) The number $10x$ is always a multiple of 10, whatever the value of x. It follows, then, that there will be no remainder when $10x$ is divided by 10.

$\therefore$ R($10x$) = 0.

(iii) In order to answer this part of the question, let us look at one or two examples:

$$R(63+21) = R(84) = 4$$
$$R(63)+R(21) = 3+1 = 4.$$

In this case it is true.

$$R(65+26) = R(91) = 1$$
$$R(65)+R(26) = 5+6 = 11.$$

In this case it is not true. It should be clear to the reader that the reason for this is that the last digits of x and y added together result in a number greater than or equal to 10. We can summarise this mathematically by the statement

$$R(x+y) = R(x)+R(y) \quad \text{if} \quad R(x)+R(y) < 10.$$

10.3 Domain and range

In order to complete the definition of a mathematical function, we need to state the set of objects that the function is going to operate on. This set of objects is called the *domain* of the function.

Consider $g: x \rightarrow x^2$ with domain $\{-1, 0, 1, 2\}$ (Fig. 10.1)
The set of values obtained from the domain is called the *range* of the function.

In this example, the range will be the set $\{0, 1, 4\}$.

Image: If we consider the subset $\{-1, 0\}$ of the same domain, this set is mapped on to the set $\{0, 1\}$. We say that $\{0, 1\}$ is the image of the set $\{-1, 0\}$.

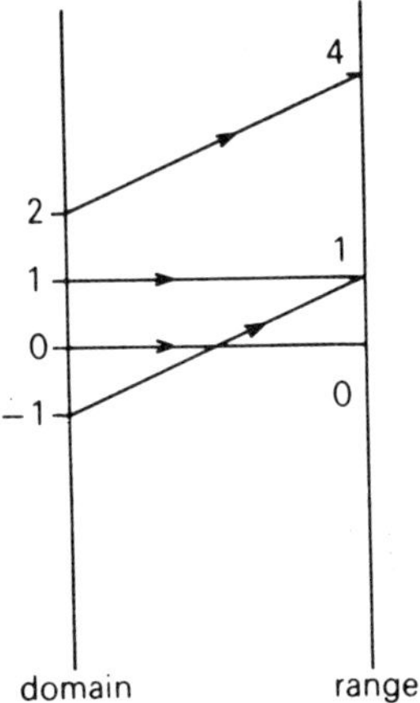

Fig. 10.1

10.4 Function of a function

Consider the following statement: If $f: x \rightarrow 2x+1$, find $f(2x)$. We have stated that the function f operates on the letter inside the brackets. We no longer have a single letter inside the bracket, but in fact a simple function (hence the phrase 'function of a function'). However, the function f will double what is inside the bracket, and add one.

$$\therefore\ f(2x) = 2 \times (2x) + 1$$
$$= 4x + 1.$$

10.5 Inverse of a function

Consider the example at the beginning of this chapter:

x		y
-2	$\longleftarrow$	2
-1	$\longleftarrow$	3
0	$\longleftarrow$	4
1	$\longleftarrow$	5
2	$\longleftarrow$	6

You will notice that we have reversed the arrows. The rule which takes us back again is called the *inverse* of the original mapping (written f^{-1}). In this case the rule is to subtract four, i.e.

$$f^{-1}: y \rightarrow y-4.$$

EXAMPLE 3

Find the inverse of the function $g: x \to 4x+3$.
This can be written $y = 4x+3$.

$$\therefore\ y-3 = 4x$$
$$\therefore\ \frac{y-3}{4} = x$$

(see 'Change of subject of formulae', chapter 5, page 71.)

$$\therefore\ g^{-1}: y \to \frac{y-3}{4}.$$

10.6 Composition of functions

If $f: x \to x+1$ and $g: x \to x^2$, then fg is called the product (or *composition*) of f and g and denotes $f(g(x))$, i.e. f applied to whatever $g(x)$ is.

$$g(x) = x^2$$
$$\therefore\ fg = f(g(x)) = f(x^2)$$
$$= x^2+1.$$

Consider $gf = g(f(x)) = g(x+1) = (x+1)^2$. We see that $fg \neq gf$. Composition of functions is *not commutative.*

EXAMPLE 4

If $f(x) = x^3-5x^2+2$, evaluate f(0) and f(1). Hence estimate a solution of the equation $x^3-5x^2+2 = 0$.

$$f(0) = 0^3-5\times 0^2+2 = 2$$
$$f(1) = 1^3-5\times 1^2+2 = -2.$$

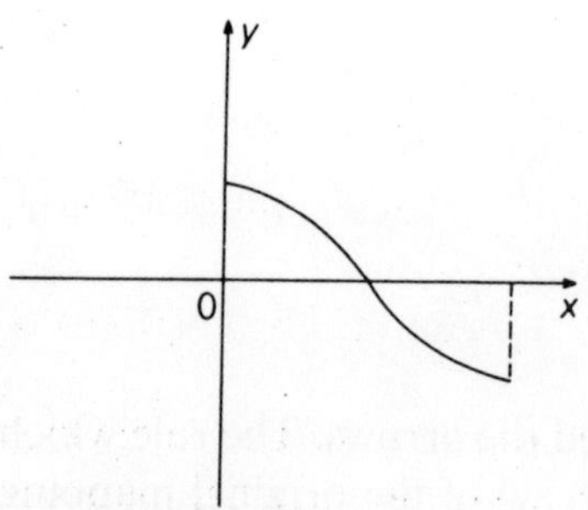

Fig. 10.2

Figure 10.2 shows a graph of this function (see chapter 11). In order to solve $x^3 - 5x^2 + 2 = 0$, we want to know where the curve cuts the x-axis. We would guess that this value is $x = \frac{1}{2}$. [Accurate methods would give the answer $x = 0.65$.]

EXAMPLE 5

If $f: x \to 3x - 1$ and $g: x \to x + 1$, find (i) $f^{-1}g^{-1}(3)$, (ii) $f^{-1}g^{-1}(3x)$.

To find f^{-1}:

$$y = 3x - 1$$
$$\therefore\ y + 1 = 3x$$
$$\therefore\ \frac{y+1}{3} = x$$
$$\therefore\ f^{-1}: x \to \frac{x+1}{3}.$$

Similarly $g^{-1}: x \to x - 1$.

(i) $g^{-1}(3) = 3 - 1 = 2$.

$$\therefore\ f^{-1}g^{-1}(3) = f^{-1}(2) = \frac{2+1}{3} = 1.$$

(ii) $g^{-1}(3x) = 3x - 1$

$$\therefore\ f^{-1}g^{-1}(3x) = \frac{(3x-1)+1}{3} = x.$$

EXERCISE 10b

1 If $f(x) = \dfrac{x+1}{x}$, find (i) $f(3)$; (ii) $f(\frac{1}{2})$; (iii) $f(-4)$.

2 $G(x)$ denotes the sum of all integers less than x which are perfect squares. Find (i) $G(25)$; (ii) $G(26)$.

3 If $g: x \to 3x + 1$, find g^{-1}.

4 If $h: x \to 2x - 1$, find $h^{-1}(-1)$.

5 If $f: x \to 2x + 3$, find t such that $f(t) = 0$.

6 $f(x)$ denotes the sum of all positive integers which do not divide into x. Find (i) $f(8)$; (ii) $f(20)$.

7 Given that $f: x \to x^2 - 14x + 33$, find (i) x if $f(x) = 0$; (ii) a value of x such that $f(x) < 0$.

8 $R(x)$ denotes the remainder when x is divided by 3. Find (i) $R(20)$; (ii) $R(20m)$; (iii) $R(21n)$, where m and n are integers.

9 Given that $m(x) = (x+1)(x-5)^2 - 28$, evaluate (i) $m(0)$; (ii) $m(1)$. Hence estimate a value of x for which $m(x) = 0$.

10 Find the inverse of the following functions:

(i) $f: x \to 1-x$ (ii) $g: x \to \frac{1}{x}$ (iii) $h: x \to x^2$

(iv) $m: x \to 1-3x$ (v) $n: x \to \frac{1}{x+1}$ (vi) $q: x \to \frac{ax+b}{cx+d}$.

HARDER EXAMPLES 10

1 Functions f and g map x on to $x+1$ and x^2+1 respectively. Show that fg maps x on to x^2+2 and find gf. What does this tell you about composition of functions?

A third function n maps x on to $b-ax$. If fgn maps x on to $x^2-6x+11$, find a and b.

2 The image of the set $\{-1, 0, 1\}$ under the function ax^2+b is $\{7, 5\}$. Find a and b.

3 The function f is defined on the domain $\{0, 1, 3, 5, 7, 9\}$ by $f: x \to$ the units digit of x^3. Draw a mapping diagram for this function.

4 $[x]$ denotes the largest integer which is less than or equal to x, i.e. $[3\frac{1}{2}] = 3$, $[5] = 5$, $[-2\frac{1}{4}] = -3$. The function $f: x \to [x]$ is defined on the domain $\{-2, 1\frac{1}{4}, -\frac{1}{2}, 0, 1, 1\frac{3}{4}, 1\frac{1}{8}, 2, 2\frac{3}{4}, 3\}$. Draw a mapping diagram for this function.

5 If $f: x \to 4x+1$ and $g: x \to x-1$, find f^{-1} and g^{-1}. What is $g^{-1}f^{-1}(3)$?

6 (i) $x * 4$ denotes the integer part when x is divided by 4, and $y \circ 7$ denotes the remainder when y is divided by 7. Calculate $81 * 4$ and $117 \circ 7$.

If $d = a+(a*4)+b \circ 7$, then the bth day of July in the year $(1900+a)$ is Saturday if $d = 0$, Sunday if $d = 1$ and so on up to Friday if $d = 6$. Calculate the day of the week on which the 16th July 1981 will fall.

(ii) Given that

$$f(x) = \frac{k}{x-1} + \frac{6}{x-2}$$

and that $f(5) = 8$, calculate the value of (a) k, (b) $f(4)$.

[C]

7 For any positive integer n, $T(n)$ is defined as the smallest multiple of 3 which is larger than or equal to n, e.g. $T(5) = 6$, $T(6) = 6$.

(i) Write down the value of $T(9)$ and that of $T(10)$.

(ii) Give the solution set of the equation $T(n) = 12$.

(iii) State whether each of the following equations is satisfied by all values of n, by some values of n, or by no values of n. Give the solution set of each equation which is satisfied by some values of n.
(a) $T(n+1) = T(n)+1$,
(b) $T(n+3) = T(n)+3$,
(c) $T(2n) = 2T(n)$.

[L]

8 The functions f, g and h map the set of all rational numbers to itself and

$$f: x \to 2$$
$$g: x \to \frac{x}{3}+1$$
$$h: x \to x^2$$

(i) Write down a formula for the mapping obtained, (a) by using g followed by f, and (b) by using f followed by h.
(ii) Express the function

$$x \to \frac{x^2+3}{3}$$

(with domain the same as f, g and h) in terms of f, g and h.
(iii) Find a function m such that $m(g(x)) = x$.

9 If $p: x \to 3x$ and $q: x \to x^2$, state which of the following are correct and which incorrect: (i) $p(5^{-1}) = \frac{3}{5}$, (ii) p and q are commutative, (iii) q^{-1} is always a function, (iv) $pq(x) = 3x^2$.

[O & C (SMP)]

10 (i) Given that $f(x)$ is defined by $f(x) = x^3 - x$, find the largest integer which is a common factor of f(3), f(4) and f(5).
(ii) $R(x)$ denotes the remainder when x is divided by 7. Write down two different positive integers, x and y, such that $R(x) = R(y)$.

[C]

11 Co-ordinates, graphs and linear programming

11.1 Plotting points

We can represent the position of a point P with respect to the x-axis ($y = 0$) and the y-axis ($x = 0$) by two numbers called the co-ordinates of P. The x-co-ordinate is the distance of P from the y-axis. The y-co-ordinate is the distance of P from the x-axis. The co-ordinates are then enclosed in brackets, i.e. the co-ordinates of P are (x, y). Note that the order is very important: $(2, 3) \neq (3, 2)$. In Fig. 11.1 the point A is $(4, 3)$; B is $(-2, 2)$; C is $(2, -2)$.

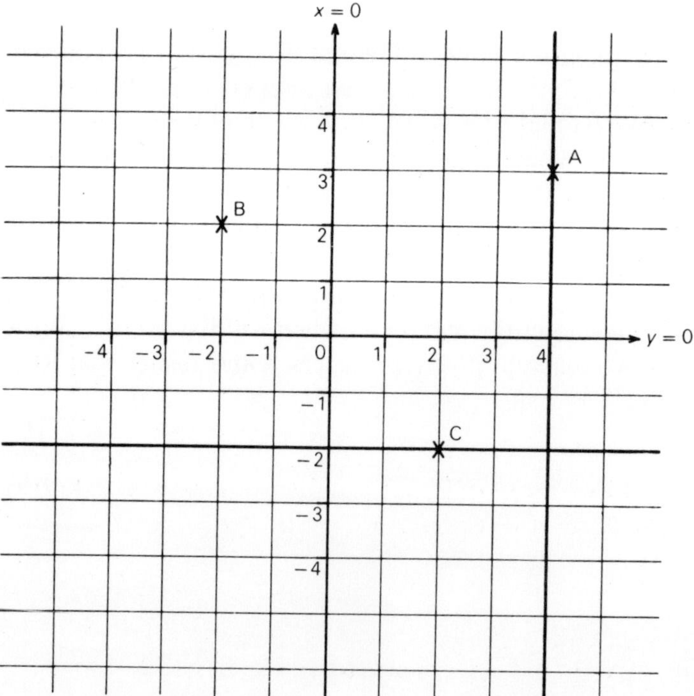

Fig. 11.1

11.2 The equation of a line

(i) *Lines parallel to the axes.* Referring to Fig. 11.1 again, consider the line through A parallel to $x = 0$. Whatever position you take on the line, the x-co-ordinate is 4. We say that its *equation* is

$$x = 4.$$

Similarly, consider a line through C parallel to $y = 0$. For every position on this line the y-co-ordinate is -2. Its equation is

$$y = -2.$$

(ii) *Lines not parallel to the axes.* The co-ordinates of the points in Fig. 11.2 are A(0, 3), B(1, 2), C(2, 1), D(3, 0).

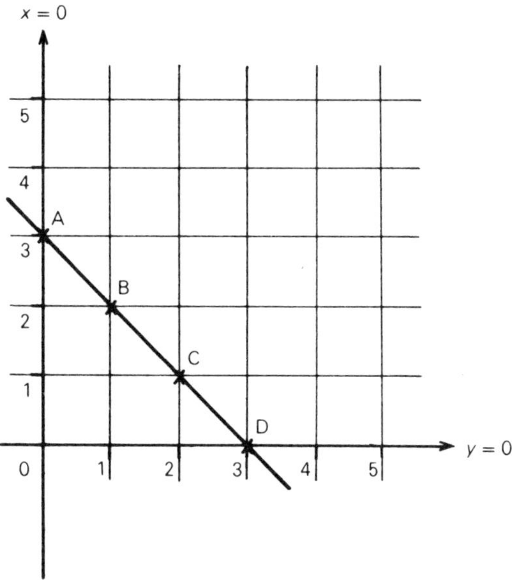

Fig. 11.2

If we add the x and y co-ordinates together we always get 3. We say that the equation of this line is

$$x + y = 3.$$

11.3 A more difficult example

The co-ordinates of the points in Fig. 11.3 are A(-4, 0), B(-2, 1), C(2, 3). The relationship here between x and y is not so obvious. It

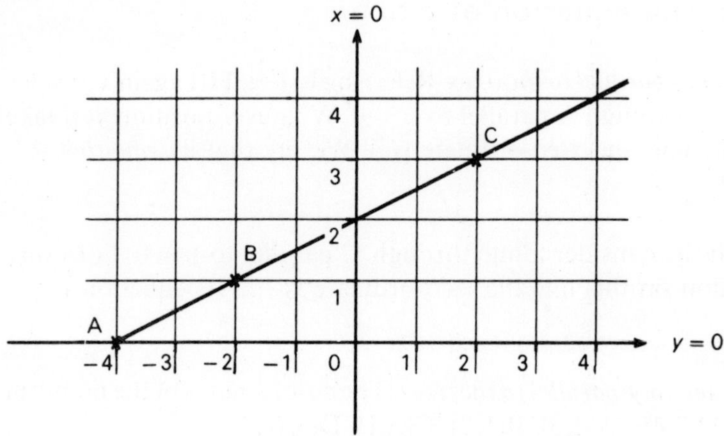

Fig. 11.3

is, in fact, that if you double the y-co-ordinate and subtract 4 you get the x-co-ordinate. Its equation is

$$x = 2y - 4 \qquad \text{or} \qquad 2y - x = 4.$$

Let us look at these two equations again:

$$x + y = 3 \qquad\qquad 2y - x = 4.$$

If we divide the first equation by 3 and the second by 4, we get

$$\frac{x}{3} + \frac{y}{3} = 1 \qquad\qquad \frac{y}{2} - \frac{x}{4} = 1.$$

The first line cuts the x-axis at 3 (the value under x in the equation) and the y-axis at 3. The second line cuts the y axis at 2 and the x-axis at -4 (note there is a negative sign in front of x).

Summary: The equation of a straight line which cuts the x axis at a and the y axis at b is

$$\frac{x}{a} + \frac{y}{b} = 1.$$

11.4 The gradient (sometimes called the slope) of a straight line

The *gradient* of a line is defined as the tangent of the angle that the line makes with the positive x-axis.

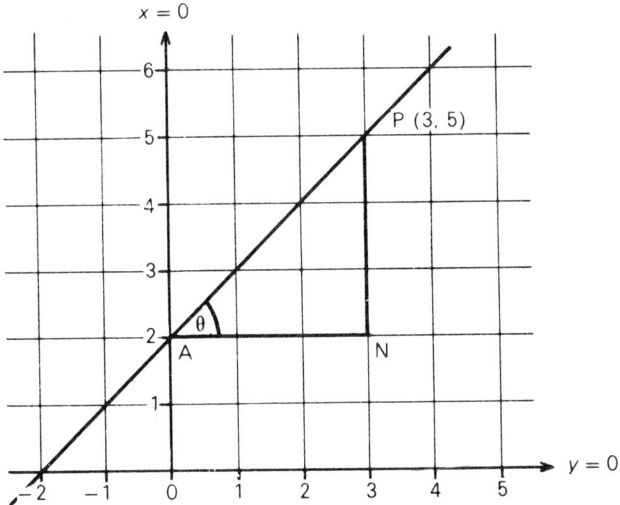

Fig. 11.4

In Fig. 11.4, the angle required is θ and

$$\tan\theta = \frac{PN}{AN} = \frac{3}{3} = 1$$

$\therefore$ the gradient of the line is 1.

If $\theta > 90°$, then the tangent becomes negative (see chapter 12).

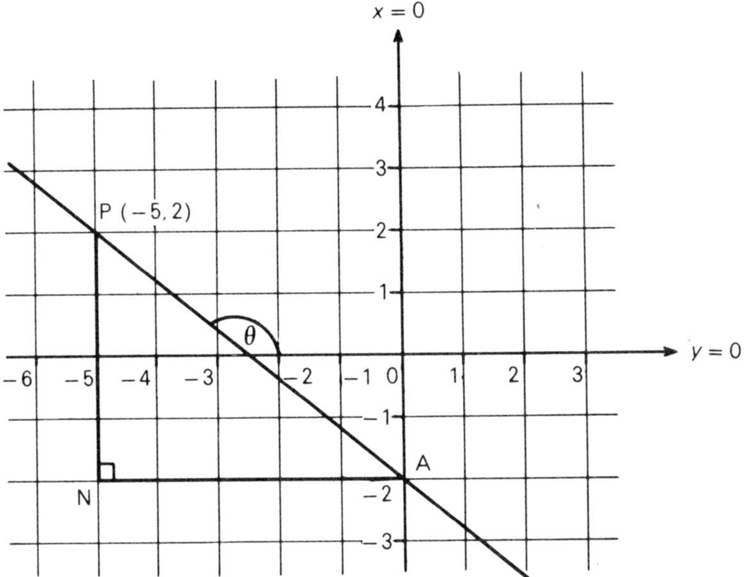

Fig. 11.5

In Fig. 11.5,

$$\tan \theta = -\frac{PN}{NA} = -\frac{4}{5}.$$

11.5 The equation of a straight line in the form $y = mx + c$

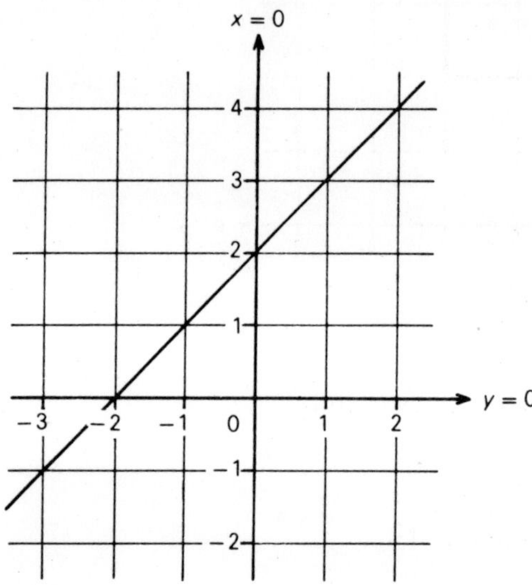

Fig. 11.6

Consider the straight line in Fig. 11.6. Section 11.3 gives its equation as:

$$\frac{x}{-2} + \frac{y}{2} = 1$$

i.e.

$$x - y = -2 \qquad \text{(multiplying by} -2)$$
$$y = x + 2.$$

We know that the slope is 1, and the point where it cuts the y-axis is 2. In the equation $y = mx + c$, m is the slope, and c the point where it cuts the y-axis.

$\therefore\ m = 1$ and $c = 2$

$\therefore$ equation of the line is $y = x + 2$.

Proof: given $y = mx + c$.
Put $x = 0$ (this is on the y-axis)

$y = m \times 0 + c = c$

$\therefore$ the line cuts the y-axis at $y = c$.

When $x = 1$, $y = m + c$.

$$\therefore \text{ slope of the line} = \frac{\text{increase in } y}{\text{change in } x} = \frac{(m+c)-c}{1} = m$$

$\therefore$ the slope of the line is m.

EXAMPLE 1

Draw a freehand sketch to show the position of the line $2y = -3x + 6$.

$$2y = -3x + 6$$
$$\therefore y = -\tfrac{3}{2}x + 3$$

$\therefore$ the slope is $-\frac{3}{2}$ and the line cuts the y-axis at 3.

The line is shown in Fig. 11.7.

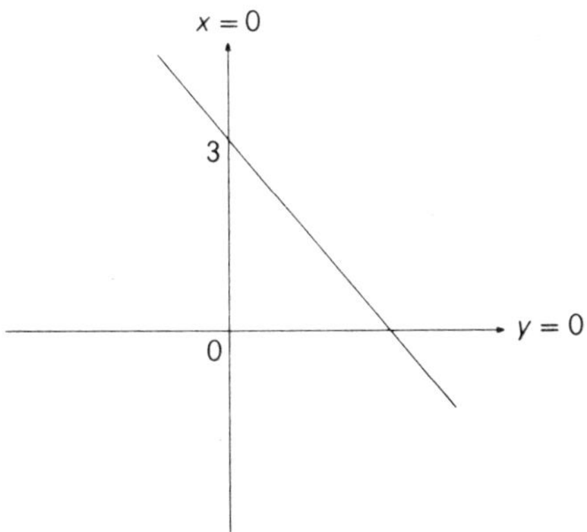

Fig. 11.7

EXERCISE 11a

Using any method you like, draw rough sketches to show the position of the following lines:

1 $\frac{x}{3} + \frac{y}{5} = 1$

2 $\frac{x}{6} - \frac{y}{2} = 1$

3 $3x + 2y = 12$

4 $4x - 5y = 20$

5 $y = 4x + 1$

6 $y = 3x - 5$

7 $y - x = 8$

8 $y - 3x + 5 = 0$

9 $x-2y+1=0$

10 $3x-y=5$

11 $\frac{1}{2}x-\frac{3}{4}y=7$

12 $2y-\frac{1}{4}x+1=0.$

11.6 Graphs and functions

The ideas that we have developed so far in this chapter are very similar to those put forward in chapter 10 on mappings and functions. In that chapter, we developed the idea of a mathematical rule assigning a value to each value of x. This is precisely what we are doing when plotting a graph.

EXAMPLE 2

Plot the graph of the function $f: x \to 2x+1$ for the domain $-4 \leq x \leq 4$. In this case, x, is mapped on to $2x+1$, i.e. $y = 2x+1$.

x	-4	-2	0	2	4
$y = 2x+1$	-7	-3	1	5	9

Note that we have set the results out in tabular form.

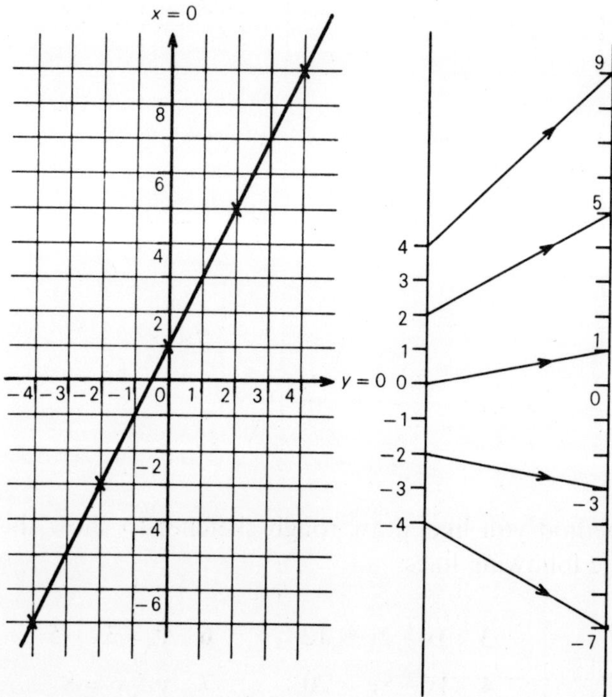

Fig. 11.8

In Fig. 11.8 we can see the two ways of representing this function: a graph, and a mapping diagram. Although we have only plotted a few points on the graph, experience of this type of function tells us we can join the points with a straight line, thus enabling us to find the value of y for any value of x in the given domain. This is obviously of considerably more use than any information we can gain from the mapping diagram.

EXAMPLE 3

Plot the graph of the function $g: x \rightarrow 3x - 2$ for $x \in \{-2, 1, 3, 0\}$

x	-2	0	1	3
$y = 3x - 2$	-8	-2	1	7

The domain this time consists only of 4 values.

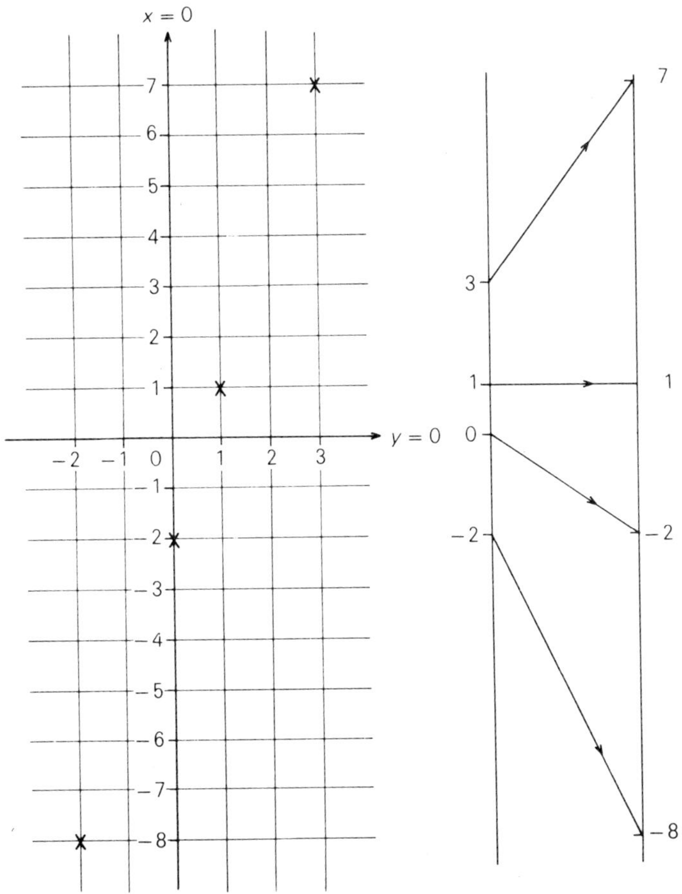

Fig. 11.9

EXAMPLE 4

Plot the graph of the function $f: x \to x^2 + 3x - 7$ for the domain $-4 \leq x \leq 2$. What is the minimum value of the function in this domain?

x	-4	-3	-2	-1	0	1	2
x^2	16	9	4	1	0	1	4
$3x$	-12	-9	-6	-3	0	3	6
-7	-7	-7	-7	-7	-7	-7	-7
$x^2 + 3x - 7$	-3	-7	-9	-9	-7	-3	3

Notice how we have set out the table. Although it appears to create extra work, breaking the function down into various parts helps to minimise mistakes.

Note: In Fig. 11.10, although the values at A and B are equal, the curve continues below AB, to give a minimum value of -9.25.

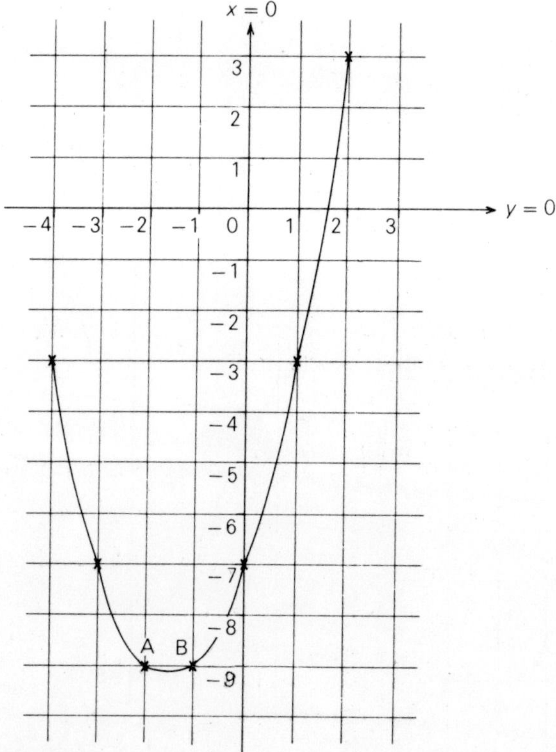

Fig. 11.10

11.7 Using graphs to solve equations

In order to solve the equation

$$x^2 - \frac{1}{x} = x^3,$$

plot the graphs of the functions

$$\text{f}: x \rightarrow x^2 - \frac{1}{x} \quad \text{and} \quad \text{g}: x \rightarrow x^3.$$

The points of intersection will give the solution of the equation. It is helpful if some idea of the answer is obtained first which enables the domain to be limited.

EXAMPLE 5

Show that a solution of the equation $x^3 - 1 = x$ lies between 1 and 2. By plotting suitable functions, find this solution. On the same diagram, draw a suitable straight line to solve the equation $x^3 + 2x + 1 = 0$.

$$x^3 - 1 = x \Rightarrow x^3 - x - 1 = 0.$$

If $x = 2$, left-hand side $= 2^3 - 2 - 1 = 5$.
If $x = 1$, left-hand side $= 1^3 - 1 - 1 = -1$.
Since the function has changed sign, the solution must be between 1 and 2. We plot the function $\text{f}: x \rightarrow x^3 - 1$ and $\text{g}: x \rightarrow x$. See Fig. 11.11. The required solution is $x = 1.32$.

In order to solve $x^3 + 2x + 1 = 0$, proceed as follows:

$$x^3 + 2x + 1 = 0 \Rightarrow x^3 - 1 = -2x - 2.$$

since we have already plotted f, we plot $\text{h}: x \rightarrow -2x - 2$. We see from Fig. 11.11 overleaf that the solution is $x = -0.45$.

EXERCISE 11b

Plot the following functions, using the domain $-3 \leq x \leq 3$ (include $x = \pm\frac{1}{2}$)

1 $\text{f}: x \rightarrow 3x - 2$

2 $\text{g}: x \rightarrow 4x + 5$

3 $\text{h}: x \rightarrow x^2 - 2$

4 $\text{k}: x \rightarrow 2x^2 - 1$

5 $\text{m}: x \rightarrow x^3 + 1$

6 $\text{n}: x \rightarrow \frac{1}{x} (x \neq 0)$

7 $\text{p}: x \rightarrow x^2 - 3x + 1$

8 $\text{t}: x \rightarrow x + \frac{1}{x} (x \neq 0)$

9 $\text{u}: x \rightarrow x^3 - 2x^2 + 1$

10 $\text{w}: x \rightarrow 1 - x - x^2$.

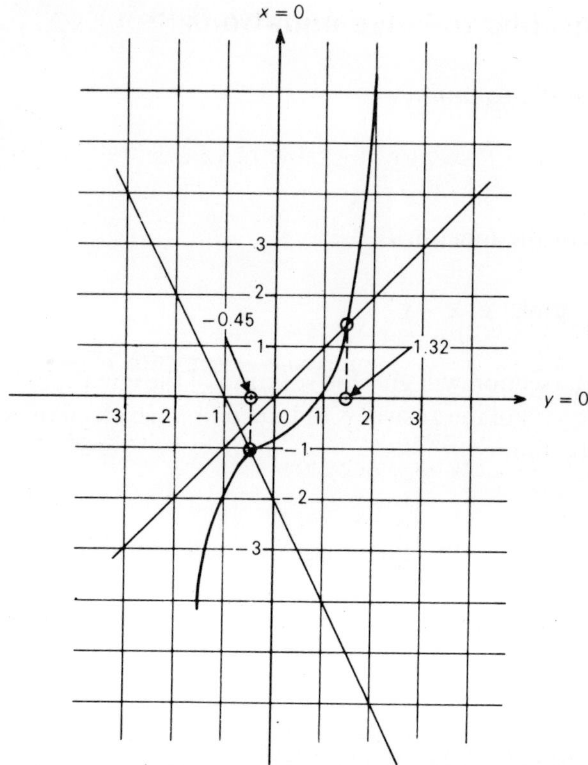

Fig. 11.11

11 Plot a graph of the function $f: x \to x + 1/x$ with domain $\{-1, 1, 2, 3, 4, 5\}$. Illustrate on a mapping diagram the image of the subset $\{-1, 1\}$.

12 Choose a suitable domain in order that the range of the function $g: x \to 2x^2 - 4$ should not be outside the limits of -6 to 10. Using this domain, plot the graph. What is the minimum value of $g(x)$ in this region?

13 Plot the graphs of the functions $f: x \to x^2 - 5$ and $g: x \to ax$ for $a = 1, 2, \frac{1}{3}$. Use the graphs you have drawn to solve the equations

(i) $x^2 - x - 5 = 0$

(ii) $x^2 - 2x - 5 = 0$

(iii) $3x^2 - x - 15 = 0$

Use a domain of $-4 \leq x \leq 4$.

14 Using your graph from question 10, plot a suitable straight line which enables you to solve the equation $2x^2 - 4x + 1 = 0$.

15 A cube of metal of side x cm has a hole of cross-sectional area 4 cm^2 drilled right through it in a direction perpendicular to one of the faces. The volume of the resulting block is $V\,\text{cm}^3$. Show that

$$V = x(x-2)(x+2).$$

Copy and complete the following table of values:

x	3	4	4.5	5	6
V			73		

and hence plot the graph of V against x.

The block is now melted down and made into a solid rectangular block whose sides are 3 cm, 3 cm and $(15-x)$ cm. Estimate the value of x by drawing a suitable straight line on the same axes.

[C]

11.8 Regions

(i) *Lines parallel to the axes.*

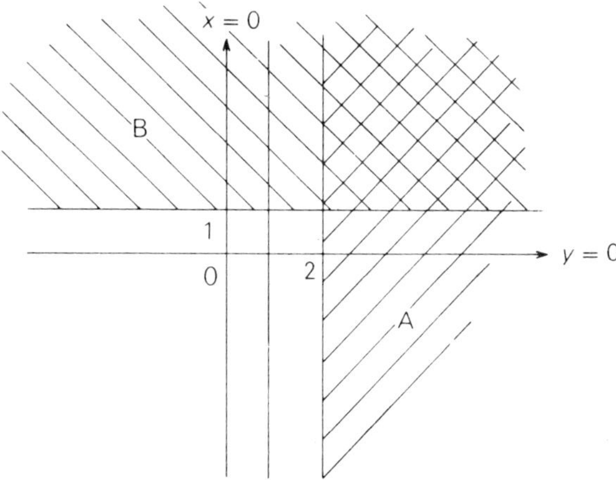

Fig. 11.12

For all the points in the region A shaded with lines from top right to bottom left, whatever the value of y, we always have $x \geq 2$. We describe region A as follows: $A = \{(x, y): x \geq 2\}$. Similarly region $B = \{(x, y): y \geq 1\}$. The region shaded with crossing lines is in fact $A \cap B$, and we have

$$A \cap B = \{(x, y): x \geq 2,\ y \geq 1\}.$$

(ii) *Lines not parallel to the axes.*

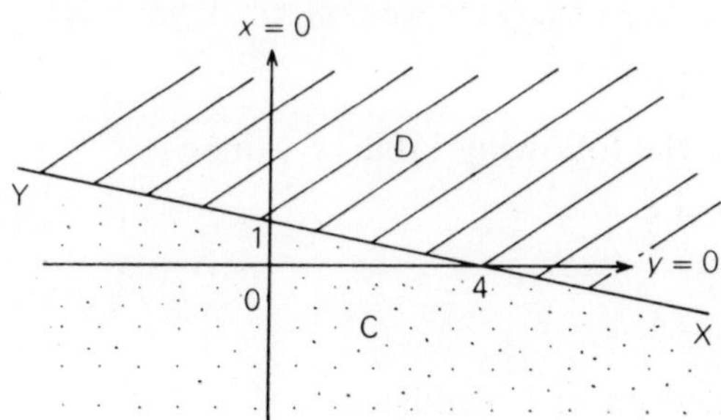

Fig. 11.13

In order to describe regions C and D we proceed as follows: The equation of YX is $y+\frac{1}{4}x = 1$, i.e. $4y+x = 4$.

Consider any point in region D, for example (3, 3). If we substitute these co-ordinates into the equation, we get $4\times 3+3 > 4$. The sign $>$ applies for every point in region D.

$\therefore$ D $= \{(x, y): 4y+x > 4\}$.

Similarly

C $= \{(x, y): 4y+x < 4\}$.

EXAMPLE 6

Draw a diagram and shade the region R given by

R $= \{(x, y): y \geq 1,\ x+y \leq 5,\ 3y-2x < 6\}$.

In order to do this, first draw a diagram and mark the lines $y = 1$, $x+y = 5$, $3y-2x = 6$. (See Fig. 11.14.)

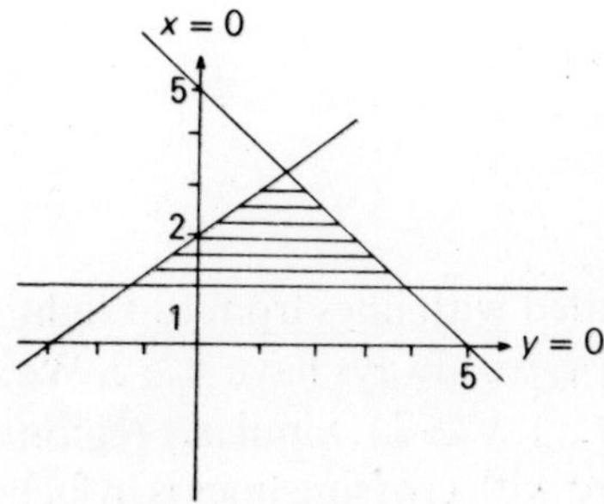

Fig. 11.14

Consider the origin (0, 0). Substituting $x = 0$ and $y = 0$ into the equations of the two sloping lines we certainly get

$$x+y = 0 < 5$$

and

$$3y-2x = 0 < 6$$

Introducing the extra condition that $y \geq 1$ the region is that which is shaded.

EXERCISE 11c

Draw diagrams to illustrate the following regions

1 $A = \{(x, y); x \leq 1, y \geq -1\}$
2 $B = \{(x, y): x \geq -3, y \leq -2\}$
3 $C = \{(x, y): x+y \leq 4, x \geq -2, y \geq 1\}$
4 $D = \{(x, y): x+y \geq -3, x \leq 2, y \leq 4\}$
5 $E = \{(x, y): x \geq y, y \leq 1\}$
6 $F = \{(x, y): 2y+x \geq 0, x \leq -3, y \geq 2\}$
7 $G = \{(x, y): y \geq x, x \geq 0, y \leq 4\}$
8 $H = \{(x; y): y+x \geq 0, y-x \leq 0\}$
9 $I = \{(x, y): y \leq 2, x \leq -y+1\}$
10 $J = \{(x, y): -1 \leq x \leq 3, -2 \leq y \leq x\}$.

11.9 Introduction to linear programming

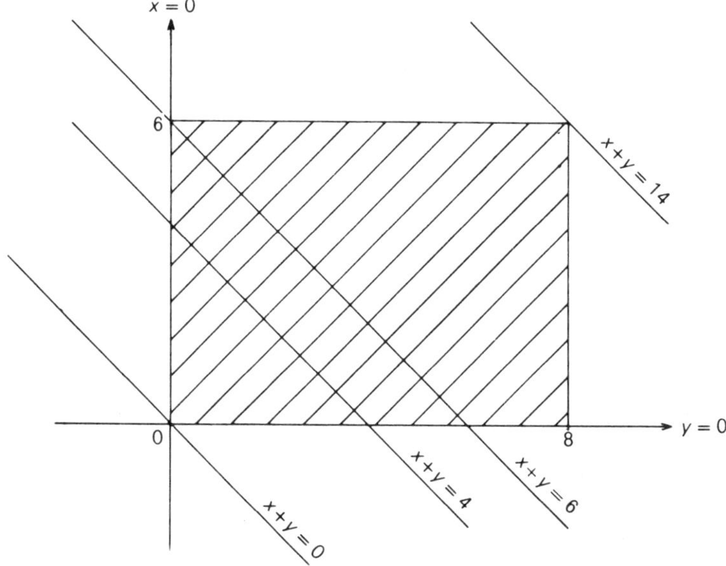

Fig. 11.15

Figure 11.15 shows the position of lines $x+y=c$, for various values of c. It can be seen that if we restrict our points (x, y) to the shaded region shown in the diagram, the smallest value of $x+y$ is 0, and the largest value of $x+y$ is 14. This idea is the basic idea in linear programming. We try to find the maximum or minimum value, of an expression of the form $ax+by$, in a given region. The process involves moving the line $ax+by=c$ (by altering c) until we have found the necessary result.

EXAMPLE 7

Find the maximum value of $2x+3y$ in the region

$$A=\{(x, y): x+y \leq 12,\ y \leq 6,\ x \geq 0,\ y \geq 0\}.$$

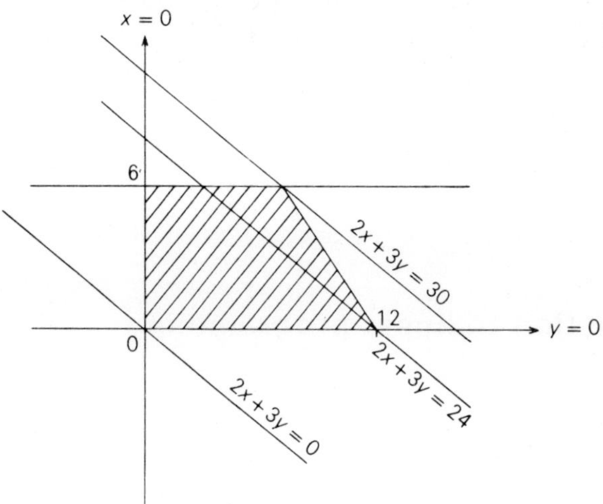

Fig. 11.16

Figure 11.16 shows that the maximum value occurs at the point (6, 6). So the maximum value of $2x+3y$ is 30.

EXAMPLE 8

A wholesaler stocks three items A, B and C. His total stock of these items is 400. His stock of A is at least 50 and of B at least 25. However, he finds it uneconomical to keep more than 80 of A or 60 of B or 100 of both taken together. He pays 50p for item A, 30p for item B and £1.20 for item C. His selling prices are 65p for item A, 40p for item B and £1.50 for item C. How can the wholesaler maximise his profits?

Let x be the number of item A stocked.
Let y be the number of item B stocked.
$\therefore$ number of item C stocked is $400-x-y$.

The conditions imposed are as follows:

$$50 \leq x \leq 80$$
$$25 \leq y \leq 60$$
$$x+y \leq 100$$

His total profit in pence is $15x+10y+30(400-x-y)$

$$= 12\,000-15x-20y.$$

This will be a maximum when $15x+20y$ is as small as possible. But

$$15x+20y = 5(3x+4y)$$
$\therefore$ $3x+4y$ has to be as small as possible.

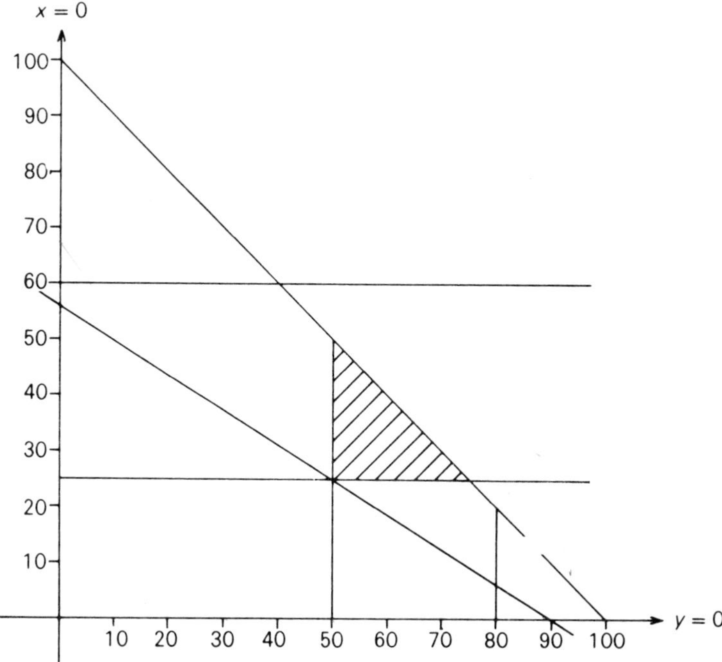

Fig. 11.17

Figure 11.17 shows that this occurs when $x = 50$, $y = 25$.

EXERCISE 11d

1 Using the regions from exercise 11c, find, if possible, the maximum and minimum values of (i) $x+2y$, (ii) $2y-x$.

2 A party of 87 children and 9 teachers are being taken on a river cruise in small boats. One type of boat takes 8 passengers and the other type 12. There must be at least one teacher in each boat. If x of the smaller boats and y of the larger boats are hired, show that $2x+3y \geq 24$, and write down a further constraint on x and y. Represent these on a linear programming diagram. If the cost of hiring the boats is £7 and £9 respectively, calculate the minimum cost of the expedition.

[SU]

3 A builder has 10 plots of land available on which to build old people's bungalows or houses. A bungalow uses half a plot and a house 1 plot. A bungalow requires 1500 man-hours in building and a house 2000 man-hours. The builder estimates he has 22500 man-hours available. If he builds x bungalows and y houses show that

$$x+2y \leq 20$$
$$3x+4y \leq 45.$$

Plot the straight lines $x+2y = 20$ and $3x+4y = 45$ on squared paper and shade in the region in which the two inequalities above are satisfied as well as $x \geq 0$ and $y \geq 0$.
(i) If he makes £800 profit on a bungalow and £1400 profit on each house, determine what values of x and y will give him the greatest profit.
(ii) In order to induce him to build old people's bungalows the local council offer him a grant so that he makes £1000 profit on each bungalow instead of £800. What values of x and y now give him the greatest profit?

[O]

HARDER EXAMPLES 11

1 The functions f and g are defined as follows:

$$f: x \to 1-2x, \quad g: x \to \frac{4}{x}.$$

(i) Find the values of (a) f(3); (b) $g(\frac{1}{4})$; (c) $gf(-1)$.
(ii) If $fg(x) = -0.6$, find the value of x.

[NI]

2 The function $f: x \to x^2+2$ has as domain the set of all integers. Are the following statements true or false?
(a) $f(-3) = 7$.
(b) The image of the set $\{-2, 2\}$ is $\{6\}$.

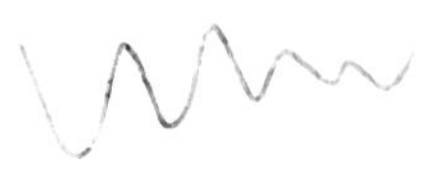

(c) The image of the domain is the same set as the domain. [AEB]

3 Using the domain $-4 \leq x \leq 4$, plot the graphs of the following functions. In each case, (a) state the greatest and least values of the function in the range; (b) by drawing a suitable straight line, solve if possible the equation $f(x) = x-1$, giving the equation that you have solved in its simplest form.

(i) $f:x \to x^2$ (ii) $f:x \to 2x^2-3$

(iii) $f:x \to \dfrac{1+x}{x}$ $(x \neq 0)$ (iv) $f:x \to x^3$

(v) $f:x \to 1-\dfrac{x^2}{x-5}$.

4 The first table gives a set of values of x and the corresponding values of y such that $y = (6/x)+x$:

x	1	2	3	4	5	6
y	7	5	5	5.5	6.2	7

Using these values of x and y and any others required, draw the graph of $y = (6/x)+x$, for the range $x = 1$ to $x = 6$.

The second table gives the values of x and the corresponding values of y such that $y = 6x-x^2$:

x	1	2	3	4	5
y	5	8	9	8	5

Using the same axes and the same scale as for the first graph, draw the graph of $y = 6x-x^2$ for the range $x = 1$ to $x = 5$.
(i) Calculate the value of y when $x = 1.25$ in each equation. State whether one solution of the equation $(6/x)+x = 6x-x^2$ is slightly greater or is slightly less than 1.25.
(ii) Use the values $x = 2.4$ and $x = 2.5$ to obtain a good estimate of the value of x for which $(6/x)+x$ has a minimum value within the given range.

[O]

5 The functions f and g are defined on the set of real numbers by

$f:x \to |x^2-4|, \quad g:x \to 4.$

Using the same scales and axes, draw the Cartesian graphs of f and g for values of x from -3 to $+3$ and show from your graphs that there are three values of x in this domain such that $f(x) = g(x)$.
Estimate these values of x correct to one decimal place.

[JMB]

6 Write down the three inequalities which define the region shown in Fig. 11.18.

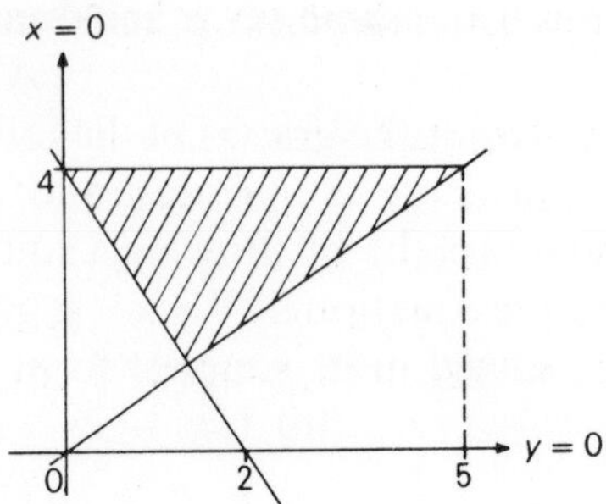

Fig. 11.18

7 On the same diagram, using a scale of 2 cm to represent 5 units on each axis and taking the x axis from -10 to $+30$, draw and clearly label the graphs of $2x+3y=60$ and $y-x=8$. On your graph, shade vertically the region in which $2x+3y<60$ and shade horizontally the region in which $y-x>8$.

Write down the co-ordinates of the points in the unshaded region (including its boundaries) for which $(x+y)$ takes the least possible value in each of the following cases:
(a) if x and y may take any real values
(b) if x and y are restricted to integer values.

8 The ordered pair (x, y) satisfies the following inequalities: $x>-4$; $x+y<0$; $y+2>x$. If A is the set of all such ordered pairs with x and y both integers, find $n(\text{A})$.

9 A firm has a fleet of vans and trucks. Each van can carry 9 crates and 3 cartons; each truck can carry 4 crates and 10 cartons. The firm has to deliver a total of 36 crates and 30 cartons. If x vans and y trucks are used to make the delivery, write down the two inequalities (other than $x \geq 0$ and $y \geq 0$) which must be satisfied by x and y. Illustrate your two inequalities graphically.

Given that the cost of using a van is exactly the same as that of using a truck, find the two pairs of values of x and y which give the smallest total cost of delivery.

[C]

10 Draw a sketch of the following curves:

(i) $f: x \to \dfrac{10}{x}$

(ii) $g: x \to x^2 + \dfrac{1}{x}$

(iii) $f: x \to 2x, \quad x \leq -1$
$g: x \to x^2-3, \quad -1 \leq x \leq 1$
$h: x \to x-3, \quad x \geq 1.$

11 Draw sketch graphs to show the relationship between the following quantities:

(i) The area of a circle (A) and its radius (r).

(ii) The length (x) and breadth (y) of a rectangle of area 10 cm^2.

(iii) Temperature in °F (F) and temperature in °C (C).

(iv) The total surface area of a cylinder (S) of height 5 cm, and its radius (R).

12

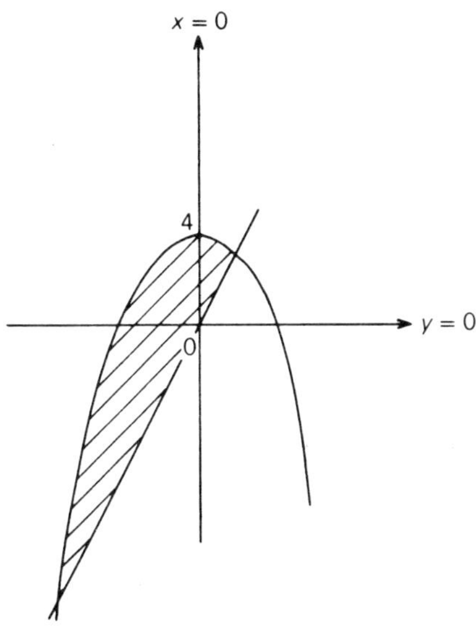

Fig. 11.19

Figure 11.19 shows the shaded region

$$A = \{(x, y): y \leq 4 - x^2,\ y \geq 3x\}.$$

If the ordered pair (a, b) lies in this region, find:

(i) the smallest value of b;

(ii) the largest value of a;

(iii) the number of points for which a and b are integers.

12 Right-angled triangles and beyond

12.1 Projection

In Fig. 12.1, the line BY is perpendicular to AX. BY meets AX at C.

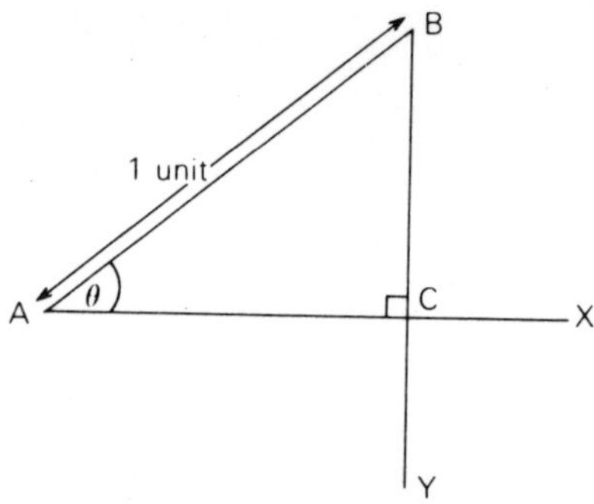

Fig. 12.1

We say that AC is the *projection* of AB in the direction AX. Similarly, BC is the projection of BA in the direction BY.

AC is said to be *adjacent* to $\angle A$.

BC is said to be *opposite* to $\angle A$.

If the length of AB is one unit, these projections have a special name:

AC = *cosine* of θ (or $\cos\theta$)
BC = *sine* of θ (or $\sin\theta$)

If θ increases, then clearly AC decreases; therefore $\cos\theta$ decreases with increasing θ.

If θ increases, then clearly BC increases; therefore $\sin\theta$ increases with increasing θ.

Since the projections are always less than AB, we have

$\cos\theta \leq 1$ (equality occurs if $\theta = 0°$)

and

$\sin\theta \leq 1$ (equality occurs if $\theta = 90°$).

The values of sin θ are tabulated. An extract is shown in Fig. 12.2. For example, to find sin 36° 45′:

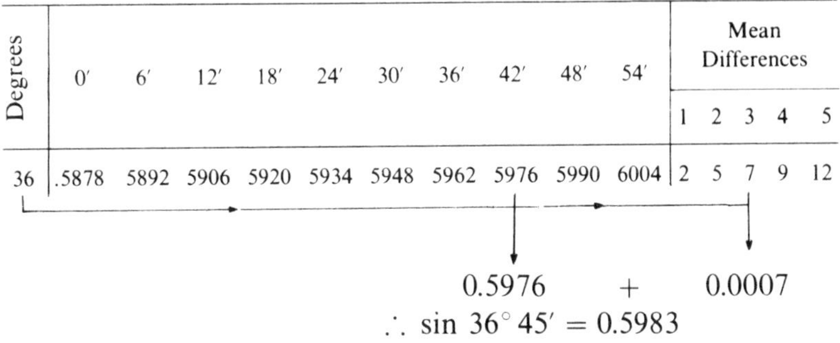

Degrees	0′	6′	12′	18′	24′	30′	36′	42′	48′	54′	Mean Differences 1	2	3	4	5
36	.5878	5892	5906	5920	5934	5948	5962	5976	5990	6004	2	5	7	9	12

0.5976 + 0.0007

$\therefore$ sin 36° 45′ = 0.5983

Fig. 12.2

In order to look up the cosine of an angle, we subtract the angle from 90° and use the sine tables.

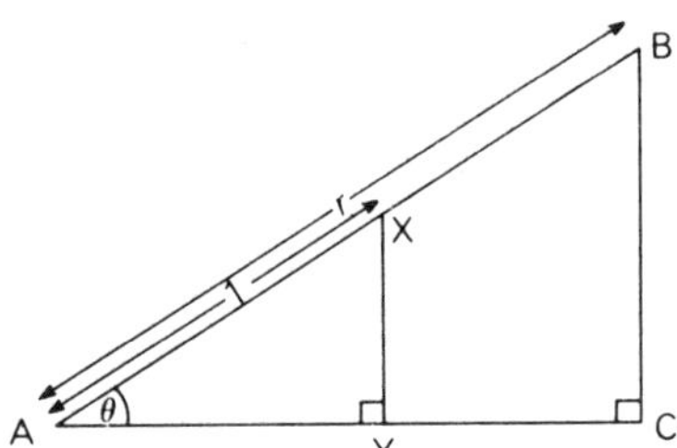

Fig. 12.3

If AB is not of unit length, then we can use a simple scale factor technique to find the projections. In Fig. 12.3, triangle ABC is obtained from triangle AXY by an enlargement centre A, scale factor r.

$$\therefore \text{AC} = r \times \text{AY}$$
$$= r \cos \theta.$$

Similarly, BC = $r \sin \theta$.

EXAMPLE 1

A car travels a distance of 500 m up a hill inclined at an angle of 20° to the horizontal. If the car was originally at sea level, find the height it is above sea level at the end of 500 m.

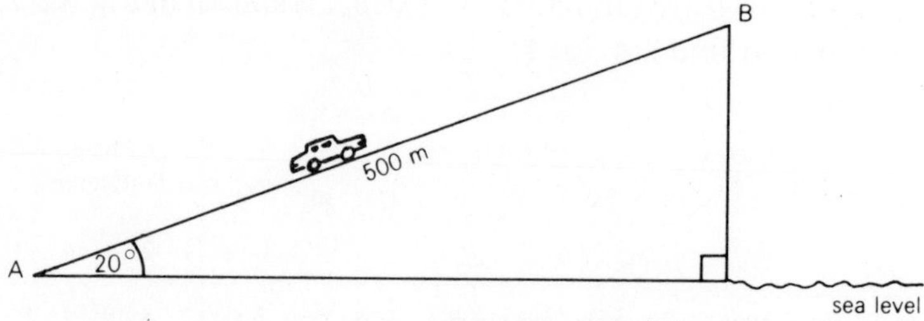

Fig. 12.4

The height above sea level is equal to BC.

$$\begin{aligned} BC &= 500 \sin 20^\circ \\ &= 500 \times 0.342 \\ &= 171 \text{ m.} \end{aligned}$$

EXAMPLE 2

The distance apart on a map of two points A and B is 750 m. It is known that AB slopes upwards at an angle of 15° to the horizontal. What is the real distance between A and B?

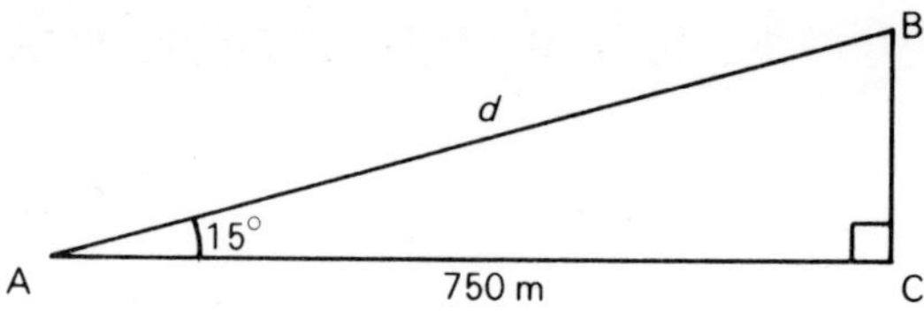

Fig. 12.5

The distance represented on the map is the projection of AB on the horizontal. (See Fig. 12.5.)

$$\therefore d \cos 15^\circ = 750$$

$$\therefore d = \frac{750}{\cos 15^\circ} = \frac{750}{\sin 75^\circ} = \frac{750}{0.9659}$$

$$\Rightarrow d = 776 \text{ m.}$$

EXAMPLE 3

In the right-angled triangle ABC, where $\angle B = 90^\circ$, AB = 3 cm and AC = 5 cm, find $\angle A$.

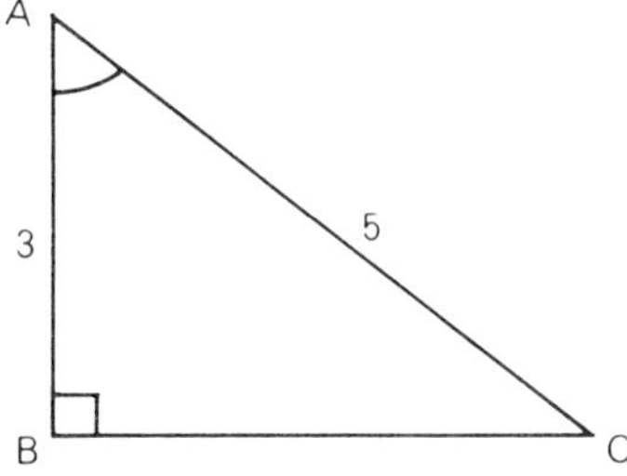

Fig. 12.6

$$3 = 5\cos A$$
$$\therefore \tfrac{3}{5} = \cos A$$
$$\therefore 0.6 = \sin(90^\circ - A).$$

In Fig. 12.2, we see that the nearest number to 0.6 is 0.6004 = sin 36° 54′. In the mean difference column, the nearest number to 4 is 5, which is 2′.

$$\therefore 0.6000 = \sin 36^\circ 52'$$
$$\Rightarrow 90^\circ - A = 36^\circ 52'$$
$$\therefore A = 90^\circ - 36^\circ 52' = 53^\circ 08'.$$

12.2 The tangent of an angle

Referring back to Fig. 12.1, we define the *tangent* of θ ($\tan\theta$) by

$$\tan\theta = \frac{\sin\theta}{\cos\theta} = \frac{BC}{AC} = \frac{(\text{side opposite } \theta)}{(\text{side adjacent to } \theta)}.$$

The tangents of all the angles between 0° and 90° are tabulated in a similar way to the sines and cosines.

EXAMPLE 4

A ladder leans against a wall. It is held in place by a piece of string tied to the bottom of the ladder, and a hook at the bottom of the wall. The string is $1\frac{1}{2}$ m in length. The top of the ladder rests on a window-sill and the ladder makes an angle of 75° with the horizontal. How far is the window-sill from the ground?

Very roughly, we have

$$\tan 75^\circ = \frac{h}{1\frac{1}{2}}$$

$$\therefore h = 1\tfrac{1}{2} \times \tan 75^\circ$$
$$= 1\tfrac{1}{2} \times 3.7321$$
$$\text{i.e. } h = 5.6 \text{ m}.$$

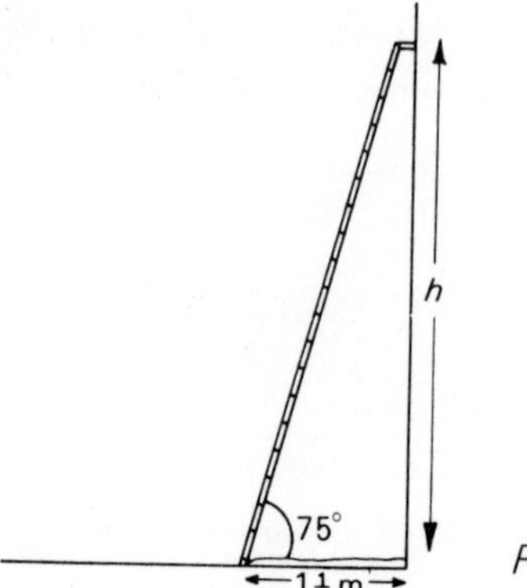

Fig. 12.7

12.3 Pythagoras' theorem

The famous Greek mathematician Pythagoras has provided us with a simple formula for finding the sides of right-angled triangles, without considering the angles.

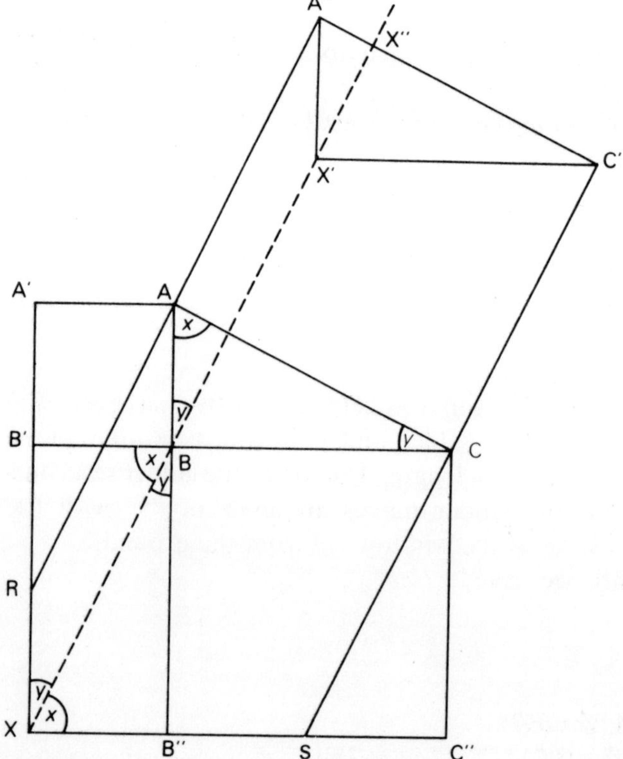

Fig. 12.8

ABC is a right-angled triangle (see Fig. 12.8). Consider squares drawn on the sides AB, BC, CA. He proved that the area of the square on AC (the hypotenuse) is equal to the sum of the area of the squares on the other two sides, i.e.

$$AC^2 = AB^2 + BC^2.$$

To prove it here, we shall use the ideas of transformation geometry developed in chapter 16.

Proof: Draw XX′ at right angles to AC.

The area of ABB′A′ = area ABXR (shear parallel to AB).

The area of BCC″B″ = area BCSX (shear parallel to CB)

ABC → BB′X (by a rotation of 90°)
∴ XB = CC′
ABC → A″X′C′ (by a translation $\overrightarrow{CC'}$)
∴ area BCSX = area BCC′X′ = area YCC′X″ (shear parallel to CC′)

Similarly,

area ABX′A″ = area ABXR = area AYX″A″ (shear parallel to AA″)

∴ area BCC″B″ + area ABB′A′ = area AA″C′C.

$$\therefore AC^2 = AB^2 + BC^2.$$

EXAMPLE 5

In triangle ABC, ∠B = 90°, AB = 5, BC = 12. Find AC.

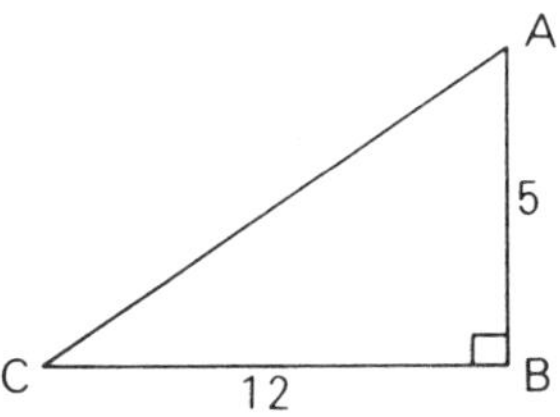

Fig. 12.9

$$AC^2 = AB^2 + BC^2$$
$$= 5^2 + 12^2 = 25 + 144 = 169$$
$$\therefore AC = 13.$$

EXAMPLE 6

In triangle RST, $\angle S = 90°$, RS = 4, RT = 5. Find ST.

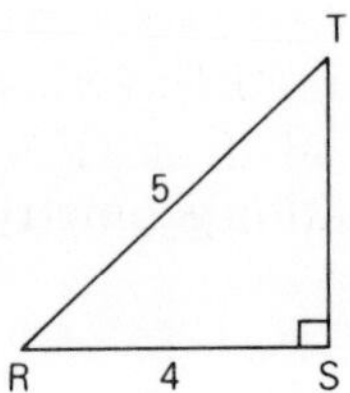

Fig. 12.10

$$RT^2 = RS^2 + TS^2$$
$$\therefore\ 5^2 = 4^2 + TS^2$$
$$\therefore\ 25 = 16 + TS^2$$
$$\therefore\ 25 - 16 = TS^2$$
$$\therefore\ 9 = TS^2$$
$$\therefore\ TS = 3.$$

EXERCISE 12

Find all the unknown sides and angles in the following triangles:

1 $\angle B = 90°$, $\angle C = 23°$, AB = 7 cm

2 $\angle A = 90°$, AB = 5 cm, BC = 13 cm

3 $\angle R = 90°$, RS = 6 cm, TR = 3 cm

4 $\angle X = 24°$, $\angle Y = 90°$, XY = 2.7 m

5 $\angle A = 23°\,15'$, $\angle B = 90°$, AC = 152.4 cm

6 $\angle P = 90°$, RQ = 6.32 cm, PQ = 4.18 cm

7 $\angle M = 90°$, MN = 1200 m, ML = 485 m

8 AC = 0.24 m, $\angle B = 90°$, AB = 0.18 m

9 $\angle D = 35°\,19'$, $\angle E = 90°$, EF = 6.28 km

10 AB = 0.032 cm, AC = 0.049 cm, $\angle A = 90°$.

11 A car is travelling along a road which is inclined at 20° to the horizontal. How long does it take the car's height above sea level to increase by 50 m, if it is travelling at 50 m/s?

12 A large radio aerial is 20 m high. It is supported by 3 stout wires which are attached to a point 2 m from the top of the aerial and fastened to points on the ground which lie on a circle of radius 5 m. If a total of 2 m of wire is required at the various fixing points, how much wire is needed?

13 An aircraft flies in the direction due north at 200 km/h from an airfield A. After 50 minutes, the pilot changes course and flies due east. After a further 25 minutes, he flies over a town (B). Find the distance from A to B.

14 A man stands on a cliff 100 m high, and observes a boat travelling away from him in a straight line. The angle of depression is 40°, and one minute later, it is 30°. How fast is the boat travelling?

15 A right-angled triangle has a hypotenuse of length $(x+h)$ cm, and the other two sides are of length x cm and $2h$ cm. If A is the set of all such triangles with x and h integers, and $1 \leq x \leq 20$, find $n(\text{A})$.

12.4 Angles of elevation and depression

Figure 12.11 illustrates the meaning of the two terms *angle of elevation* (measured upwards from the horizontal), and *angle of depression* (measured downwards from the horizontal). We see that:

the angle of depression from the man = angle of elevation from the boat.

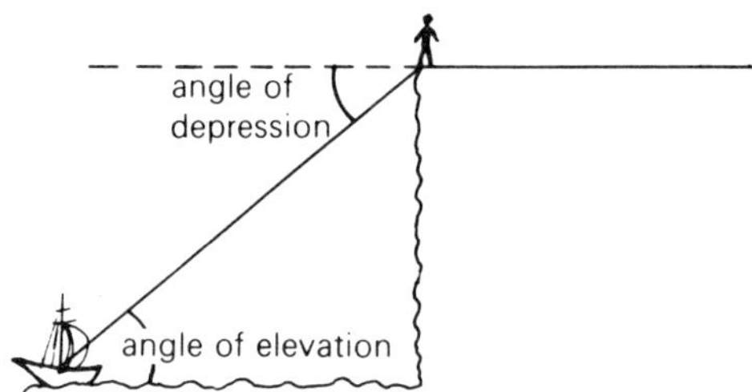

Fig. 12.11

EXAMPLE 7

A man is standing on the top of a cliff 150 m high. He observes a boat travelling at 40 km/h leaving the base of the cliff and he watches it for 1 minute moving in a straight line. What is the angle of depression of the boat after this time?

Before we can find x, we need to know the distance (d) the boat has travelled. In metres

$$d = \frac{40\,000}{360} = \frac{2\,000}{3}$$

$$\therefore \tan x = \frac{150}{2000/3} = \frac{150 \times 3}{2000} = \frac{9}{40} = 0.225$$

$$\therefore \text{angle of depression} = 12^\circ\,41'.$$

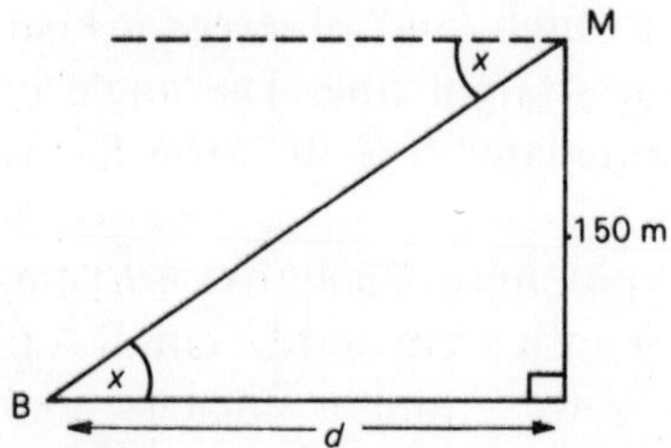

Fig. 12.12

12.5 Angles greater than 90°

We can extend the definition of sine, cosine and tangent to angles greater than 90°.

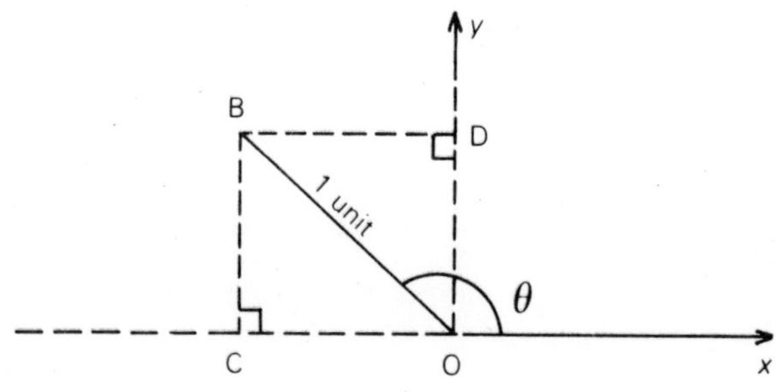

Fig. 12.13

In Fig. 12.13, we define the projection of OB in the direction OX to be $-$OC.

$$\therefore \cos\theta = -\text{OC} = -\cos(180^\circ - \theta).$$

The projection in the direction of OY is OD

$$\therefore \sin\theta = \text{OD} = \text{BC} = \sin(180^\circ - \theta).$$

For example, $\cos 127^\circ = -\cos 53^\circ$; $\sin 146^\circ = \sin 34^\circ$.

In Fig. 12.14, we see how the sine and cosine vary between 0° and 360°. Between 180° and 270°:

$$\sin 231^\circ = -\sin(231^\circ - 180^\circ) = -\sin 51^\circ$$
$$\cos 247^\circ = -\cos(247^\circ - 180^\circ) = -\cos 67^\circ.$$

Between 270° and 360°:

$$\sin 310^\circ = -\sin(360^\circ - 310^\circ) = -\sin 50^\circ$$
$$\cos 325^\circ = \cos(360^\circ - 325^\circ) = \cos 35^\circ.$$

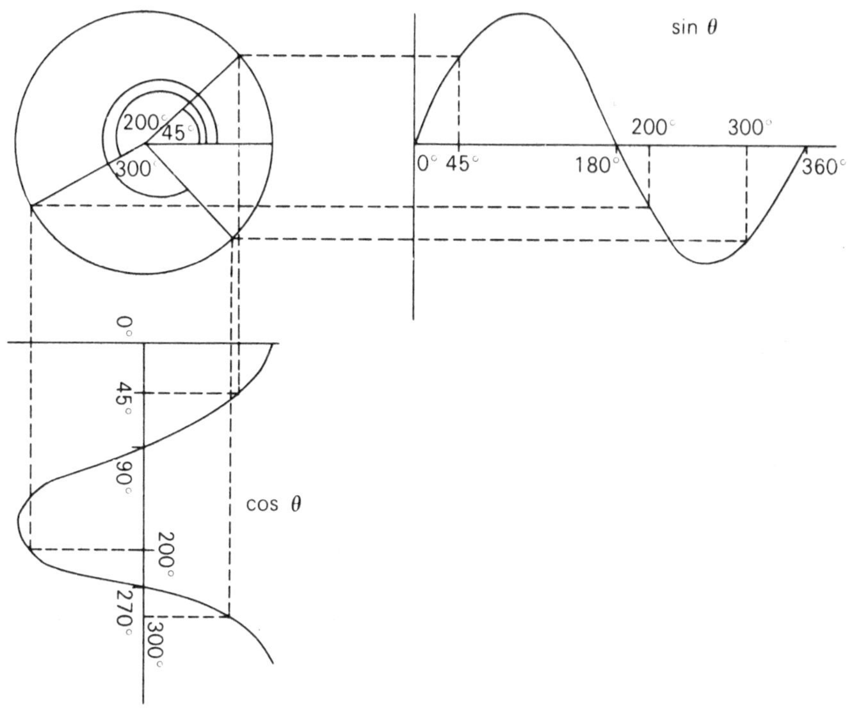

Fig. 12.14

HARDER EXAMPLES 12

1

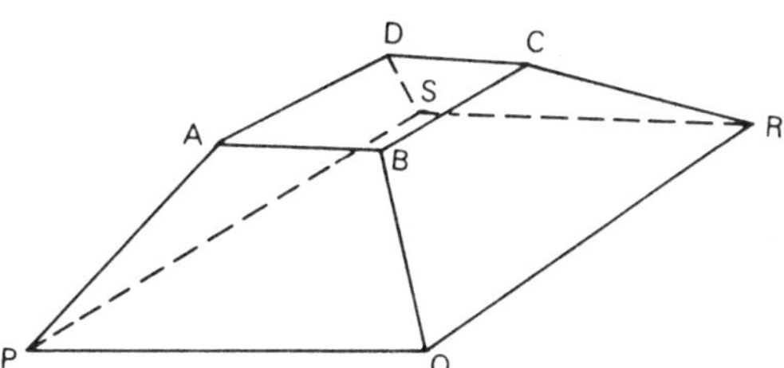

Fig. 12.15

The top and bottom of a pedestal are horizontal squares of side 2 metres and 4 metres respectively. The centres of the squares are in the same vertical line and the corresponding sides of the squares are parallel. The length of each sloping edge is 2 metres. Calculate

(i) the length of each of the diagonals AC and PR,
(ii) the angle which the sloping edge AP makes with the diagonal PR,
(iii) the perpendicular distance between the two planes ABCD and PQRS,
(iv) the angle which a sloping face makes with the plane PQRS.

2 [NI]

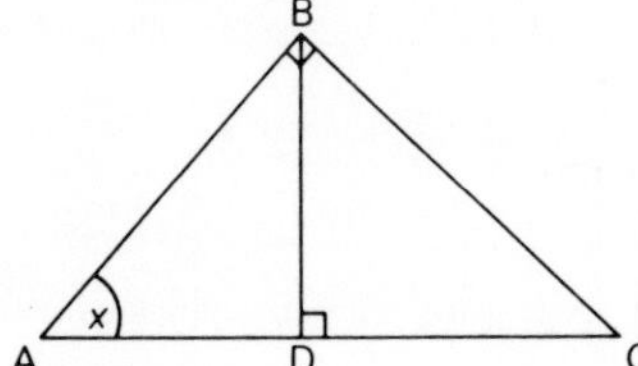

Fig. 12.16

In Fig. 12.16 ABC is a right-angled triangle, and $\angle A = x^\circ$. Mark another angle in the diagram equal to x°, and by expressing $\tan x^\circ$ in two different ways, prove that $BD^2 = AD.DC$.

3 Find all the angles between 0° and 360° which satisfy the equations:
(i) $\sin x = \frac{1}{2}$; (ii) $\cos x = -0.866$.

4 A man in a boat observes the top of a cliff at an angle of elevation of 35°. He looks on a map and notices that the height of the cliff is 180 m. How far is he from the cliff?

5 Two axes OX and OY are at right angles. A thin rod 20 cm long moves with one end of the rod on OX and the other end on OY. Initially the rod makes an angle of 63° with XO. Calculate the distances of the ends of the rod from O.
The end of the rod on OX is moved 2.9 cm nearer O. Calculate
(i) the angle the rod makes with OX in the new position,
(ii) the distance the other end moves along OY. [O]

6

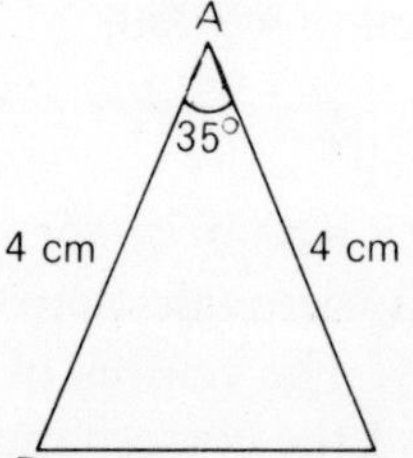

Fig. 12.17

In Fig. 12.17 ABC is an isosceles triangle. Use right-angled triangles only to find BC.

7 A spider sets out to walk from the bottom corner of a room measuring 3 m by 4 m, which is $2\frac{1}{2}$ m high, to his web, which is at the furthest corner of the room from his present position. What is the shortest distance for his journey?

8

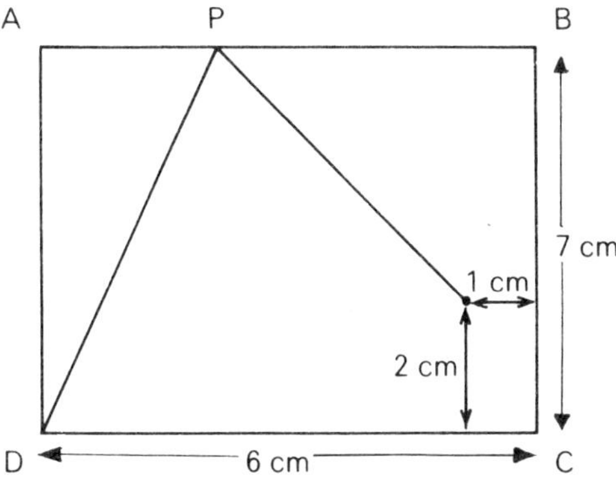

Fig. 12.18

In Fig. 12.18 ABCD is a rectangle, and Q is a point distant 1 cm from BC and 2 cm from DC. P is any point on AB. What is the smallest value of DP + PQ?

9 A sphere whose centre is O and whose radius is 10 cm floats in water. The water-surface meets the sphere in a circle of radius 6 cm, and the points A and B are on this circle. Calculate

(i) the distance of O from the centre of the circle,
(ii) the largest possible value of the angle AOB.

[L]

10 A pyramid stands on a square base ABCD of side 10 cm. Its apex is V and its triangular faces (e.g. VAB) are all equilateral triangles. M is the mid-point of VB. Show that VA is inclined at 45° to the base and calculate

(i) the inclination of the plane VAB to the base,
(ii) the angle between the opposite faces VAB and VCD,
(iii) the angle AMC.

[O]

11 Figure 12.19 shows a cube of side x cm. P is a point on the top edge B′C′, such that B′P = a cm. If y is the angle that AP makes with the horizontal, show that

$$\tan y = \frac{x}{\sqrt{x^2+a^2}}.$$

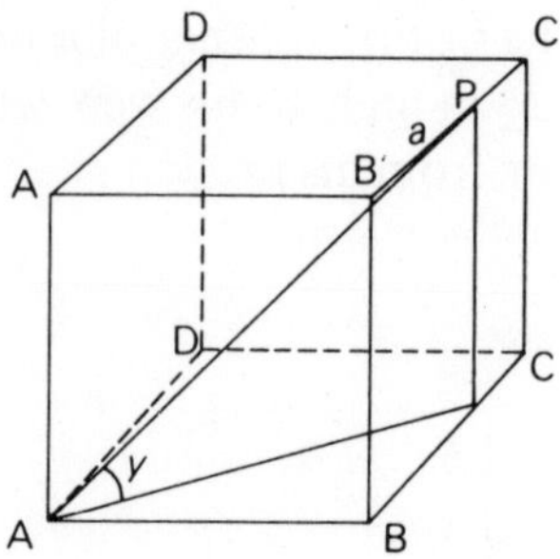

Fig. 12.19

12 AB is a base-line 20 km long, in a direction due north, from the ends of which the points C, D are observed, the angles being as shown in Fig. 12.20. (All the points A, B, C, D are in a horizontal plane.) M is the point of AB which is nearest to D. Prove that AM = 5.00 km and MD = 8.66 km. Hence or otherwise find the distance (to 3 significant figures) and the bearing (to nearest $\frac{1}{2}$ degree) of D from C.

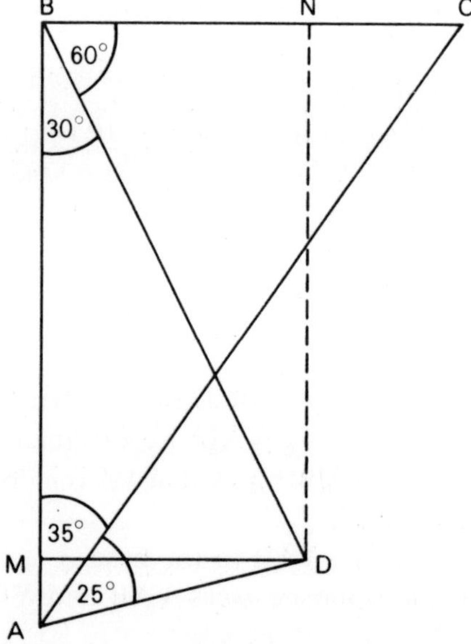

Fig. 12.20

The northerly construction line DN is suggested as an aid to the calculation.

[O]

13 Variation and proportion

13.1 Introduction

Consider the following situation which occurs frequently in our everyday lives.

John goes into a shop to buy some nails. He sees that they cost 22p for 100. For the job he is doing he requires 250. If he had some mathematical training he would think that the quantity of nails he requires is $2\frac{1}{2}$ times the basic unit, therefore the cost will be $2\frac{1}{2}$ times the cost of the basic unit. Total cost $= 2\frac{1}{2} \times 22 = 55$p.

He has used the fact that the cost is *directly proportional to* or *varies as* the quantity, written C (cost) $\propto$ N (number of nails). The sign for proportional ($\propto$) should be removed as soon as possible to avoid confusion. Let us now solve the problem again using this new idea. The proportional sign is replaced by an equality sign and a constant multiplying one side.

$$\therefore\ C \propto N \quad \text{becomes} \quad C = kN$$

We do not yet know the value of k. However, we know that if $N = 100$, $C = 22$.

$$\therefore\ 22 = k \times 100$$

$$\therefore\ k = \frac{22}{100} = \frac{11}{50}.$$

We now have a simple equation to represent our problem, i.e.

$$C = \frac{11}{50}N.$$

$\therefore$ if $N = 250$,

$$C = \frac{11}{\cancel{50}_1} \times {}^5\cancel{250} = 55$$

$\therefore$ the cost is 55p.

13.2 Graphical representation

In chapter 11, we saw that a function of the form $f: x \to kx$ is a straight line through the origin. It follows, then, that any two quantities which are directly proportional can be represented by a straight line graph.

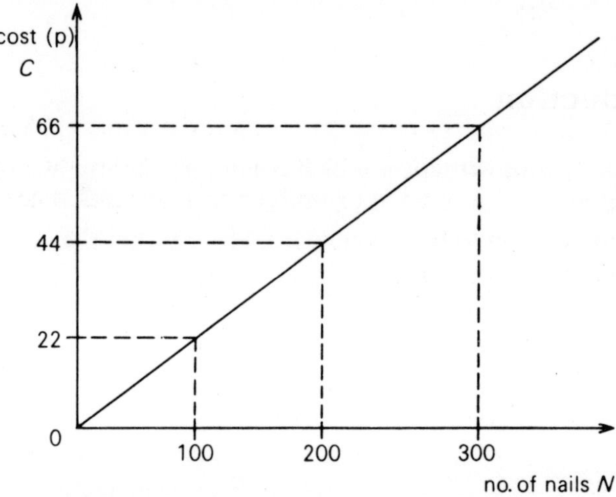

Fig. 13.1

Figure 13.1 shows the graph for the problem above. It can be used to read off the cost of any quantity.

EXERCISE 13a

1 In the following questions, y is directly proportional to x. Use the information given to write down the equation relating y and x, and find the value of y corresponding to the given value of x.

(i) $y = 4$ when $x = 3$; $x = 5$
(ii) $y = 10$ when $x = 10$; $x = 8$
(iii) $y = 25$ when $x = 10$; $x = 8$
(iv) $y = 5$ when $x = 100$; $x = 2$
(v) $y = \frac{1}{4}$ when $x = \frac{1}{8}$; $x = \frac{3}{4}$
(vi) $y = a$ when $x = b$; $x = 3b$
(vii) $y = 30$ when $x = a$; $x = 2a$
(viii) $y = 0.01$ when $x = 0.02$; $x = 0.3$
(ix) $y = 0.15$ when $x = 60$; $x = 14$
(x) $y = a + b$ when $x = a - b$; $x = a^2 - b^2$.

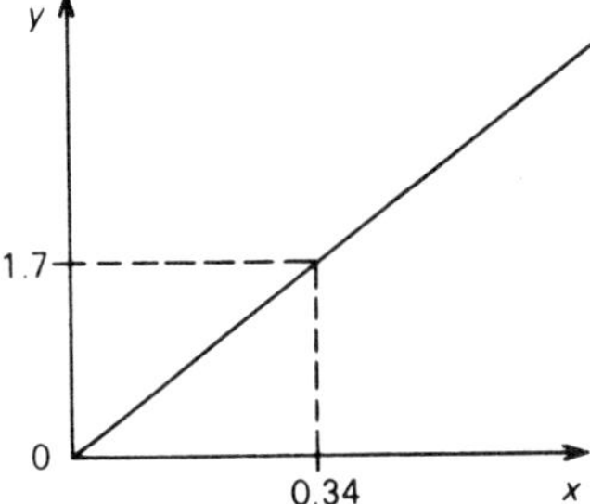

Fig. 13.2

2 Figure 13.2 shows the graph for two quantities x and y which are directly proportional. Find the equation of the line and find x when $y = 1.3$ (give your answer to 1 decimal place).

13.3 Other types of variation

Consider the following problem. The radius of a circle is 7 cm. If we take $\pi = \frac{22}{7}$ we find that the area of the circle is

$$\frac{22}{\not{7}_1} \times 7^{\not{2}} = 154 \text{ cm}^2$$

If we now double the radius, does the area become $2 \times 154 = 308$ cm^2? The answer is no, because if we double the radius it becomes 14 cm, therefore the area is

$$\frac{22}{7} \times 14^2 = 616 \text{ cm}^2.$$

In this case we say that the area (A) is proportional to (or varies as) the square of the radius (r), because we have to double r^2, not r, to double the area.

i.e. $A \propto r^2$
$\therefore\ A = kr^2$ [in fact we know that $k = \pi$].

EXAMPLE 1

The radius of a circle is 5 cm. Find by how much the radius has to be increased in order to treble the area. [Do not assume a value for π.]

We know that $A = kr^2$.
$\therefore$ if $r = 5$,
$A = 25k$ (i)

If A becomes $3A$, then

$3A = kr^2$ (ii)

(i) $\times 3 \Rightarrow 3A = 75k$ (iii)

From (iii) and (ii),

$$75k = kr^2$$
$$r^2 = 75$$
$$r = \sqrt{75}.$$

Hence the radius has to be increased by $(\sqrt{75} - 5)$ cm.

13.4 Inverse proportion

If $y \propto 1/x^n$, we say that y is *inversely proportional* to x^n and the equation that we work from is $y = k/x^n$.

EXAMPLE 2

It is known that the force between two atoms (F) is inversely proportional to the square of their distance (d) apart. If $F = 10^{-2}$ when $d = 10^{-5}$, find F when $d = 4 \times 10^{-5}$.

$$F \propto \frac{1}{d^2} \Rightarrow F = \frac{k}{d^2}$$

$$10^{-2} = \frac{k}{(10^{-5})^2} = \frac{k}{10^{-10}}$$

$$\therefore\ k = 10^{-10} \times 10^{-2} = 10^{-12}$$

$$\therefore\ F = \frac{10^{-12}}{d^2}.$$

When $d = 4 \times 10^{-5}$,

$$F = \frac{10^{-12}}{(4 \times 10^{-5})^2} = \frac{10^{-12}}{16 \times 10^{-10}}$$

$$F = \frac{10^{-2}}{16} = 6.25 \times 10^{-4}.$$

EXERCISE 13b

1 Repeat question 1 of exercise 13a given that y is proportional to x^2.

2 A is inversely proportional to r^2. If $A = 10$ when $r = 7$, find A when $r = 3.5$.

3 Given that H varies as the cube of v, and that $H = 100$ when $v = 5$, find H when $v = 3$.

4 V varies directly as p, and inversely with q. This means that $V = kp/q$. If $V = 11$ when $p = 9$ and $q = 2$, find k and the value V when (i) $p = 3$ and $q = 1$, (ii) $p = 2q$.

5 H varies directly with p and inversely with the square of T. If $H = 5$ when $p = 7$ and $T = 12$, find T when $H = p = 3$.

13.5 Similar figures

(i) *Areas.* Figure 13.3 shows a square of side 3 cm. Its area is 9 cm^2. Consider this in relationship to the unit square, shown shaded. We have increased the sides of this square by a factor $(3)^2 = 9$.

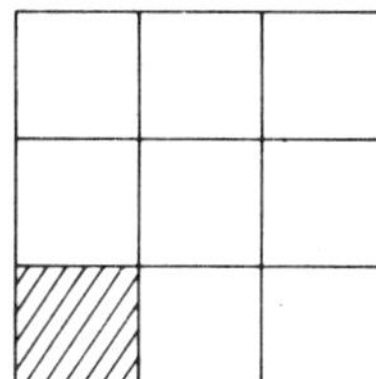

Fig. 13.3

In general, for any two similar shapes, if corresponding lengths are multiplied by a scale factor r, then the area is multiplied by r^2.

$$\therefore \frac{\text{new area}}{\text{old area}} = r^2.$$

EXAMPLE 3

A map has a scale of 1 : 1000. On the map, the area of a pond is 0.25 cm^2. What is the true area of the pond? Express your answer in m^2.

The scale factor is 1000.
Areas are multiplied by a factor $(1000)^2 = 10^6$.

$$\therefore \text{the true area} = 0.25 \times 10^6 = 2.5 \times 10^5 \text{ cm}^2.$$

Now 1 m = 100 cm.

$$\therefore 1 \text{ m}^2 = 10^4 \text{ cm}^2$$

$$\therefore \text{area} = \frac{2.5 \times 10^5}{10^4} = 25 \text{ m}^2.$$

EXAMPLE 4

In Fig. 13.4 the triangles ACB and AED are similar. Find the ratio of the area of triangle ACB to that of the trapezium BCED.

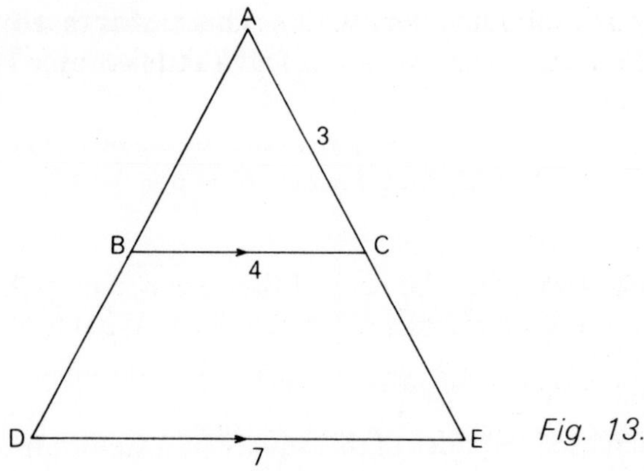

Fig. 13.4

Let the area of triangle ABC = x.
Let the area of triangle ADE = y.
Since BC : DE = 4 : 7, the scale factor for the length of the triangle is $\frac{7}{4}$.
Hence, areas are multiplied by a factor $(\frac{7}{4})^2 = \frac{49}{16}$.

$$\therefore\ y = \frac{49x}{16}$$

Now the area of the trapezium is $y - x$

$$= \frac{49x}{16} - x = \frac{33x}{16}.$$

$\therefore$ the required ratio is $x : \frac{33x}{16}$

i.e. 16 : 33.

(ii) *Volumes*. In Fig. 13.5 the unit cube has its sides doubled. It can be seen that the volume of the larger cube is 8 $\text{cm}^3 = (2)^3\ \text{cm}^3$.

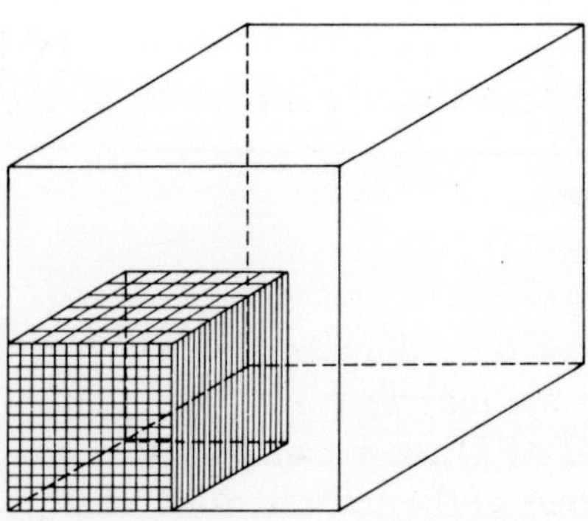
Fig. 13.5

In general, if the corresponding lengths of similar shapes are multiplied by a scale factor r, the volumes are multiplied by r^3.

$$\therefore \frac{\text{new volume}}{\text{old volume}} = r^3.$$

EXAMPLE 5

A cylinder has height h cm, and the area of the base is 6 cm^2. A similar cylinder has base area h^2 cm^2, and height 3 cm. What is h?

Volume of the first cylinder $= 6 \times h = 6h\ cm^3$.
Volume of the second cylinder $= h^2 \times 3 = 3h^2$ cm^3.
The lengths of the sides are in the ratio $h:3$.

$\therefore$ the volumes are in the ratio $h^3:27$

$$\therefore 6h:3h^2 = h^3:27$$

$$\text{i.e. } \frac{6h}{3h^2} = \frac{h^3}{27} \Rightarrow \frac{2}{h} = \frac{h^3}{27}$$

$$\therefore 54 = h^4$$

$$\Rightarrow h = 2.71.$$

HARDER EXAMPLES 13

1 It is given that x is both directly proportional to y and inversely proportional to the square of z. Complete the table of values:

x	y	z
3	1	2
2		3
1	3	

[C]

2 Given that y is inversely proportional to x^2 and that $x = 3$ when $y = 5$, find the value of y when $x = 5$.

[C]

3 Some corresponding values of s and t are given in the following table:

s	60	80	150	200
t	20	15	8	6

State which one or more of the following statements are possibly correct and which are incorrect:

(i) $s = 3t$,

(ii) t is proportional to s,
(iii) $s \times t = 1200$,
(iv) t is inversely proportional to s.

4 Given that y varies as x^n, write down the value of n in each of the following cases:
(i) y is the area of a circle of radius x,
(ii) y is the volume of a cylinder of given base area and height x,
(iii) y and x are the sides of a rectangle of given area.

5 Z and t are connected by a formula of the form $Z = at^n$, where a is a constant and n a positive integer. From the table below deduce the values of a and n. Also complete the table.

t	1	2	3	4	5	6
Z	0.5	4	13.5			

6 In triangle ABC, X is the midpoint of AB, and Y the midpoint of AC. Z is the point on XY such that XZ = 2ZY. Find the ratio of the area of triangle AYZ to that of the trapezium XYCB.

7 H consists of the sum of two quantities, one of which varies as the square of r; the other is inversely proportional to r. If $H = 1$ when $r = 3$, $H = 2$ when $r = 4$, find H when $r = 5$.

8 Two spheres have volumes in the ratio 8 : 27. What is the ratio of their surface areas?

9 Two similar cylinders have volumes in the ratio $a^3 : b^3$. The total surface area of the smaller cylinder is 20 cm^2. What is the total surface area of the larger?

10 A scale model of a hut is a cuboid of volume 12 cm^3. The height of the model is 2 cm, and the height of the real building is 25 m. What is the area of the floor of the building?

14 Straight lines and circles

14.1 Parallel lines

In Fig. 14.1, FC and GE are parallel lines (indicated by the arrows). Three angles have been marked which are equal. Let us look at the reasons for this.

$\angle BDE \rightarrow \angle ABC$ by a translation parallel to AH.

We say that $\angle ABC$ and $\angle BDE$ are *corresponding* angles.
If AB is rotated 180° about B it becomes BD; if BC is rotated 180° about B it becomes BF.

$\therefore \angle ABC \rightarrow \angle DBF$ by a rotation of 180° about B.

We say that $\angle ABC$ and $\angle FBD$ are *vertically opposite*.
$\angle FBD$ and $\angle BDE$ are called *alternate* angles.

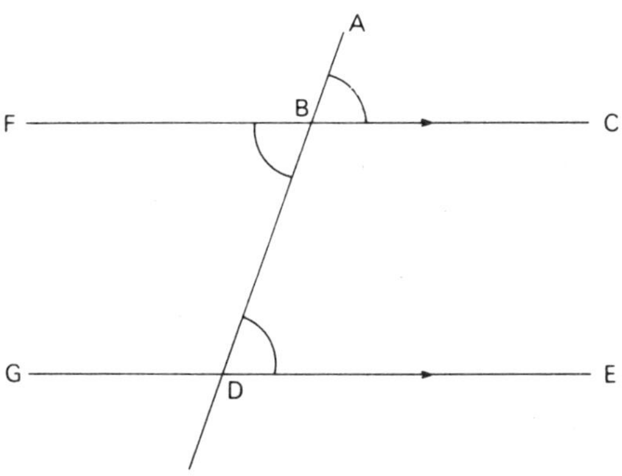

Fig. 14.1

14.2 The angles of a triangle

If $\angle x$ is translated to C and $\angle z$ translated to C, then clearly

$$x + y + z = 360°$$

but

$x = 180° - B$, $y = 180° - C$ and $z = 180° - A$

$\therefore (180° - A) + (180° - B) + (180° - C) = 360°$

$\therefore A + B + C = 180°$.

The angles of a triangle add up to 180°. Also the exterior angles of a triangle add up to 360°. A proof of this follows in the next section.

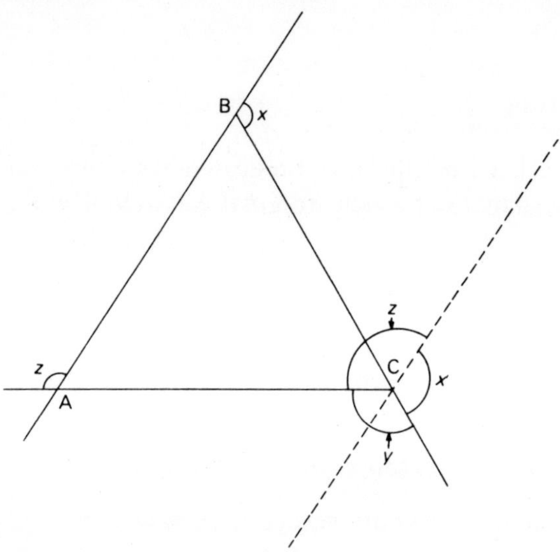

Fig. 14.2

14.3 Angles of a polygon

Look at the hexagon (six sides) in Fig. 14.3. Each exterior angle is translated to B and it can be seen that

$$a + b + c + d + e + f = 360°$$

Clearly this works however many sides there are. We have the result then that:

The exterior angles of a polygon add up to 360°.

Each interior angle A is related to its exterior angle a, by $a = 180° - A$. If we add all n of these together, we get

$$(a + b + c + \ldots) = 180n° - (A + B + C + \ldots)$$
$$\therefore 360° = 180n° - (A + B + C + \ldots)$$
$$\therefore (A + B + C + \ldots) = 180n° - 360° = 90(2n - 4)°$$

We have the result that:

The sum of the interior angles of an n-sided polygon is $90(2n-4)$ degrees

or $(2n-4)$ right angles.

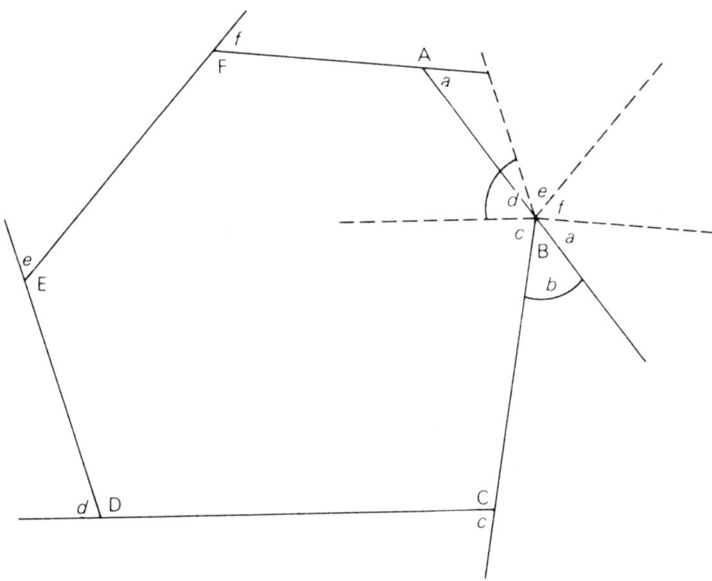

Fig. 14.3

EXAMPLE 1

Find the angles marked with a letter in Fig. 14.4.

$$\angle ABC = \angle PAB \text{ (alternate angles)}$$
$$\therefore\ 12^\circ + a = 70^\circ$$
$$\Rightarrow a = 58^\circ$$
$$\angle CBA + \angle BAC + \angle ACB = 180^\circ \text{ (angles of a triangle)}$$
$$\therefore\ 70^\circ + b + 50^\circ = 180^\circ$$
$$\Rightarrow b = 60^\circ$$
$$\angle ADB = \angle CBD \text{ (alternate angles)}$$
$$\therefore\ c = a = 58^\circ$$
$$\angle AED + \angle EAD + \angle EBA = 180^\circ \text{ (angles of a triangle)}$$
$$\therefore\ \angle AED + b + 12^\circ = 180^\circ$$
$$\Rightarrow \angle AED = 180^\circ - 12^\circ - 60^\circ = 108^\circ$$
$$\angle DEC = \angle AED \text{ (vertically opposite)}$$
$$d = 108^\circ.$$

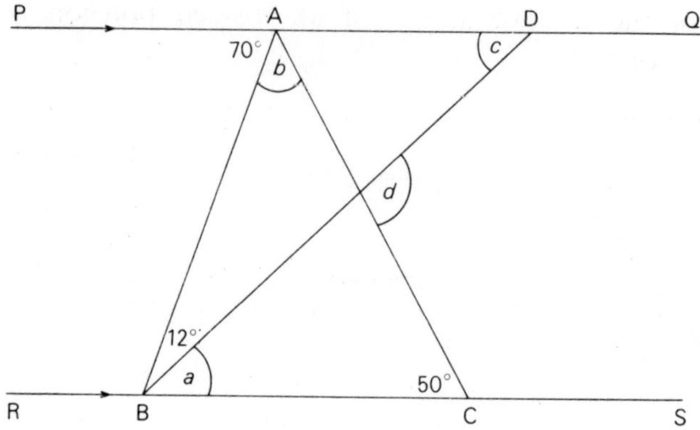

Fig. 14.4

EXAMPLE 2

The interior angles of a polygon are $2x°$, $(3x+15)°$, $x°$, $(3x-40)°$, $(x+25)°$. Find x and the exterior angles of the polygon.

The interior angles of a five-sided figure add up to $(2 \times 5 - 4)$ right angles

$$= 6 \times 90° = 540°$$

$$\therefore\ 2x + (3x+15) + x + (3x-40) + (x+25) = 540$$

$$\Rightarrow 10x = 540$$

$$\Rightarrow x = 54$$

$\therefore$ the interior angles are 108°, 177°, 54°, 122°, 79°.

To find the exterior angles, each interior angle is subtracted from 180° to give: 72°, 3°, 126°, 58°, 101°.

EXERCISE 14a

Find the angles marked with a letter in the following diagrams:

1 **2**

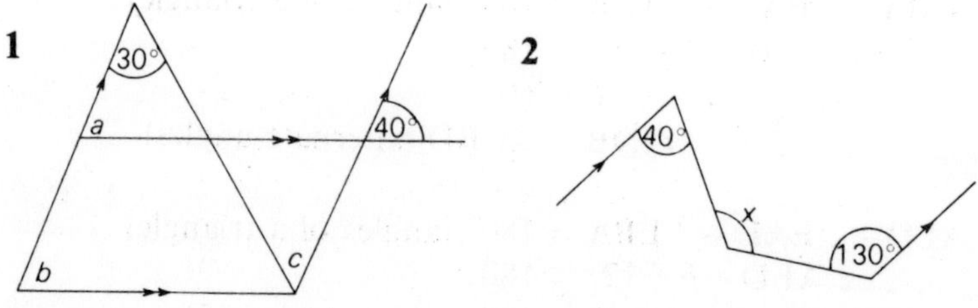

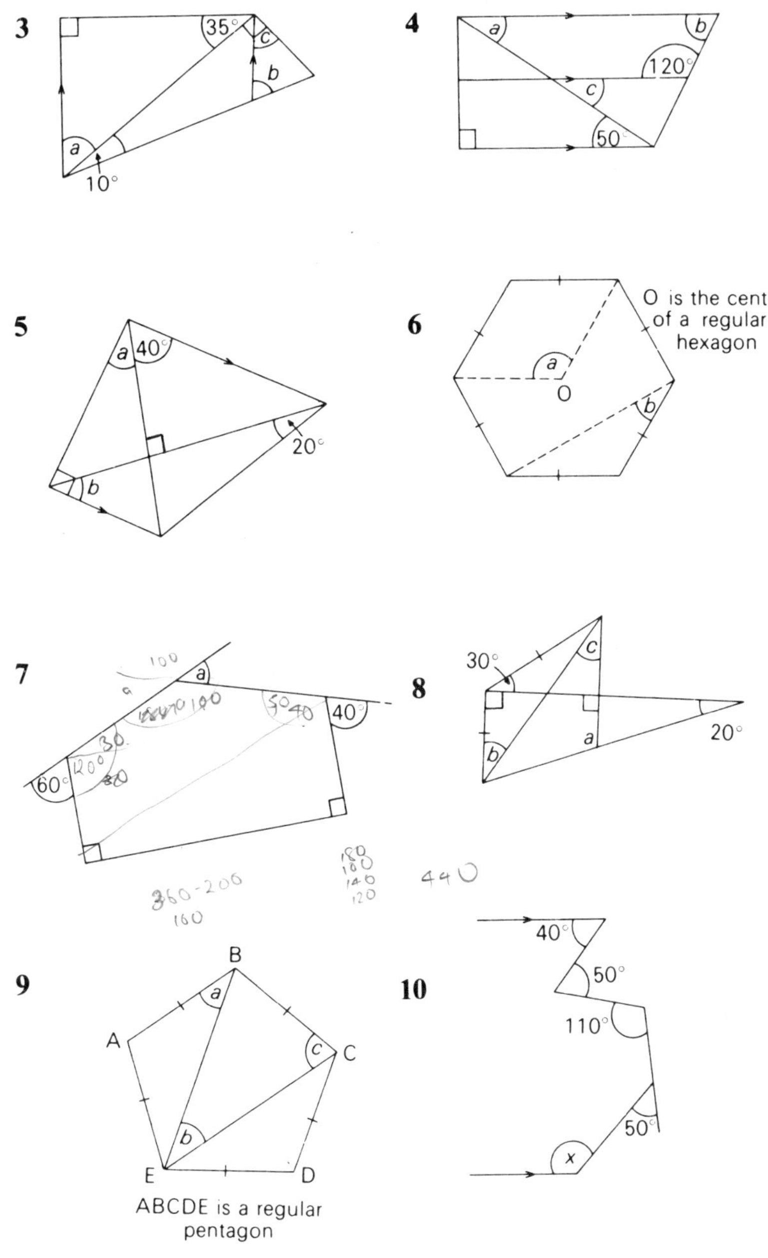
3
35°
c
b
a
10°
4
a
b
120°
c
50°
5
a
40°
20°
b
6
O is the centre of a regular hexagon
a
O
b
7
a
40°
60°
8
30°
c
b
a
20°
9
B
A
C
E
D
a
c
b
ABCDE is a regular pentagon
10
40°
50°
110°
50°
x

11 The interior angles of a pentagon are $3x^\circ$, $(2x+15)^\circ$, $4x^\circ$, $(x-20)^\circ$, and x°. Find x.

12 The interior angle of a regular polygon is three times the exterior angle. How many sides has the polygon?

14.4 Symmetry

(1) *Reflection.*

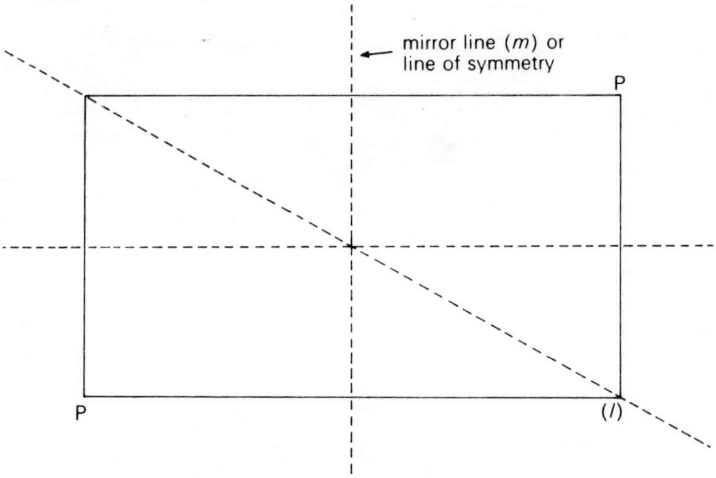

Fig. 14.5

Consider the rectangle drawn in Fig. 14.5. The line m divides the rectangle into parts which are exactly the same. We say that m is the *line of symmetry* (or mirror line) of the rectangle.

The line l, however, although dividing the rectangle into two equal parts, is not a line of symmetry. If the rectangle is folded along the line l, P′ will not be on top of P. It can be seen that the rectangle has 2 lines of symmetry.

(2) *Rotation.*
Figure 14.6 shows a regular hexagon (all sides equal). If the hexagon is rotated by 60° about a line through O, at right angles to the plane of the paper, then $A \to B$ and $B \to C$. The final position of the hexagon is the same as before, with all the letters having moved round one vertex. This process can be repeated six times in all, before A returns to its original position.

We say that the order of rotational symmetry is 6.

(*Note:* in 3 dimensions, we would have *planes* of symmetry.)

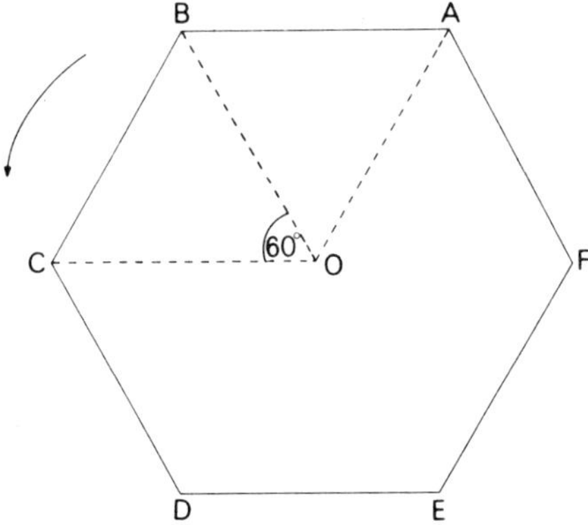

Fig. 14.6

14.5 Geometrical properties and definitions of simple plane figures

(1) *Isosceles triangle.*

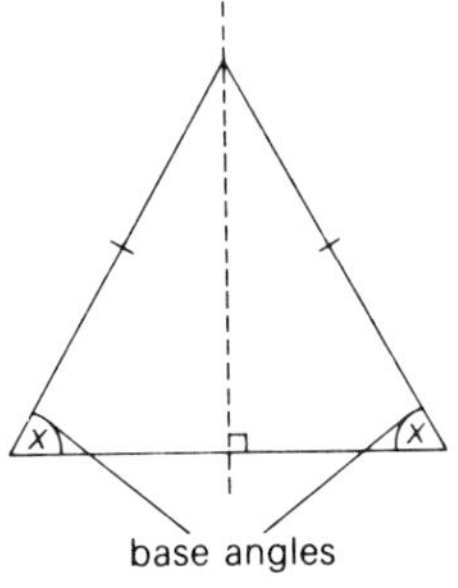

Base angles are equal and two sides are equal. Questions involving isosceles triangles are often simplified if considered as two right-angled triangles made from the line of symmetry.

(2) *Equilateral triangle.*

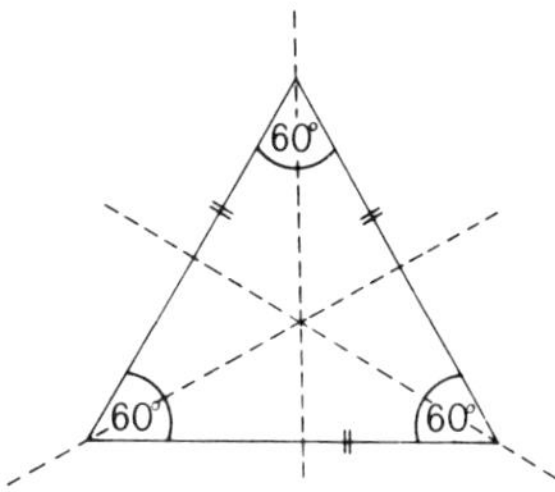

All angles are equal to 60°, and the triangle has three axes of symmetry. The order of rotational symmetry is 3.

(3) *Obtuse-angled triangle.* One angle greater than 90°.

(4) *Square.*

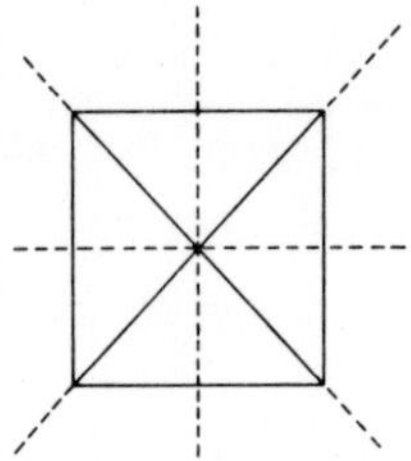

All sides are equal in length, and all angles are 90°. The diagonals are equal in length and cross at 90°. There are 4 axes of symmetry and the order of rotational symmetry is 4.

(5) *Rectangle.*

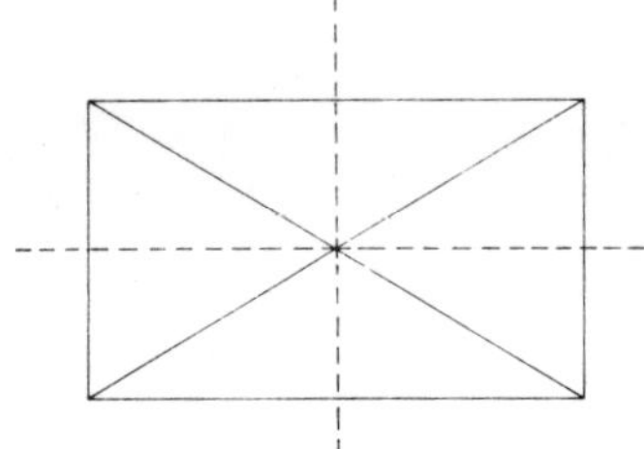

All angles are 90°. The diagonals are equal in length and cross in the centre. There are 2 axes of symmetry, and the order of rotational symmetry is 2.

(6) *Parallelogram.*

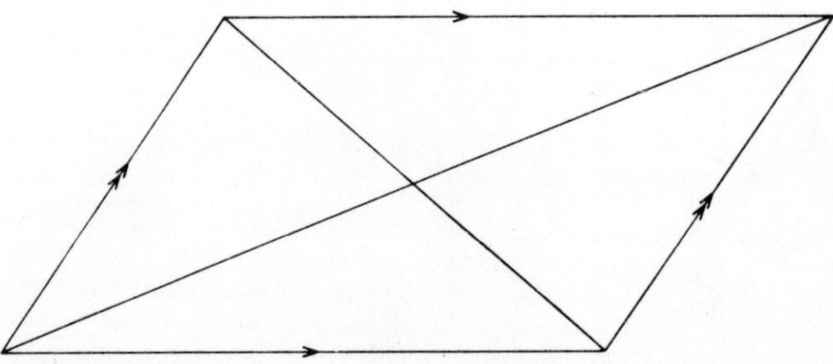

Opposite sides are parallel and equal in length. The diagonals cross at their centres. The figure has no lines of symmetry, but the order of rotational symmetry is 2.

(7) *Rhombus.*

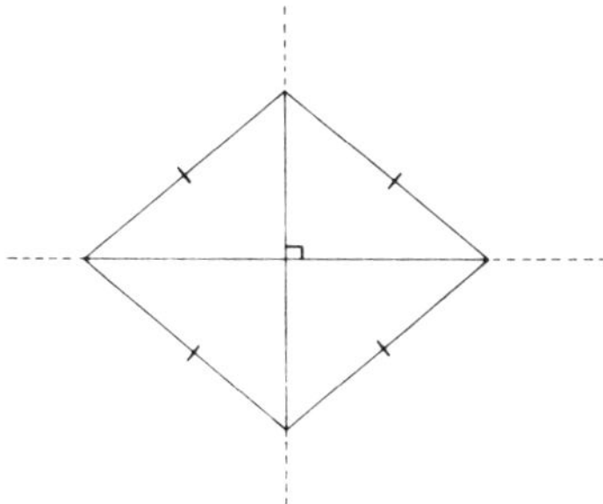

A parallelogram with all sides equal. The diagonals bisect each other at 90°. There are 2 axes of symmetry. The order of rotational symmetry is 2.

(8) *Kite.*

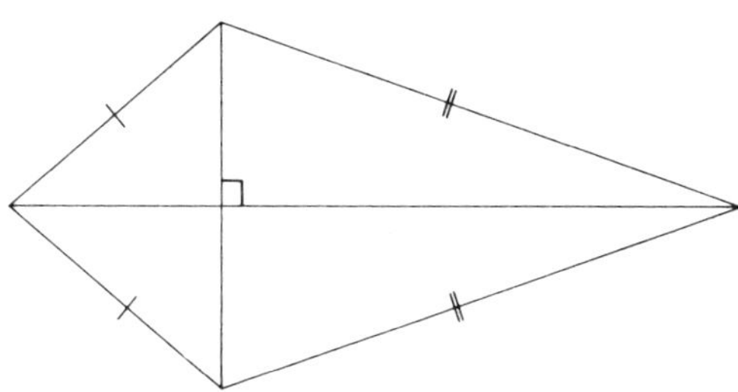

Two pairs of adjacent sides are equal. The diagonals cross at 90°.

(9) *Trapezium.*

Two opposite sides are parallel. In general, there is no symmetry.

(10) *Regular polygons.* All sides and angles are equal.

Pentagon: five sides, five lines of symmetry, order of rotational symmetry 5.

Octagon: eight sides, eight lines of symmetry, order of rotational symmetry 8.

14.6 The circle

Figure 14.7 shows the various commonly-used terms which refer to a circle. ∠AOB is the angle subtended at the centre of the circle by the chord AB.

Note: AOB is an isosceles triangle.

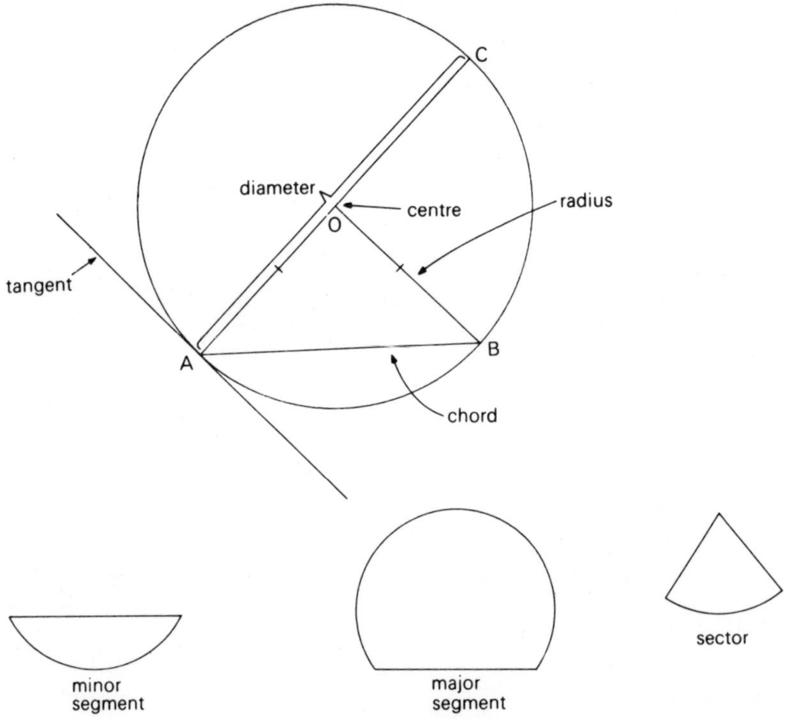

Fig. 14.7

14.7 The symmetry properties of the circle

(1) The angle between a tangent TP and the radius OT at the point of contact (with the circle) is always 90°. See Fig. 14.8.

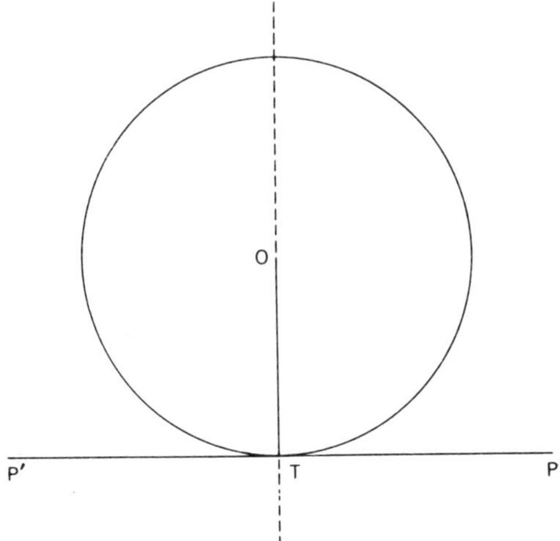

Fig. 14.8

This follows because the diagram is symmetrical about OT.

(2)

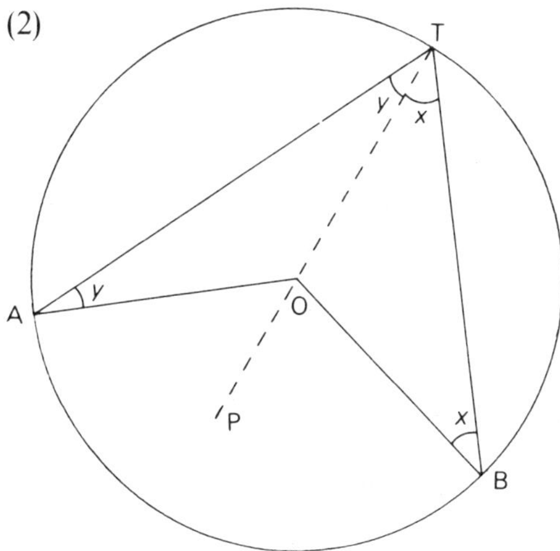

Fig. 14.9

The angle ∠AOB subtended at the centre by two points on the circumference of the circle A and B is twice the angle subtended by these same two points A and B at any point T on the circle, i.e. ∠ATB.

∠AOB = 2 × ∠ATB

TBO and AOT are isosceles triangles.

$\therefore \angle TOB = 180° - 2x$ and $\angle AOT = 180° - 2y$

$\therefore \angle BOP = 180° - (180° - 2x)$ and $\angle AOP = 180° - (180° - 2y)$

$= 2x$ $= 2y$

$\therefore \angle AOB = 2x + 2y = 2(x + y) = 2 \times \angle ATB.$

(3)

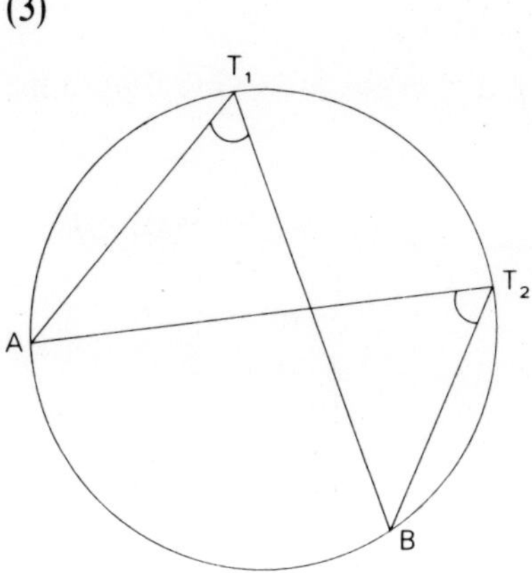

Fig. 14.10

Angles in the same segment of a circle which have the same base points A and B, must be equal, because they make the same angle at the centre of the circle.

$\therefore \angle AT_1B = \angle AT_2B$

EXAMPLE 3

In Fig. 14.11, prove that $\angle ABC + \angle ADC = 180°$.

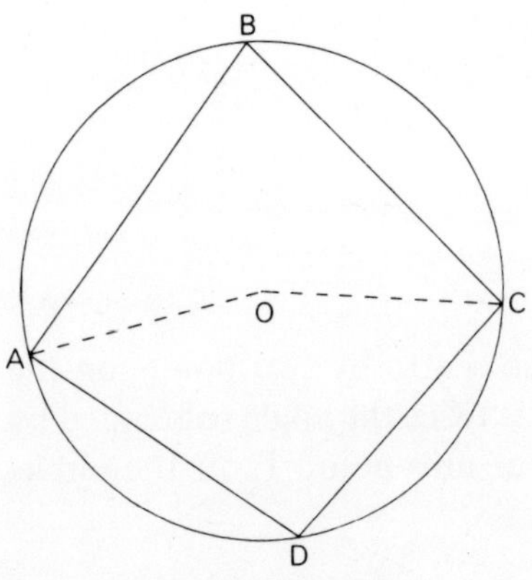

Fig. 14.11

In order to prove this, we need to draw the lines AO, OC.

$$\begin{aligned} \angle ABC &= \tfrac{1}{2}\angle AOC \text{ (obtuse)} \\ \angle ADC &= \tfrac{1}{2}\angle AOC \text{ (reflex, i.e.} > 180^\circ) \\ \therefore \angle ABC + \angle ADC &= \tfrac{1}{2}(\angle AOC \text{ (obtuse)} + \angle AOC \text{ (reflex)}) \\ &= \tfrac{1}{2} \times 360^\circ = 180^\circ \end{aligned}$$

EXERCISE 14b

Find all the angles marked with a letter in the following diagrams:

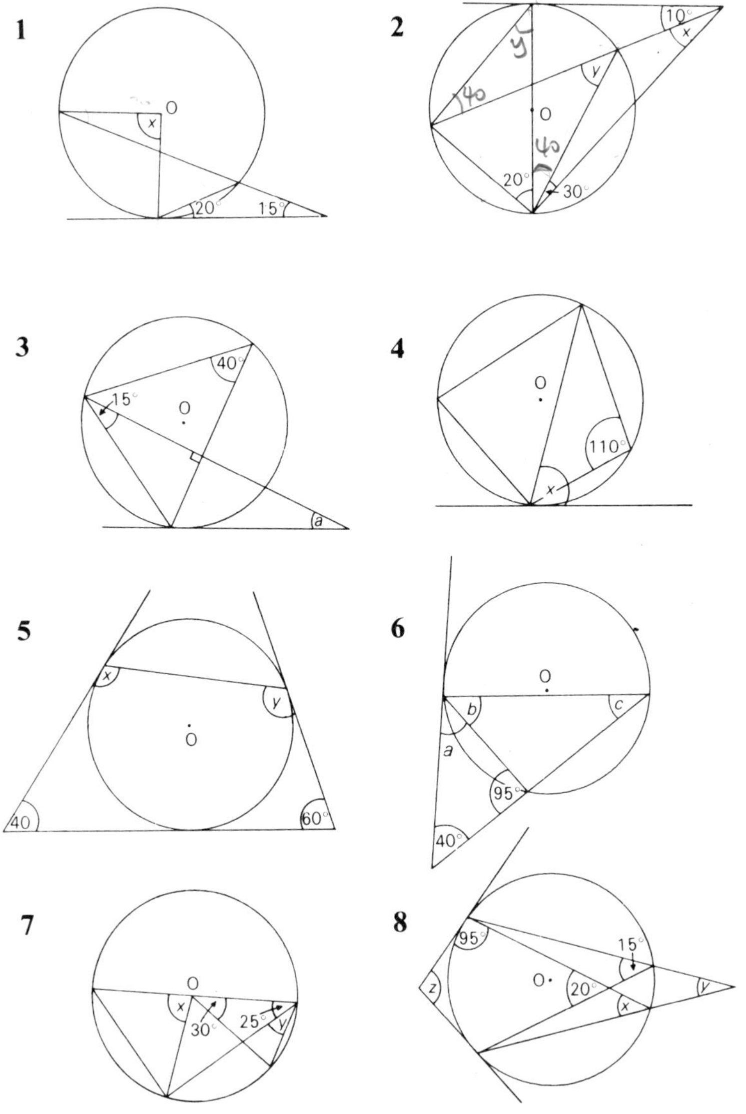

14.8 Locus problems

The word *locus* is the name given to the path traced out by a point which moves subject to certain conditions.

(i) The locus of a point P which moves so that its distance from A is constant is a circle, centre A, in two dimensions, or a sphere in three dimensions.

(ii) The locus of a point P which moves so that it is always the same distance from two fixed points A and B is a line in two dimensions (the perpendicular bisector of AB), or a plane in three dimensions.

(iii) The locus of a point P which moves so that it subtends a constant angle at a fixed line AB, is an arc of a circle in two dimensions, or part of a sphere in three dimensions. See Fig. 14.12.

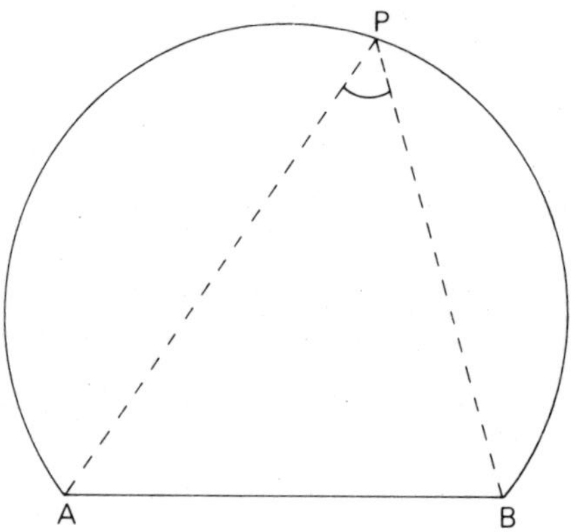

Fig. 14.12

If ∠APB is constant, then P always lies in the same segment of a circle in two dimensions.

14.9 Regions

A and B are fixed points. P is a variable point which satisfies the following condition:

$PA \geq PB.$

We can show that P lies in the shaded region shown in Fig. 14.13. In three dimensions, P would lie in that half of space to the right of the dividing plane.

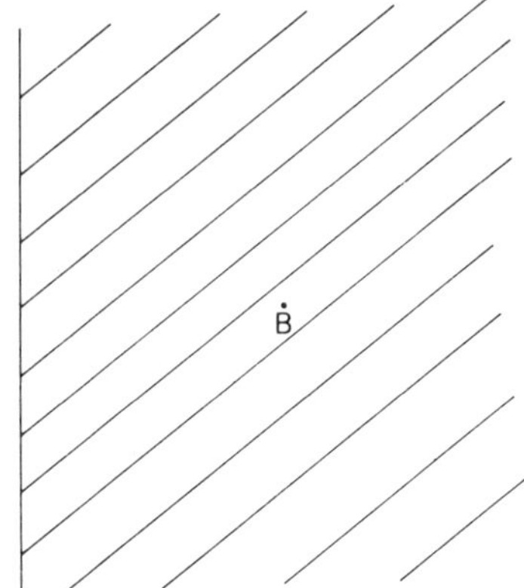

Fig. 14.13

EXAMPLE 4

In Fig. 14.14, ABCD is a square.
$X = \{P : P$ is on the opposite side of DB to A, and $\angle BPD \geq 90°\}$.
$Y = \{P : P \in S'\}$, where $S = \{$points inside the square$\}$, and
$= \{$points inside the circle$\}$. Shade $X \cap Y$.

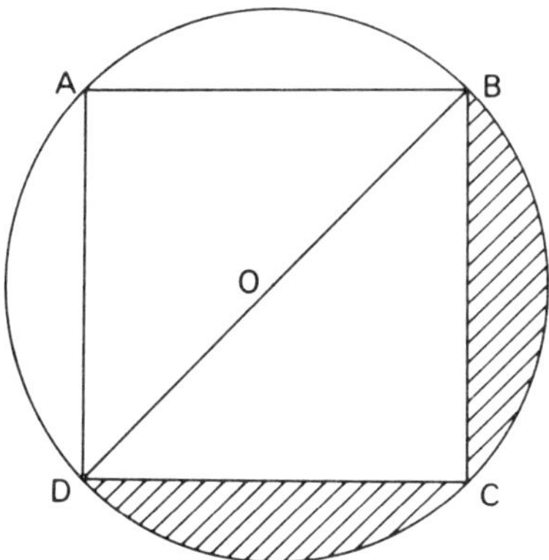

Fig. 14.14

On the circle, $\angle BPD = 90°$, therefore $X = \{$points inside the semi-circle not containing A$\}$. Y is the set of points outside the square and inside the circle. Hence, $X \cap Y$ is the shaded region as shown.

There are too many examples to include every possibility here, but these three examples should enable you to attempt the following questions. Remember, you could be in three dimensions.

EXERCISE 14c

In the following questions, P is a variable point, and all other points are fixed. Find the locus of P if:

1 $\angle PAB = 30°$.
2 The perpendicular distance of P from a fixed line is 4 cm.
3 The perpendicular distance of P from two non-intersecting lines is equal.
4 $PA + PB = 16$ cm ($AB = 12$ cm).
5 $PA \leq 5$ cm.
6 The distance of P from a fixed plane is never greater than 4 cm.
7 The distance of P from a fixed line is never less than 10 cm.
8 $\angle APB = 90°$.
9 $PA = 2PB$ (try to sketch it).
10 $\angle PAB \leq 90°$.

HARDER EXAMPLES 14

1 In Fig. 14.15, find a and b.

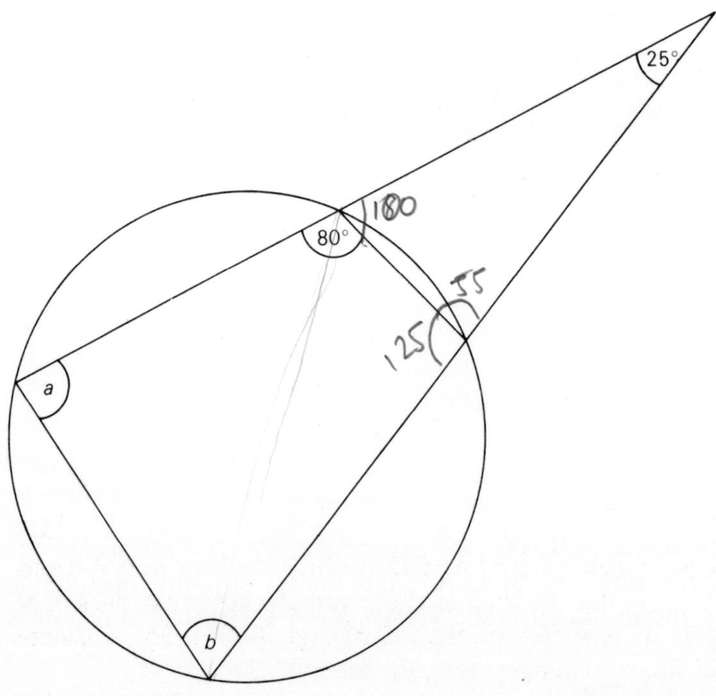

Fig. 14.15

2 In Fig. 14.16, the angles are as stated and the marked lines are parallel. Calculate the values of x and y.

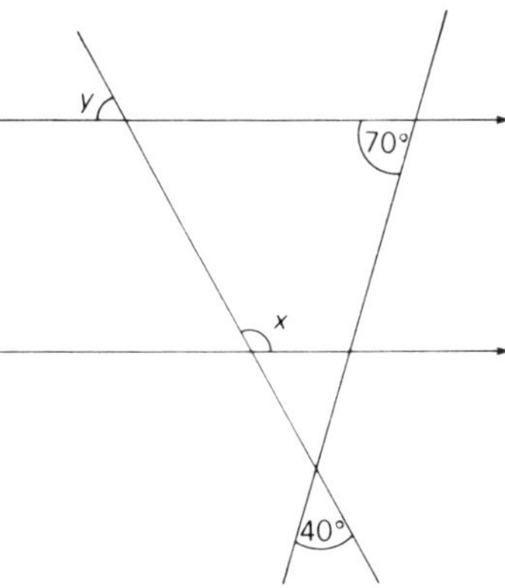

Fig. 14.16

[SU]

3 In Fig. 14.17, the diameter AB of a circle is produced to a point P and PT is a tangent. Given that $\angle TPB = x$ and $\angle TAB = 2x$, find (i) $\angle BTP$ in terms of x, (ii) the value of x.

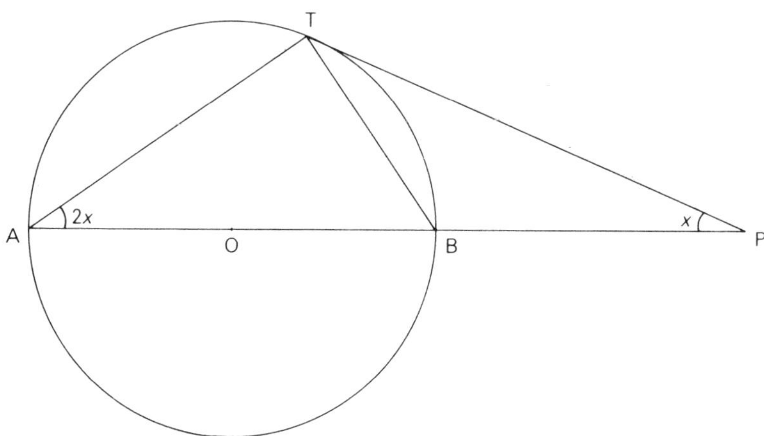

Fig. 14.17

[C]

4 Given a quadrilateral ABCD in which the lengths of AB and BC are equal, the lengths of CD and DA are equal, the size of angle BCA $= 50°$ and the size of angle BDA is $35°$, calculate the size of (i) angle ABC, (ii) angle CAD.

5 TA is a tangent to a circle. (A is on the circle.) AB is a chord of the circle and O is the centre of the circle. If $\angle BAT = 50°$, find $\angle BOA$.

6 Draw a four-sided plane figure which has one line of symmetry.

7 In Fig. 14.18, O is the centre of the circle and BTC is a tangent. Given that $\angle BAT = x°$ and $\angle CAT = 2x°$, name three other angles of this figure each equal to $2x°$.

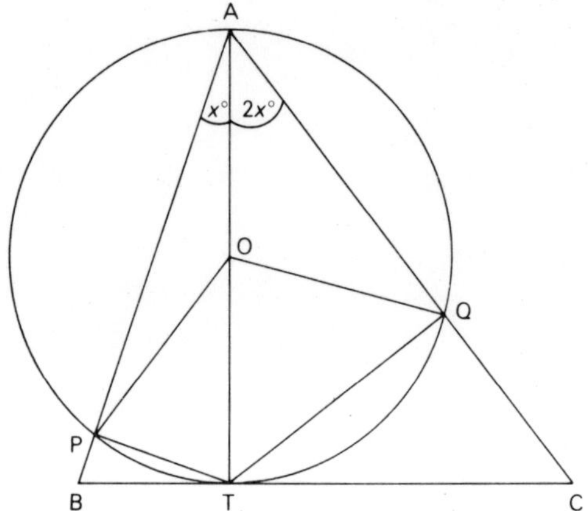

Fig. 14.18

8 In Fig. 14.19, TA is the tangent to the circle at A. Prove that $\angle TAB = \angle ACB$. (Hint: Draw a line through the centre of the circle and A.)

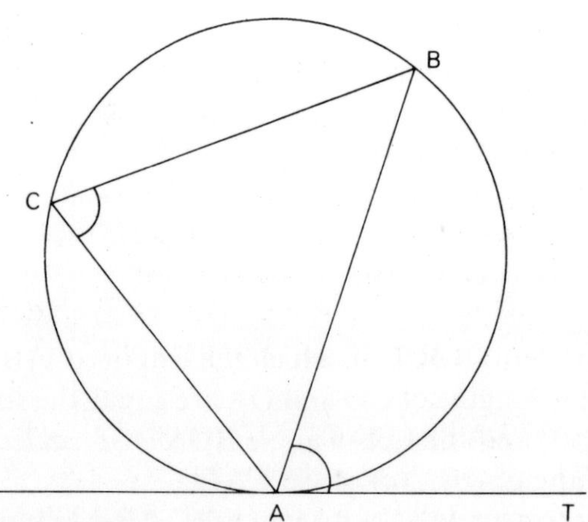

Fig. 14.19

9 A regular polygon has exterior angles of x°. Find, in terms of x,

(i) the interior angle;
(ii) the number of sides;
(iii) the number of lines of symmetry;
(iv) the order of rotational symmetry of the polygon.

10 In Fig. 14.20, PQ is the tangent at P to the circle. Write down the sizes of the angles marked x, y and z.

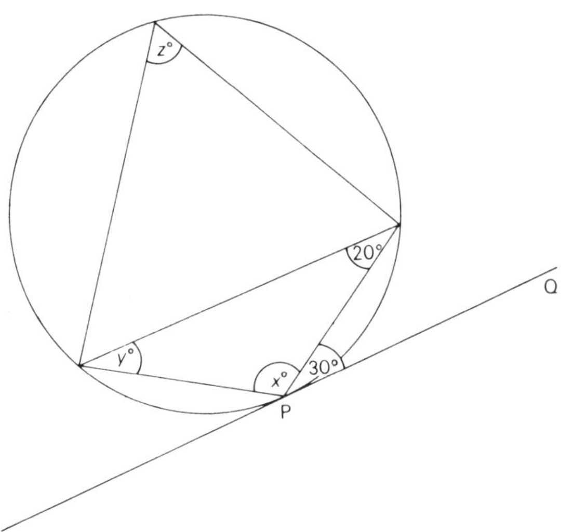

Fig. 14.20

11 Two points P_1 and P_2 are fixed in a plane, P_3 is the mid-point of P_1P_2, and P is a variable point in that plane. Give a geometric interpretation of the following sets:

$S_1 = \{P: \text{angle } P_1PP_2 = 90^\circ\}$
$S_2 = \{P: PP_1 = P_1P_3\}$
$S_3 = \{P: PP_1 = PP_3\}$

Show that $S_1 \cap S_2$ is a subset of S_3.
If $P_1 = (0, 0)$ and $P_2 = (4, 0)$, sketch the set S_1 and S_2 and calculate the co-ordinates of the members of $S_1 \cap S_2$.

[AEB]

12 The points X, Y, Z lie on the circumference of a circle whose centre is O. The perpendiculars from O to YZ, ZX, XY are of lengths p, q, r respectively. Show that if $pqr = 0$, then XYZ is a right-angled triangle.

(i) Mark two points, P and Q, 4 cm apart. Draw the locus

{R : PQR is a right-angled triangle}. (This locus consists of three parts, each of which is a straight line or a circle.) Shade as much as lies on your paper of the locus {S : PQS is an obtuse-angled triangle}.

(ii) Mark two points, M and N, 4 cm apart. Draw the locus {T : MNT is an isosceles triangle}. (Locus T also has three parts, each of which is a straight line or a circle.)

[L]

15 Vectors

15.1 Definition

A vector is a quantity which has magnitude and direction. Vector quantities are of the utmost importance in mathematics, and occur in a great variety of situations. In this book, the theory will be developed using the ideas of co-ordinates.

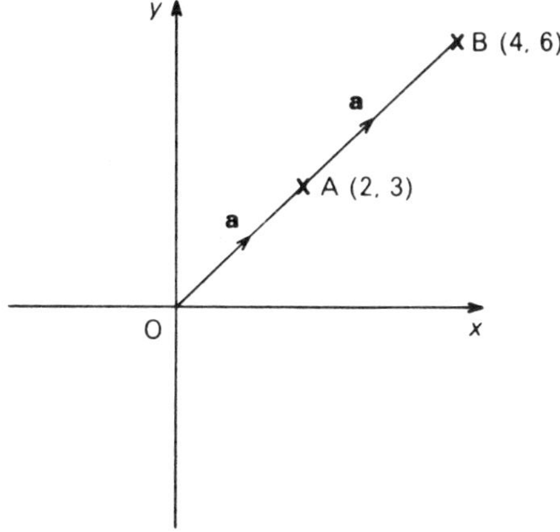

Fig. 15.1

In Fig. 15.1, the co-ordinates of A are (2, 3). If we wanted to describe the instruction 'move from O to A in a straight line' we would say

$$\overrightarrow{OA} = \begin{pmatrix} 2 \\ 3 \end{pmatrix}.$$

$\overrightarrow{OA}$ stands for 'O to A (the direction of the arrow) in a straight line'. This is known as a displacement or translation. It is represented as a 2×1 matrix, the top number being the x-displacement and the bottom number the y-displacement.

15.2 Multiple of a vector

In Fig. 15.1 again, if on reaching A we make the same displacement again to reach B, we would be at the point (4, 6).

If we denote the vector $\begin{pmatrix} 2 \\ 3 \end{pmatrix}$ by **a** and the vector $\overrightarrow{OB}$ by **b**, then we have the following:

$$\overrightarrow{OB} = \begin{pmatrix} 4 \\ 6 \end{pmatrix}$$

which is just **a** followed by **a** i.e. 2**a**. But

$$\mathbf{a} = \begin{pmatrix} 2 \\ 3 \end{pmatrix}$$

$$\therefore \mathbf{b} = 2 \times \begin{pmatrix} 2 \\ 3 \end{pmatrix}.$$

Now

$$\begin{pmatrix} 4 \\ 6 \end{pmatrix} = 2 \times \begin{pmatrix} 2 \\ 3 \end{pmatrix}.$$

Therefore, to obtain a multiple of a vector, we simply multiply the elements of the vector by that multiple.

$$\text{If } \mathbf{a} = \begin{pmatrix} 2 \\ 3 \end{pmatrix}, \text{ then } -3\mathbf{a} = \begin{pmatrix} -6 \\ -9 \end{pmatrix}.$$

This displacement is in the opposite direction, and so a negative multiple rotates the direction of the vector through 180°.

15.3 Addition and subtraction of vectors

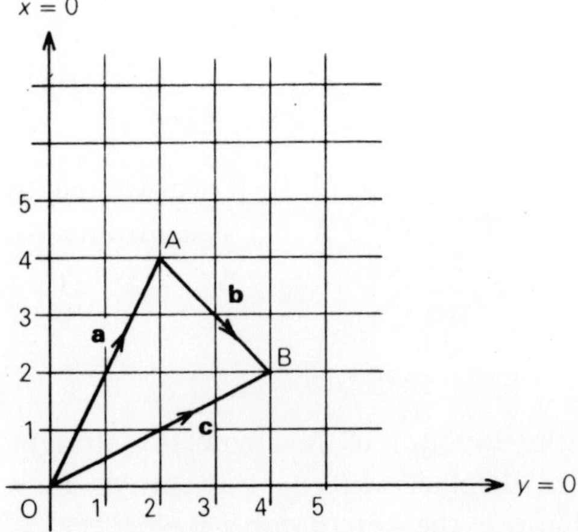

Fig. 15.2

In Fig. 15.2,

$$\overrightarrow{OA} = \begin{pmatrix} 2 \\ 4 \end{pmatrix} = \mathbf{a}$$

$$\overrightarrow{AB} = \begin{pmatrix} 2 \\ -2 \end{pmatrix} = \mathbf{b}.$$

If we follow **a** by **b** we arrive at the point B (4, 2). We could also have arrived at B by the displacement

$$\mathbf{c} = \overrightarrow{OB} = \begin{pmatrix} 4 \\ 2 \end{pmatrix}.$$

We say that **c** is the sum or resultant of **a** and **b**.

$$\therefore\ \mathbf{a}+\mathbf{b} = \mathbf{c} \qquad \text{i.e.} \begin{pmatrix} 2 \\ 4 \end{pmatrix} + \begin{pmatrix} 2 \\ -2 \end{pmatrix} = \begin{pmatrix} 4 \\ 2 \end{pmatrix}.$$

To achieve the answer, we have simply added the corresponding elements of the vector.

Subtraction follows in a similar way. If

$$\mathbf{d} = \mathbf{a} - \mathbf{b}$$

then

$$\mathbf{d} = \begin{pmatrix} 2 \\ 4 \end{pmatrix} - \begin{pmatrix} 2 \\ -2 \end{pmatrix} = \begin{pmatrix} 0 \\ 6 \end{pmatrix}.$$

We have subtracted the corresponding elements. Figure 15.3 shows us that this is the same as adding $-\mathbf{b}$.

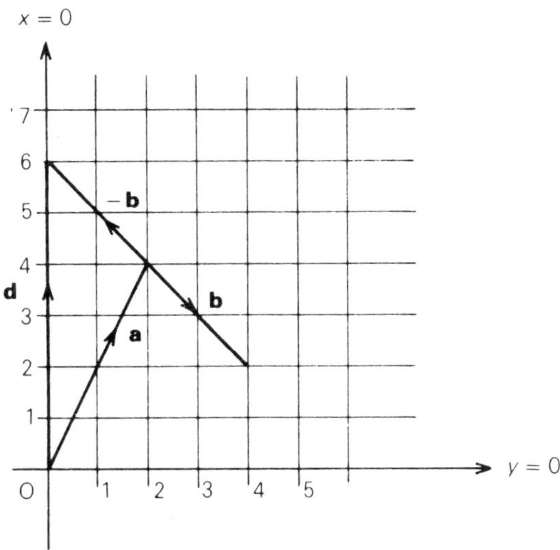

Fig. 15.3

EXAMPLE 1

If $\mathbf{a} = \begin{pmatrix} 3 \\ -2 \end{pmatrix}$ and $\mathbf{b} = \begin{pmatrix} 4 \\ 1 \end{pmatrix}$, evaluate (i) $3\mathbf{a}+2\mathbf{b}$, (ii) $\mathbf{a}-4\mathbf{b}$.

(i) $3\mathbf{a} = \begin{pmatrix} 9 \\ -6 \end{pmatrix}$ and $2\mathbf{b} = \begin{pmatrix} 8 \\ 2 \end{pmatrix}$.

$$\therefore\ 3\mathbf{a}+2\mathbf{b} = \begin{pmatrix} 9 \\ -6 \end{pmatrix} + \begin{pmatrix} 8 \\ 2 \end{pmatrix} = \begin{pmatrix} 17 \\ -4 \end{pmatrix}.$$

(ii) $4\mathbf{b} = \begin{pmatrix} 16 \\ 4 \end{pmatrix}$.

$$\therefore\ \mathbf{a}-4\mathbf{b} = \begin{pmatrix} 3 \\ -2 \end{pmatrix} - \begin{pmatrix} 16 \\ 4 \end{pmatrix} = \begin{pmatrix} -13 \\ -6 \end{pmatrix}.$$

EXERCISE 15a

If $\mathbf{a} = \begin{pmatrix} 1 \\ 2 \end{pmatrix}$, $\mathbf{b} = \begin{pmatrix} 3 \\ -1 \end{pmatrix}$ and $\mathbf{c} = \begin{pmatrix} 4 \\ 0 \end{pmatrix}$, evaluate the following:

1 $\mathbf{a}+\mathbf{b}$
2 $\mathbf{a}+\mathbf{c}$
3 $\mathbf{b}+\mathbf{c}$
4 $\mathbf{b}-\mathbf{c}$
5 $\mathbf{c}-\mathbf{a}$
6 $\mathbf{a}-\mathbf{b}$
7 $\mathbf{a}+\mathbf{b}+\mathbf{c}$
8 $2\mathbf{a}-\mathbf{b}$
9 $\mathbf{a}-3\mathbf{c}$
10 $3\mathbf{a}-2\mathbf{b}$
11 $5\mathbf{a}-4\mathbf{b}+\mathbf{c}$
12 $\mathbf{a}-5\mathbf{b}-\mathbf{c}$
13 $3\mathbf{a}-4\mathbf{b}+5\mathbf{c}$
14 $\mathbf{a}+\mathbf{c}-\mathbf{b}$
15 $3\mathbf{a}-\mathbf{b}+6\mathbf{c}$
16 $3\mathbf{c}-4\mathbf{a}-\mathbf{b}$
17 $2\mathbf{a}+6\mathbf{c}-3\mathbf{b}$
18 $9\mathbf{b}-8\mathbf{c}+\mathbf{a}$
19 $3\mathbf{b}+7\mathbf{c}-2\mathbf{a}$
20 $\mathbf{a}-7\mathbf{b}+\mathbf{c}$.

15.4 Length or modulus of a vector

$|\mathbf{a}|$ stands for the length or modulus of the vector **a**.

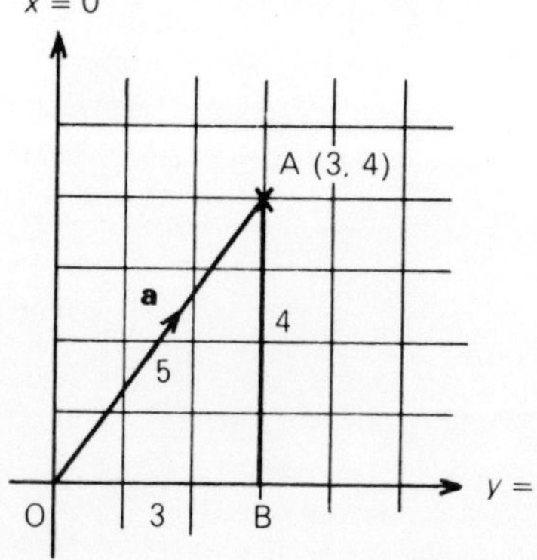

Fig. 15.4

In Fig. 15.4, since A is the point (3, 4), then

$$\mathbf{a} = \begin{pmatrix} 3 \\ 4 \end{pmatrix}.$$

We can obtain the length of OA by Pythagoras' theorem.

$$OA^2 = 3^2 + 4^2 = 25.$$
$$\therefore\ OA = 5$$
$$\text{so } |\mathbf{a}| = 5.$$

In general of course the length will not be a whole number. If

$$\mathbf{a} = \begin{pmatrix} x \\ y \end{pmatrix}$$

then

$$|\mathbf{a}| = \sqrt{x^2 + y^2}.$$

EXERCISE 15b

Work out the lengths of the vectors in exercise 15a, leaving your answers in surd form.

15.5 Geometrical proofs using vectors

What can you say about the vectors $\mathbf{a} + \mathbf{b}$ and $\frac{3}{4}\mathbf{a} + \frac{3}{4}\mathbf{b}$?
The answer is simply that they are parallel, because $\frac{3}{4}\mathbf{a} + \frac{3}{4}\mathbf{b}$ is the same as $\frac{3}{4}(\mathbf{a} + \mathbf{b})$, and in section 15.2 we saw that a multiple of a vector does not change its direction.

EXAMPLE 2

ABC is a triangle. X is the mid-point of AB, and Y is the mid-point of BC. Prove that XY is parallel to AC and half of its length.

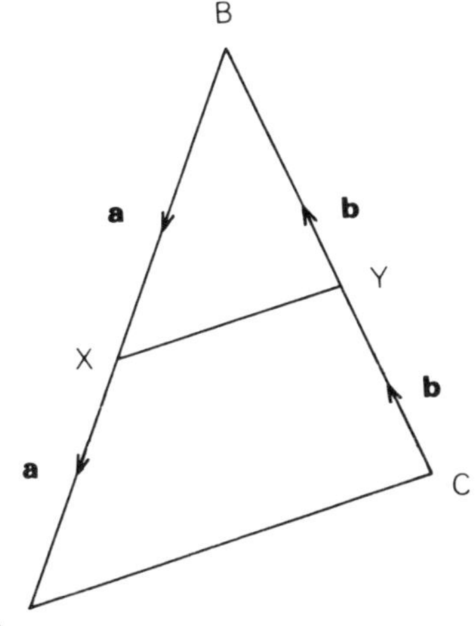

Fig. 15.5

If Fig. 15.5, if we let $\overrightarrow{BX} = \mathbf{a}$, then $\overrightarrow{BA} = 2\mathbf{a}$. If we let $\overrightarrow{YB} = \mathbf{b}$, then $\overrightarrow{CB} = 2\mathbf{b}$.

$$\overrightarrow{YX} = \overrightarrow{YB} + \overrightarrow{BX} = \mathbf{b} + \mathbf{a}$$

$$\overrightarrow{CA} = \overrightarrow{CB} + \overrightarrow{BA} = 2\mathbf{b} + 2\mathbf{a} = 2(\mathbf{b} + \mathbf{a})$$

We see that $2\,\overrightarrow{YX} = \overrightarrow{CA}$; therefore YX is parallel to CA and is half its length.

EXAMPLE 3

ABCD is a parallelogram. E is the mid-point of BC, and AE is produced to F where AE = EF. Show that DCF is a straight line.

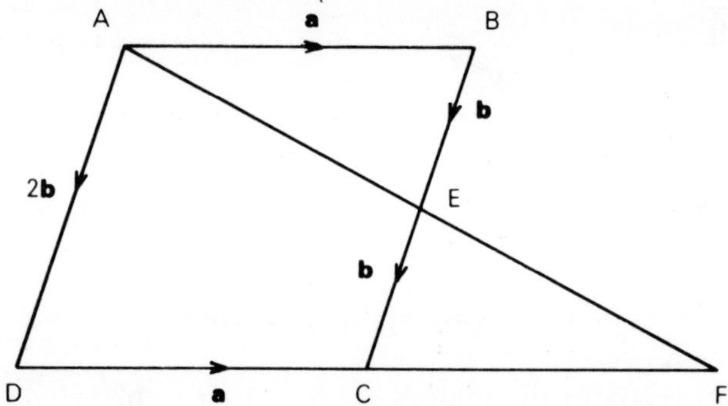

Fig. 15.6

In order to prove this, we have to show that $\overrightarrow{CF}$ is a multiple of $\overrightarrow{DC}$. In Fig. 13.6, let $\overrightarrow{AB} = \overrightarrow{DC} = \mathbf{a}$ (since ABCD is a parallelogram). Let $\overrightarrow{\mathbf{BE}} = \mathbf{b}$.

$$\therefore\ \overrightarrow{BC} = 2\,\mathbf{b} \text{ (since E is the mid-point of BC)}$$

and also

$$\overrightarrow{AD} = 2\mathbf{b}.$$

Now

$$\overrightarrow{AE} = \overrightarrow{AB} + \overrightarrow{BE} = \mathbf{a} + \mathbf{b}$$

$$\therefore\ \overrightarrow{EF} = (\mathbf{a} + \mathbf{b})$$

$$\overrightarrow{CF} = \overrightarrow{CE} + \overrightarrow{EF} = -\mathbf{b} + (\mathbf{a} + \mathbf{b}) = \mathbf{a}$$

$\therefore \overrightarrow{CF} = \overrightarrow{DC}$

so DCF is a straight line.

EXERCISE 15c

1 ABC is a triangle. X is the mid-point of AB, and Y is the mid-point of BC. XY is produced to T so that XY = YT. Prove that XBTC is a parallelogram.

2 In the triangle ABC, AC is produced to D so that $\overrightarrow{CD} = 2\overrightarrow{AC}$. E is the point on AB such that $\overrightarrow{AE} = \frac{1}{3}\overrightarrow{AB}$. Prove that ECDB is a trapezium.

3 ABCD is a parallelogram. X is the mid-point of DA and Y is the mid-point of CB. Z is the point on CB such that $\overrightarrow{CZ} = \frac{1}{4}\overrightarrow{CB}$. If DZ is produced to T such that DZ = ZT, prove that XYT is a straight line.

4 ABCDEF is a regular hexagon. G is the mid-point of CD, and FG is produced to H so that G is the mid-point of FH. AC is produced to K so that AC = CK, and the point L is constructed such that $\overrightarrow{KL} = \overrightarrow{FA}$. Prove that AL is equal and parallel to FH.

15.6 Locus problems using vectors

The modulus notation defined in section 15.4 can be used to represent simple curves.

Let O be a fixed point and P a variable point. If $\overrightarrow{OP} = \mathbf{r}$ and we are told that $|\mathbf{r}| = 4$, what is the locus of P?

$|\mathbf{r}| = 4$ tells us that the distance OP is always 4 units; therefore the locus of P is a circle, with centre O.

EXAMPLE 4

A and B are points with position vectors **a** and **b** with respect to an origin O. A variable point P with position vector **p** moves so that:

$$|\mathbf{p} - \mathbf{a}| = |\mathbf{p} - \mathbf{b}|.$$

Describe the locus of P.

$|\mathbf{p} - \mathbf{a}|$ stands for the distance between P and A; therefore P moves so that it is equidistant from A and B. Hence, P lies on the perpendicular bisector of AB. (See Fig. 15.7.)

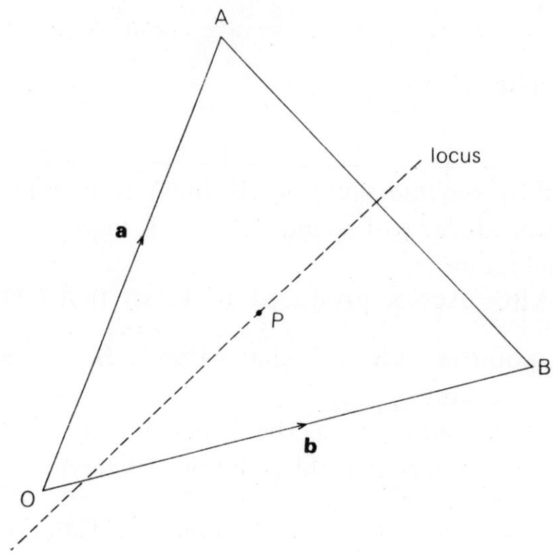

Fig. 15.7

EXERCISE 15d

1 OABC is a quadrilateral. $\overrightarrow{OA} = \mathbf{a}$, $\overrightarrow{AB} = \mathbf{b}$, and $\overrightarrow{BC} = \mathbf{c}$. Write down the following vectors in terms of **a**, **b** and **c**:

(i) $\overrightarrow{AC}$ (ii) $\overrightarrow{OC}$ (iii) $\overrightarrow{OB}$.

2 RSTU is a rectangle. The diagonals of the rectangle meet in M. If $\overrightarrow{MR} = \mathbf{a}$ and $\overrightarrow{MS} = \mathbf{b}$, write down in terms of **a** and **b**:

(i) $\overrightarrow{RS}$ (ii) $\overrightarrow{ST}$ (iii) $\overrightarrow{US}$.

3 ABC is a triangle, and X is the mid-point of BC. If $\overrightarrow{CX} = \mathbf{a}$ and $\overrightarrow{BA} = \mathbf{b}$, write down in terms of **a** and **b**:

(i) $\overrightarrow{XB}$ (ii) $\overrightarrow{XA}$ (iii) $\overrightarrow{AC}$.

4 If $\mathbf{a} = \begin{pmatrix} 1 \\ 3 \end{pmatrix}$, $\mathbf{b} = \begin{pmatrix} 2 \\ 0 \end{pmatrix}$ and $\mathbf{c} = \begin{pmatrix} 3 \\ -1 \end{pmatrix}$, evaluate:

(i) $\mathbf{a}+\mathbf{b}$ (ii) $\mathbf{a}-\mathbf{b}$ (iii) $\mathbf{b}-\mathbf{c}$
(iv) $2\mathbf{b}+\mathbf{c}$ (v) $\mathbf{a}+\mathbf{b}+\mathbf{c}$ (vi) $\mathbf{a}-\mathbf{b}-2\mathbf{c}$
(vii) $3\mathbf{a}-4\mathbf{b}$ (viii) $\frac{1}{4}\mathbf{a}-\mathbf{b}$ (ix) $\frac{1}{2}\mathbf{a}-\mathbf{b}$
(x) $2\mathbf{a}-\frac{1}{2}\mathbf{b}-\frac{1}{4}\mathbf{c}$

5 X is the point (4, 1), Y is the point $(-2, -1)$ and Z is the point $(5, -3)$.
Express, as a 2×1 column vector, the following:
(i) $\overrightarrow{XY}$ (ii) $\overrightarrow{YZ}$ (iii) $\overrightarrow{XZ}$.

6 O(0, 0), A(3, 1) and B(2, −1) are three vertices of a quadrilateral OABC. If $\overrightarrow{BC} = \begin{pmatrix} 3 \\ -2 \end{pmatrix}$, what are the co-ordinates of C? Find $\overrightarrow{AC}$.

7 If $2\begin{pmatrix} x \\ y \end{pmatrix} + 3\begin{pmatrix} 2x \\ y \end{pmatrix} = \begin{pmatrix} 16 \\ -20 \end{pmatrix}$, find x and y.

8 If O is the origin, and $\overrightarrow{OX} = \mathbf{x}$, draw a diagram and mark the point Y such that $3\mathbf{x} - 2\mathbf{y} = 0$.

9

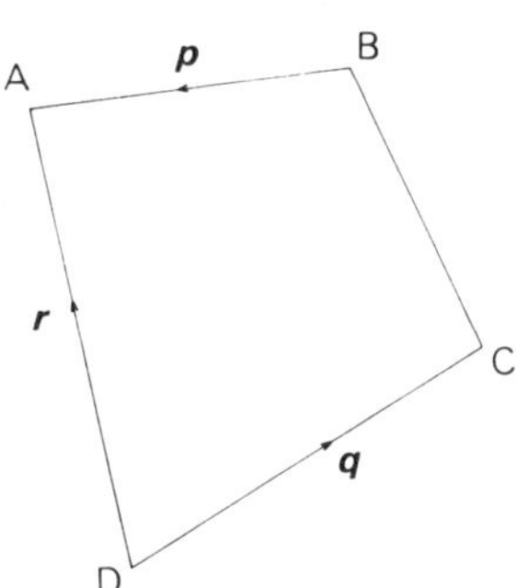

Fig. 15.8

This question refers to Fig. 15.8. Write down in terms of **p**, **q**, **r**:

(i) $\overrightarrow{DB}$ (ii) $\overrightarrow{AC}$ (iii) $\overrightarrow{CB}$.

10 **a** and **b** are two non-parallel vectors. If $(m+n+1)\mathbf{a} - (3n+2)\mathbf{b} = 0$, find m and n.

11 O is the origin, and A is a fixed point with position vector **a**. **b** is a vector which has a variable direction, but its modulus is 3 units. A point P with position vector **p** is such that $\mathbf{p} = \mathbf{a} + \mathbf{b}$. Describe the locus of P as fully as you can.

12 ABCD is a quadrilateral. A, B, D have co-ordinates (0, 2), (2, 5), (8, 0) respectively. If $\overrightarrow{AD} = 2\overrightarrow{BC}$, find the co-ordinates of C.

AC is produced to P such that $\overrightarrow{CP} = 2\overrightarrow{AC}$. Find the co-ordinates of P, and $|\overrightarrow{BD}|$.

HARDER EXAMPLES 15

1 **a** is the displacement from the point (2, 3) to the point (5, −2). **b** is the displacement from the point (1, 2) to the point (2, 3). To what point will the displacement $\mathbf{b} - 3\mathbf{a}$ move the point (1, 2)?

2 Illustrate clearly on a diagram, the points with position vectors $\overrightarrow{OC} = \mathbf{a}+\mathbf{b}$ and $\overrightarrow{OD} = 2\mathbf{b}-\mathbf{a}$. If E is the mid-point of CD, find the position vector of E and show that O, B and E lie in a straight line.

3 The position vectors of two points A and B relative to an origin O are **a** and **b** respectively, as illustrated in Fig. 15.9.
(i) The position vector of a point P is $\mathbf{a}+h\mathbf{b}$ where $h \geq 0$. Draw accurately on the grid provided the locus of P.
(ii) The position vector of a point Q is $\mathbf{a}+k(\mathbf{a}+\mathbf{b})$ where $k \geq 0$. Draw accurately on the grid provided the locus of Q. Label your answers clearly.

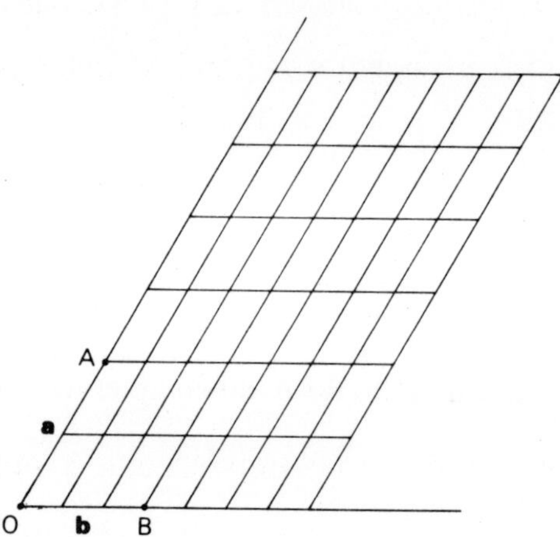

Fig. 15.9

[C]

4 Write the statements $x-4y = 2m$ and $2y-x = n$ as a vector equation.

5 The vector **a** represents a displacement of 8 km due west; the vector **b** represents a displacement of 5 km on a bearing of 030°.
(i) Using a scale of not less than 1 cm to represent 1 km, construct a carefully labelled figure and from it measure, as accurately as possible, the magnitude and direction of the vector **c**, where $\mathbf{c} = \mathbf{a}+\mathbf{b}$.
(ii) Write down the magnitude and direction of the vector $2\mathbf{c}-2\mathbf{b}$.
(iii) A displacement due north is represented by the vector $\mathbf{a}+k\mathbf{b}$, where k is a constant. Find, by calculating or by accurate drawing and measurement, both the value of k and the magnitude of the vector $\mathbf{a}+k\mathbf{b}$.

[NI]

6 A is the point (1, 2), B the point (3, 1) and C the point (3, 6). X is a point such that $\overrightarrow{CX} = 2\overrightarrow{AB}$. Find the co-ordinates of X.

7 Find, by any method, the acute angle between the vectors

$$\mathbf{a} = \begin{pmatrix} 2 \\ 4 \end{pmatrix} \text{ and } \mathbf{b} = \begin{pmatrix} -1 \\ 3 \end{pmatrix}.$$

8 (i) The vectors **p** and **q** are at right angles to each other and $|\mathbf{p}| = 2|\mathbf{q}| = 2$. Given that $\mathbf{r} = \mathbf{p} + \mathbf{q}$ and that $\mathbf{s} = -\frac{1}{2}\mathbf{p} + 2\mathbf{q}$, find by calculation

(a) $|\mathbf{r}| - |\mathbf{s}|$,

(b) the angle between **r** and **s**.

(ii) ABCDEF is a regular hexagon. Given that $\overrightarrow{AB} = \mathbf{a}$ and $\overrightarrow{BC} = \mathbf{b}$, find expressions for $\overrightarrow{AD}$ and $\overrightarrow{AF}$ in terms of **a** and/or **b**.

If $|\mathbf{a}| = 10$, calculate $|\mathbf{a} + \mathbf{b}|$ giving your answer correct to one decimal place.

9 In the triangle OAB, $\overrightarrow{OA} = \mathbf{a}$ and $\overrightarrow{OB} = \mathbf{b}$. L is a point on the side AB, M is a point on the side OB, and OL and AM meet at S. It is given that AS = SM and OS/OL = $\frac{3}{4}$. Given also that OM/OB = h and AL/AB = k,

(i) express the vectors $\overrightarrow{AM}$ and $\overrightarrow{OS}$ in terms of **a**, **b** and h,

ii) express the vectors $\overrightarrow{OL}$ and $\overrightarrow{OS}$ in terms of **a**, **b** and k. Find h and k, and hence find the values of the ratios OM/MB and AL/LB.

10 If $\mathbf{x} = \begin{pmatrix} 3 \\ 1 \end{pmatrix}$ and $\mathbf{y} = \begin{pmatrix} a \\ 3 \end{pmatrix}$ are at right angles, find a.

11 A vector of length 10 units lies along the line whose equation is $3x - 4y + 6 = 0$. If one end of the vector is the point $(0, \frac{3}{2})$, what are the possible co-ordinates of the other end?

16 Transformation geometry

16.1 Translation

In Fig. 16.1, each element of the shape A has been subject to a simple displacement, which maps the complete shape to B. We can represent this translation as a vector (see chapter 15).

In this case, A → B under the translation $\begin{pmatrix} 3 \\ 2 \end{pmatrix}$.

Similarly, B → C under the translation $\begin{pmatrix} 2 \\ -7 \end{pmatrix}$.

We will denote by $\mathbf{T}_1$ the translation $\begin{pmatrix} 3 \\ 2 \end{pmatrix}$ and by $\mathbf{T}_2$ the translation $\begin{pmatrix} 2 \\ -7 \end{pmatrix}$.

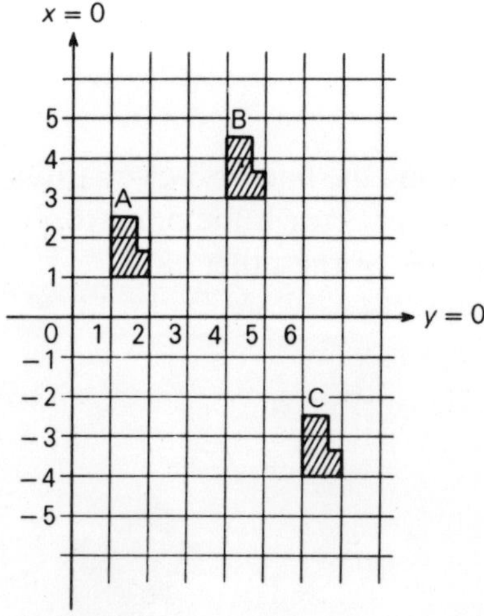

Fig. 16.1

What do we mean by $\mathbf{T}_2\mathbf{T}_1$?
In this book, we will let $\mathbf{T}_2\mathbf{T}_1$ denote the operation of translation $\mathbf{T}_1$, followed by $\mathbf{T}_2$.
(*Note:* the second transformation is carried out first.)
Look again at $\mathbf{T}_2\mathbf{T}_1(\mathrm{A})$:

$$\mathbf{T}_1(\mathrm{A}) = \mathrm{B}$$
$$\mathbf{T}_2\mathbf{T}_1(\mathrm{A}) = \mathbf{T}_2(\mathrm{B}) = \mathrm{C}.$$

Now C can be obtained directly from A by a single translation, denoted by $\mathbf{T}_3$.

$$\therefore\ \mathbf{T}_3 = \mathbf{T}_2\mathbf{T}_1.$$

But

$$\mathbf{T}_3 = \begin{pmatrix} 5 \\ -5 \end{pmatrix} = \begin{pmatrix} 3 \\ 2 \end{pmatrix} + \begin{pmatrix} 2 \\ -7 \end{pmatrix}$$

Therefore, to combine translations we add the components, as for vectors. We can prove the general result: the composition of two translations is always a translation.

16.2 Reflection

In Fig. 16.2, each element of A is mapped on to each element of B by a reflection in the x-axis, a reflection being such that PA = AP′ (see Fig. 16.2).

We can represent a mathematical reflection by a matrix. Let P(x, y) be any point. Consider the matrix

$$\begin{pmatrix} 1 & 0 \\ 0 & -1 \end{pmatrix}.$$

In order to find how this transforms P, we write the co-ordinates as a column vector

$$\begin{pmatrix} x \\ y \end{pmatrix}.$$

We pre-multiply this by the matrix to obtain

$$\begin{pmatrix} 1 & 0 \\ 0 & -1 \end{pmatrix}\begin{pmatrix} x \\ y \end{pmatrix} = \begin{pmatrix} x \\ -y \end{pmatrix}$$

$\therefore$ P′ is the point $(x, -y)$.

The x-co-ordinate remains the same, and the y-co-ordinate has been reflected in the x-axis to become $-y$.

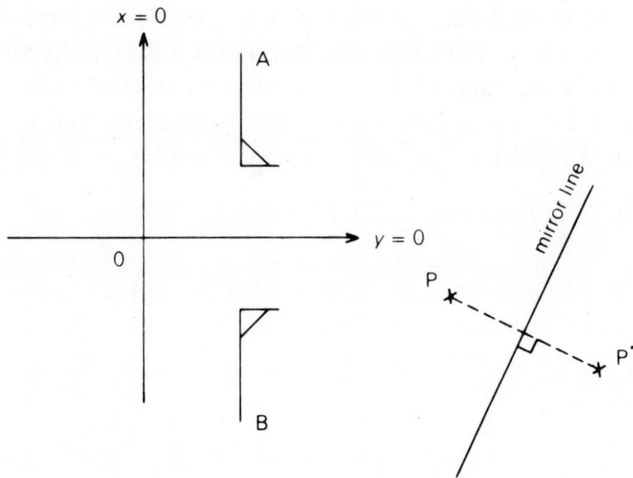

Fig. 16.2

$\therefore \begin{pmatrix} 1 & 0 \\ 0 & -1 \end{pmatrix}$ represents reflection in the x-axis.

Similarly, $\begin{pmatrix} -1 & 0 \\ 0 & 1 \end{pmatrix}$ represents reflection in the y-axis.

Note: if $\mathbf{M}$ is a reflection, $\mathbf{M}^2 = \mathbf{I}$ (the identity matrix).

16.3 Rotation

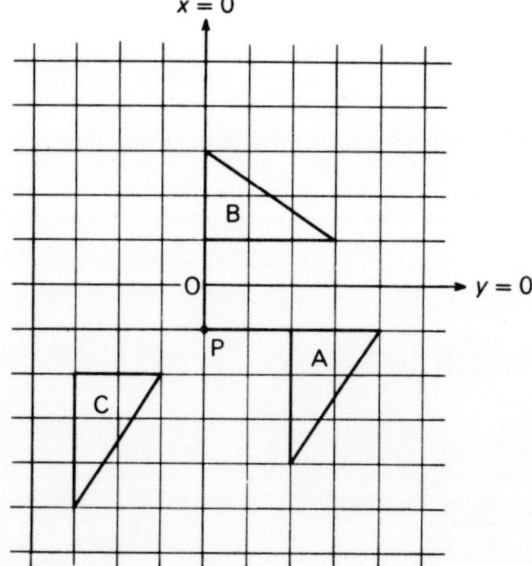

Fig. 16.3

In Fig. 16.3, A, B and C are congruent triangles (exactly the same shape). If you imagine A attached to the point P(0, − 1) by a rigid line which is rotated through +90° (anti-clockwise), then A → B.

Although A and C are congruent, it is not possible to find a point about which the triangle A can be rotated so that A → C. In fact, A → C by a translation $\begin{pmatrix} -5 \\ -1 \end{pmatrix}$.

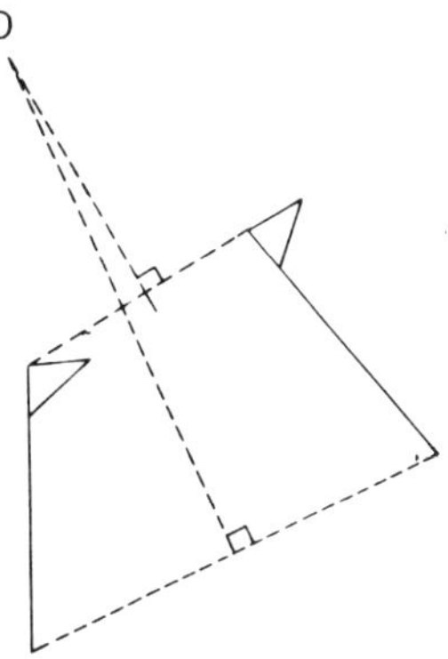

Fig. 16.4

Figure 16.4 shows how the centre of rotation can be constructed, if it exists. O is the intersection of the two perpendicular bisectors of lines joining pairs of corresponding points. (*Note:* all lines are rotated by the angle of rotation. See Fig. 16.5.)

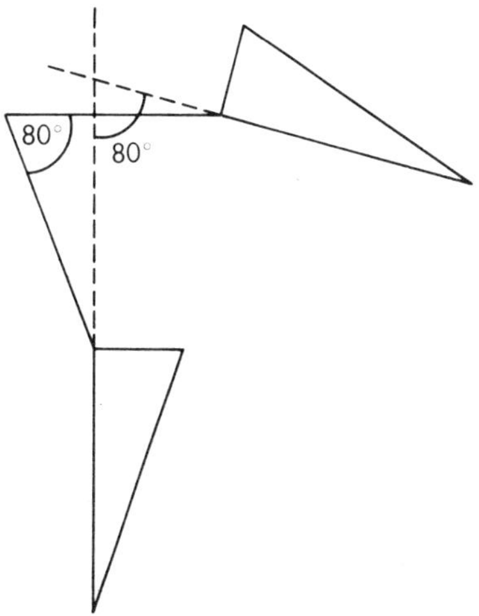

Fig. 16.5

A clockwise rotation will be taken as negative. An anti-clockwise rotation will be taken as positive.
A rotation followed by a rotation can be replaced by a single rotation.

16.4 Glide reflection

A glide reflection is the name given to a single transformation which consists of a reflection followed by a translation, or *vice versa*. (See Fig. 16.6.)

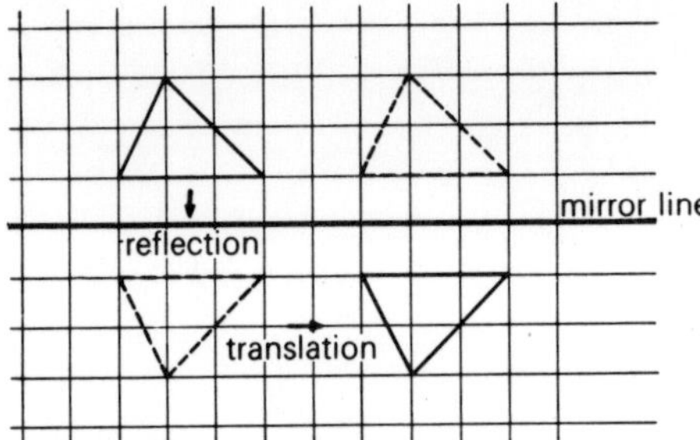

Fig. 16.6

16.5 Isometries

These transformations—reflection, rotation, translation and glide reflection—leave the shape of the transformed figure unchanged. For this reason, they are called isometries. We can prove that a reflection followed by a reflection (in a different mirror line) can usually be replaced by a rotation, unless the mirror lines are parallel.

16.6 Dilatation

In Fig. 16.7, we say that OA′B′ is the enlargement of OAB, centre O, and scale factor 2 (since OA′ = 2OA).

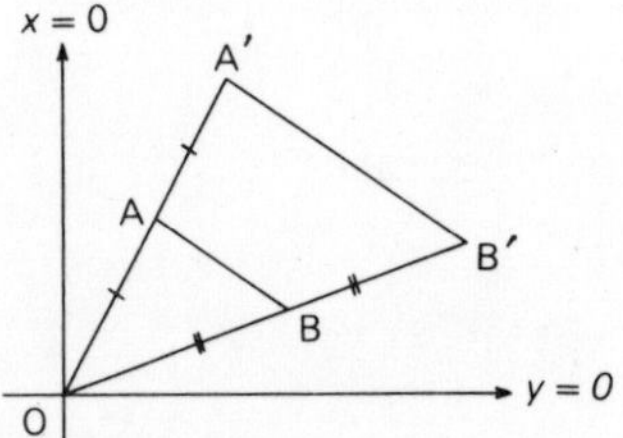

Fig. 16.7

In Fig. 16.8, A′B′C′ is an enlargement, centre O, scale factor $-\frac{1}{2}$, of ABC. All lengths are halved, and the negative sign shows that the shape is on the opposite side of O. (*Note:* all corresponding lengths are multiplied by the scale factor.)

The matrix $\begin{pmatrix} k & 0 \\ 0 & k \end{pmatrix}$ represents an enlargement, centre O, scale factor k.

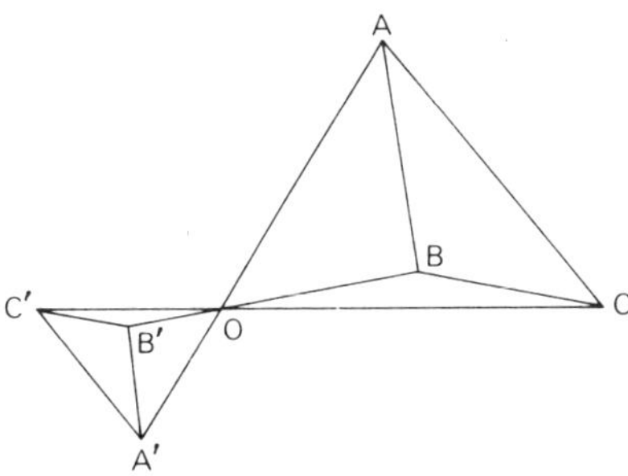

Fig. 16.8

16.7 Shearing

In Fig. 16.9, the triangle OBA → OB′A by a shear. The displacement in the direction of the x-axis increases proportionately with the distance from the x-axis.

B(3, 4) → B′(11, 4).

In terms of a matrix equation, we have

magnitude of the shear ↓ (the 2)

$$\begin{pmatrix} 1 & 2 \\ 0 & 1 \end{pmatrix}\begin{pmatrix} 3 \\ 4 \end{pmatrix} = \begin{pmatrix} 11 \\ 4 \end{pmatrix}.$$

always 1 → (the 1 in the top left)

always 0 in x-direction ↑ (the 0 in the bottom left)

We say that the magnitude of the shear is 2, because the shift in the x-direction is twice the distance from the x-axis.

A shear in the direction of the y-axis will be of the form $\begin{pmatrix} 1 & 0 \\ k & 1 \end{pmatrix}$

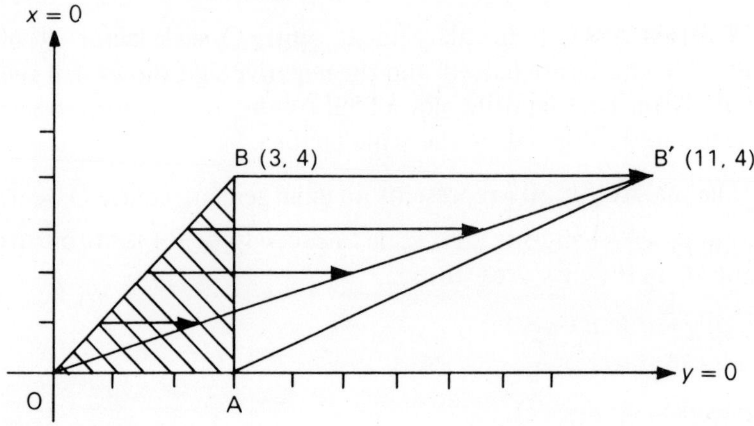

Fig. 16.9

16.8 Transforming shapes

To find the shape into which A(2, 0), B(3, 1), C(2, 4) is transformed by the matrix

$$\begin{pmatrix} 1 & 2 \\ 1 & 1 \end{pmatrix},$$

we proceed as follows. Write all the co-ordinates in one matrix, i.e.

$$\begin{pmatrix} 2 & 3 & 2 \\ 0 & 1 & 4 \end{pmatrix}.$$

Then pre-multiply by the matrix.

$$\begin{pmatrix} 1 & 2 \\ 1 & 1 \end{pmatrix}\overset{\text{A}\quad\text{B}\quad\text{C}}{\begin{pmatrix} 2 & 3 & 2 \\ 0 & 1 & 4 \end{pmatrix}} = \overset{\text{A}'\quad\text{B}'\quad\text{C}'}{\begin{pmatrix} 2 & 5 & 10 \\ 2 & 4 & 6 \end{pmatrix}}$$

See Fig. 16.10.

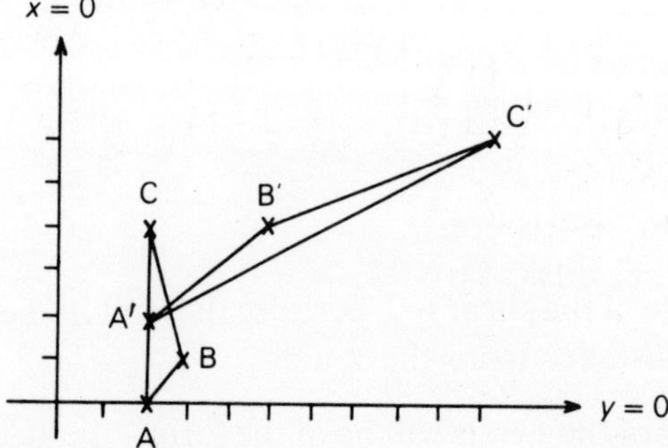

Fig. 16.10

16.9 Areas

If a shape is transformed by a matrix

$$\begin{pmatrix} a & b \\ c & d \end{pmatrix},$$

its area will change. It is possible to prove that, if A is the old area and A' is the new area, then

$$A' = A \times (ad - bc).$$

EXAMPLE 1

The triangle ABC, where A is (1, 3), B is (2, 1) and C is (4, 6), is transformed by the matrix

$$\begin{pmatrix} 3 & 1 \\ 2 & 6 \end{pmatrix}$$

into A′B′C′. Find the area of A′B′C′.

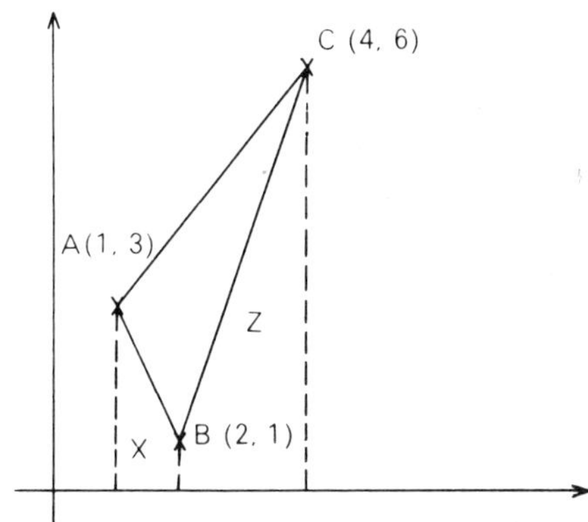

Fig. 16.11

In order to find the area of ABC, we find the area of the large trapezium, and subtract X and Y (see Fig. 16.11).

$$\text{Area ABC} = 3 \times \frac{(6+3)}{2} - 1 \times \frac{(3+1)}{2} - 2 \times \frac{(1+6)}{2} = 4\tfrac{1}{2}.$$

$$\begin{aligned} \text{Area A'B'C'} &= (ad - bc) \times 4\tfrac{1}{2} \\ &= (3 \times 6 - 2 \times 1) \times 4\tfrac{1}{2} = 72. \end{aligned}$$

If we consider the matrix

$$\begin{pmatrix} 1 & k \\ 0 & 1 \end{pmatrix},$$

$1 \times 1 - k \times 0 = 1$; hence areas are unaltered by a shear. It follows that the area of a parallelogram (a sheared rectangle) is equal to that of a rectangle on the same base with the same height.

EXERCISE 16a

1 By considering the effect on the unit square (0, 0), (0, 1), (1, 1), (1, 0), describe as accurately as you can the effect of the following matrices:

(i) $\begin{pmatrix} 0 & 1 \\ 1 & 0 \end{pmatrix}$ (ii) $\begin{pmatrix} 0 & -1 \\ 1 & 0 \end{pmatrix}$ (iii) $\begin{pmatrix} 0 & 1 \\ -1 & 0 \end{pmatrix}$

(iv) $\begin{pmatrix} 2 & 1 \\ 0 & 2 \end{pmatrix}$ (v) $\begin{pmatrix} -1 & 0 \\ 2 & -1 \end{pmatrix}$

2 If $\mathbf{M}_1$ is the matrix for reflection in the x-axis, and $\mathbf{M}_2$ is the matrix for reflection in the y-axis, find $\mathbf{M}_1\mathbf{M}_2$ and $\mathbf{M}_2\mathbf{M}_1$. What do these two matrices represent?

3 The unit square is transformed by means of the matrix

$$\begin{pmatrix} 2 & -1 \\ 3 & 4 \end{pmatrix}.$$

Describe in general terms the effect of this matrix.

4 The triangle A(1, 1), B(2, 3), C(4, 1) is transformed by means of the matrix

$$\begin{pmatrix} 1 & 3 \\ 2 & 4 \end{pmatrix}$$

into A′B′C′. Find the area of A′B′C′.

5 The rectangle A(2, 0), B(2, 4), C(5, 4), D(5, 0) is transformed by means of the matrix

$$\begin{pmatrix} \frac{1}{2} & \frac{1}{3} \\ 1 & \frac{2}{3} \end{pmatrix}.$$

Illustrate the result.

16.10 Proofs without using co-ordinates

We can use the methods of transformation geometry to prove results without the use of co-ordinates. The following example should indicate how the methods are applied.

EXAMPLE 2

In Fig. 16.12, ABCD and BEFG are squares. K is the mid-point of AB, and H is the mid-point of BG. Prove that HF is perpendicular to DK.

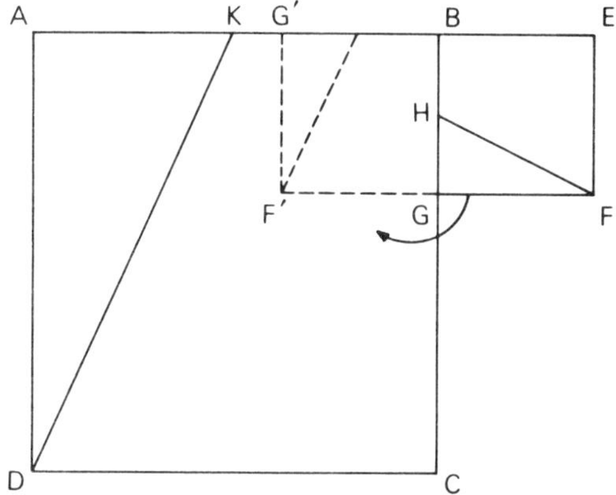

Fig. 16.12

BEFG → BGF′G′ by a rotation of $-90°$.
BGF′G′ → BCDA by an enlargement, centre B.
Therefore, the line HF has been rotated by 90° and enlarged to become KD. Hence, HF ⊥ DK.

EXERCISE 16b

1 Prove that the diagonals of a rhombus bisect each other at 90°.
2 Prove that the diagonals of a parallelogram bisect each other.
3 ABCD is a square. E is the mid-point of AB. DE is produced to meet CB at F. Prove that AD = BF.
4 PQRS is a parallelogram. T is the mid-point of QR, and PT produced meets SR produced at V. Prove that PRVQ is a parallelogram.
5 WXYZ is a parallelogram. M is the mid-point of XY. WM is produced to N, so that WM = MN. Prove that ZYN is a straight line.
6 Prove that if in triangle ABC, ∠A = ∠B, then AC = BC.

HARDER EXAMPLES 16

1 On squared paper, draw triangle ABC where A is (0, 1), B is (3, 5) and C is (1, 5), using a scale of 2 cm for one unit on both axes and allowing space on your paper for x to vary from -3 to 6, and y to vary from -5 to 5. **M** is the reflection in the line

$x = 0$, and **T** is the translation determined by the vector $\begin{pmatrix} 6 \\ 0 \end{pmatrix}$.

ABC is mapped on to $A_1B_1C_1$ by **M**, and $A_1B_1C_1$ is mapped on to $A_2B_2C_2$ by **T**. Draw these triangles in your diagram and find from your diagram the transformation **K** such that **K** = **TM**.

Triangle ABC is mapped on to $A_3B_3C_3$ by the transformation S which is determined by the matrix $\begin{pmatrix} -1 & 0 \\ 0 & -1 \end{pmatrix}$.

Draw triangle $A_3B_3C_3$ in your diagram and describe the transformation **S**.

[W]

2 The triangle ABC is mapped under the transformation
$x' = 3x$
$y' = 3y$
on to the triangle A′B′C′. Given that the vertices of ABC are (2, 1), (4, 1) and (4, 3) respectively, find the vertices of the triangle A′B′C′. Find the ratio of the area of triangle A′B′C′ to the area of triangle ABC and show that the triangles are similar.
Triangle ABC is mapped under the transformation
$x'' = 3y$
$y'' = 3x$
on to triangle A″B″C″. Show that triangles A″B″C″ and A′B′C′ are congruent.

[JMB]

3 A triangle is formed by the points A(2, 0), B(5, 0) and C(3, 2). The matrix $\begin{pmatrix} 2 & 3 \\ -1 & 0 \end{pmatrix}$ transforms this triangle into the triangle A′B′C′. Find the area of the triangle A′B′C′.

4 (i) Taking 2 cm (or 1 inch) as unit, draw on squared paper, axes showing values of x from -5 to $+2$ and y from -1 to $+3$. Mark in the points O(0, 0), A(1, 0), B(1, 1), C(0, 1), P(1, 2) and Q(0, 2).
(ii) **S** is the shear with invariant line $y = 0$ which maps (1, 1) on to $(-1, 1)$. Mark the images S(O), S(A), etc., of all the six given points under the shear S using the letters O′, A′, etc. Show the form of the image of the rectangle OAPQ by lightly shading within its outline.
(iii) **T** is the shear with invariant line $y = 1$ which maps (0, 0) on to $(-2, 0)$. Mark the images **T**(O′), **T**(A′), **T**(P′), **T**(Q′) of

the points O′, A′, P′, Q′, under the shear **T** by the letters O″, A″, P″, Q″, and show clearly the outline of the figure which they form.

(iv) State what conclusions you reach about the nature of the transformation **TS**, giving your reasons.

5 On graph paper, with the scale of x ranging at least from 0 to 15 and the scale of y ranging at least from -4 to 5, draw the quadrilateral whose vertices are O(0, 0), A(1, 0), B(3, 1) and C(0, 1).

Find and draw the image A′B′C′O of ABCO under the transformation whose matrix is

$$\mathbf{M} = \begin{pmatrix} 4 & 3 \\ 3 & -4 \end{pmatrix}.$$

Explain why OBB′ is a straight line.

Find the value of n so that the transformation whose matrix is

$$\begin{pmatrix} n & 0 \\ 0 & n \end{pmatrix}$$

transforms A′B′C′O into a reflection of ABCO. Draw this reflection and find the equation of its mirror line. Hence describe briefly the transformation corresponding to **M**.

[L]

6 The point P whose co-ordinates are (x, y) is mapped on to P′ by a reflection in the y-axis. P′ is mapped on to P″ by a reflection in the x-axis.

(i) Find the matrix $\mathbf{M}_1$ which maps P on to P′.

(ii) Find the matrix $\mathbf{M}_2$ which maps P′ on to P″.

Write down in terms of $\mathbf{M}_1$ and $\mathbf{M}_2$, the matrix which transforms P into P″. Evaluate this matrix.

7 Figure 16.13 shows a tessellation of the plane by equilateral triangular tiles. A right-angled triangle of cardboard is of such a shape that it can cover exactly half of any tile: its initial position is shown marked t in the diagram. The operation of reflecting t in its shortest side is written **A**, and starting with the given position t the result is **A**(t), which is marked with the figure 1. Reflection in the second side is denoted by **B**, so that **B**(t) is 2; reflection in the hypotenuse is **C** so that **C**(t) is 3. Position 4 is **B**(3) = **BC**(t), so that the operation carrying t to 4 is **BC**.

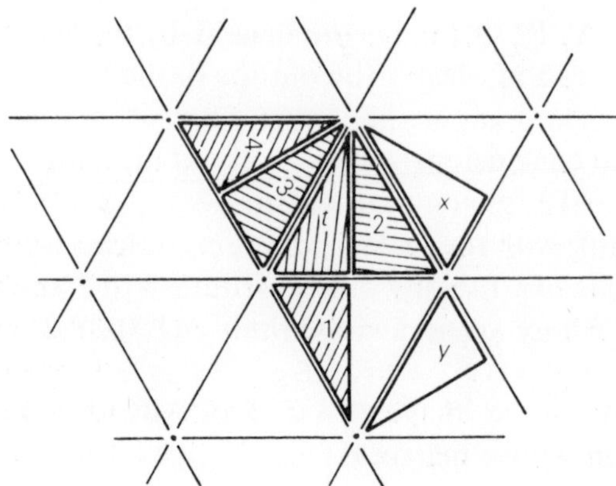

Fig. 16.13

(i) Show that operations **AB** and **BA** are the same. Describe this combined operation in simple geometrical terms.
(ii) Describe the combined operation **BC** geometrically. Show that **BC** ≠ **CB**. Simplify **C**(**CB**), (**CC**)**B** and **BC**(**CB**). Use **I** to denote the identity operation if it is required in your work.
(iii) Obtain expressions, as combinations of **A**, **B** and **C**, for operations which carry (a) t to x, (b) t to y, (c) x to t.

[O]

8 On squared paper, draw triangle ABC where A is (2, 0), B is (3, 4) and C is (0, 1), using a scale of 1 cm for one unit on both axes and allowing space on your paper for x to vary from -4 to 7 and y to vary from -4 to 6. **R** is the anti-clockwise rotation of 90° about the origin and **T** is the translation determined by vector $\begin{pmatrix} 7 \\ 3 \end{pmatrix}$. ABC is mapped on to $A_1B_1C_1$ by **R**, and $A_1B_1C_1$ is mapped on to $A_2B_2C_2$ by **T**. Draw these triangles in your diagram and find the transformation **Q** such that **Q** = **TR**.

Draw the triangle on to which triangle ABC is mapped by the transformation **K**, determined by the matrix $\begin{pmatrix} 1 & 0 \\ 0 & -1 \end{pmatrix}$, and describe the geometrical effect of the transformation **K**.

9 PQ is a diameter of a circle. C is the mid-point of one of the semi-circular arcs into which PQ divides the circumference, and X is any point on the other semi-circular arc. **R** is the rotation with centre C which maps P on to Q, and **R**(X) = X′.
(i) State the angle of the rotation **R**.

(ii) Explain why (a) PX is perpendicular to QX′, (b) PX is perpendicular to XQ.
What do these facts show about X, Q, X′?
(iii) Show that $(X'X)^2 = 2(CX)^2$.

[O & C (SMP)]

10 ABC is an equilateral triangle of side 3 cm, the corners being lettered anti-clockwise in alphabetical order. ABC is mapped on to $A_1B_1C_1$ by an anti-clockwise rotation of 120° about A. $A_1B_1C_1$ is then mapped on to $A_2B_2C_2$ by an anti-clockwise rotation of 120° about B. Finally, $A_2B_2C_2$ is mapped on to $A_3B_3C_3$ by an anti-clockwise rotation of 120° about C. Construct the complete figure accurately and state clearly a single geometrical transformation which maps ABC on to $A_3B_3C_3$.

[C]

11 The vertices of a triangle ABC have co-ordinates A(0, 0), B(3, 0) and C(0, 4). Under a rotation the triangle is mapped on to the triangle A′B′C′ whose vertices have co-ordinates A′(9, 0), B′(9, 3) and C′(5, 0). Using graph paper, plot these points and draw the triangles. Construct R, the centre of rotation, and describe clearly your construction.

Prove that $\angle ARA' = 90°$ and that the y-co-ordinate of R is $4\frac{1}{2}$.

[C]

12 (i)

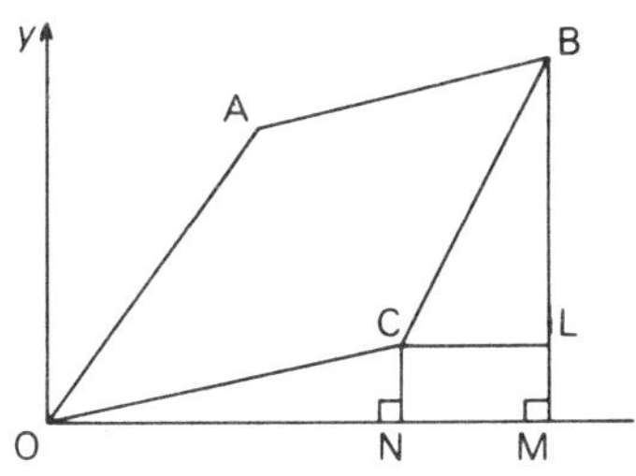

Fig. 16.14

The co-ordinates of the points O, B and C in the parallelogram OABC are (0, 0), (9, 7) and (6, 2) respectively (see Fig. 16.14). Calculate
(a) the co-ordinates of A,
(b) the areas of the right-angled triangles OCN, OBM and CBL,
(c) the area of the parallelogram.
(ii) Given any parallelogram PQRS, describe completely a simple transformation which will map triangle RSP on to triangle PQR.

[C]

17 Statistics

Statistics deals with the collection, presentation, analysis and interpretation of numerical data.

There are a number of ways of visually presenting data once they have been collected. Here are some of the methods, with examples.

17.1 Bar charts

EXAMPLE 1

The Mathemarua national football team played 24 representative games last year. Out of these games they won 7, lost 10 and drew 5, and 2 games were abandoned.

To draw a bar chart of this information, each category (win, lose, draw, abandon) is represented by a bar proportional in length to the number in that category. (See Fig. 17.1.)

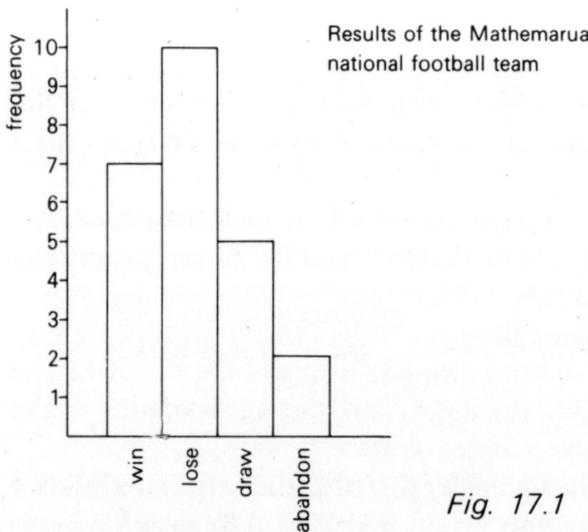

Fig. 17.1

'Frequency' refers to the number in a particular category.

17.2 Pie charts

The above information can also be represented in a pie chart. This consists of a circle divided into sectors representing categories, such that the size of each sector is proportional to the number in that category. In the previous example, the total number of games was 24. Dividing this number into 360° gives the angle representation of one game, that is, 15°. Therefore, for 7 games, the angle of the sector is $7 \times 15° = 105°$.

result of match:	win	lose	draw	abandon	total
frequency	7	10	5	2	24
degrees	105°	150°	75°	30°	360°

Fig. 17.2

EXERCISE 17a

1 Of 120 pupils, 30 prefer French, 40 prefer German, 50 prefer Spanish. Illustrate this information by (i) a bar chart, (ii) a pie chart.

2 The numbers of newspapers sold daily in Stenworth are:

Daily Mail	360	*Sun*	470
Guardian	80	*Daily Express*	400
The Times	70	*Daily Mirror*	420

Draw (i) a bar chart, (ii) a pie chart stating clearly the angles for each sector.

3 The strength of the armed forces in Mathemania is as follows:

Navy	Army	RAF	Reserve
1700	14 000	4800	3500

Represent this information in (i) a bar chart, (ii) a pie chart. What percentage of the armed forces are on reserve?

4 1620 candidates took an examination, the results of which were graded 1, 2, 3 and Fail. If a pie chart was drawn, how many students failed if the angle representing 'Fail' was 78°?

5 The local supermarket in Stenworth sells five brands of washing powder. In one week the following number of boxes were sold:

brand of washing powder:	A	B	C	D	E
number of boxes:	130	78	265	126	221

Represent this information on a pie chart, marking in the sizes of the angles of the various sectors.

17.3 The average

Consider the following three statements.

John Smythe, a local Stenworth cricketer:
'My batting average this year is 24.7.'
Mr Dickens, owner of local shoe shop:
'The most popular size of shoe I sell is size 8.'
Polly Andrews, a second form pupil:
'I only hope that I can finish in the top half of the class this year.'

All three of these people are referring to different forms of average. John Smythe's 'average' is known as the *arithmetic mean* (or *mean*). To obtain this he adds up the total number of runs scored and divides by the number of innings.

EXAMPLE 2

Find the mean of the following figures: 3, 9, 11, 28, 37, 20.

$$\text{Mean} = \frac{3+9+11+28+37+20}{6}$$

$$= \frac{108}{6}$$

$$= 18.$$

Mr Dickens, however, is concerned with the most popular or most common size of shoe. This 'average' is known as the *mode* (or *modal value*).

EXAMPLE 3

Find the mode (modal value) of these numbers: 2, 3, 3, 4, 4, 4, 4, 5, 5, 5, 6, 6, 7, 7.

Mode = 4.

Polly Andrews' 'average' is the mark which divides the top half of the class from the bottom. This is known as the *median.*

EXAMPLE 4

Find the median of the following figures: 3, 7, 2, 9, 8, 11, 16. Rearranging:

2, 3, 7, 8, 9, 11, 16.
↓
middle term

The median is the middle term.

$\therefore$ median = 8.

EXAMPLE 5

Find the median of the following figures: 5, 1, 25, 73, 2, 67. Rearranging:

2, 5, 11, ↓ 25, 67, 73

$$\text{middle term: } \frac{11+25}{2}.$$

$$\text{Median} = \frac{11+25}{2} = 18\cdot$$

EXAMPLE 6

Find the mean of the following figures: 505, 511, 525, 502, 517.

$$\text{Mean} = \frac{505+511+525+502+517}{5}$$

$$= \frac{(500+5)+(500+11)+(500+25)+(500+2)+(500+17)}{5}$$

$$= \frac{2500+5+11+25+2+17}{5}$$

$$= \frac{2500}{5} + \frac{5+11+25+2+17}{5}$$

$$= 500 + \frac{60}{5}$$

$$= 512.$$

Sometimes it is difficult to find the different forms of average from raw data, and it is helpful to tabulate the data in a frequency table.

EXAMPLE 7

Find (i) the mode, (ii) the median, (iii) the mean of the following set of figures:

3, 7, 9, 2, 4, 7, 5, 4, 5, 7,
8, 9, 7, 2, 3, 6, 5, 8, 3, 2,
4, 6, 8, 7, 3, 5, 5, 5, 6, 3.

(1) number	(2) tally	(3) frequency	(4) number × frequency
2	111	3	6
3	~~1111~~	5	15
4	111	3	12
5	~~111*1~~ 1	6	30
6	111	3	18
7	~~1111~~	5	35
8	111	3	24
9	11	2	18
	total:	30	158

(i) Clearly the mode = 5.
(ii) The starred position in the tally column is the median. Therefore the median = 5 (not 5.5).
(iii) To calculate the mean, we construct a further column to the table, headed 'number × frequency', and the mean is found by dividing the total of column 3 into column 4:

$$\text{mean} = \frac{158}{30}$$

$$= 5\tfrac{4}{15}.$$

EXERCISE 17b

1 Calculate the means of the following:
(i) 11, 6, 8, 3, 7

(ii) 14, 9, 13, 2, 5, 8
(iii) 3.1, 1.9, 1.2, 2.4, 1.5
(iv) 4, −2, −5, 1, 3
(v) −0.8, 0.2, 1.1, −2.6, 0.1, 0.4
(vi) $\frac{1}{2}, \frac{1}{2}, \frac{1}{4}, 1\frac{1}{4}, \frac{3}{4}, \frac{1}{2}$.

2 Find the mean of the figures: 8, 1, 0, 3, 15, 6.
Deduce the mean of the figures: 58, 51, 50, 53, 65, 56.

3 Find the mean of the figures: −3, 3, −11, −5, 7, −1.
Deduce the mean of the figures: 217, 223, 209, 215, 227, 219.

4 Calculate the means of the following:
(i) 79, 85, 81, 78, 82, 88, 80
(ii) 52.3, 52.1, 52.9, 52.7, 51.8, 51.2
(iii) 5248, 5253, 5249, 5243, 5256, 5240, 5251
(iv) −0.95, −0.88, −0.91, −0.84, −0.95, −0.88.

5 Find the median of the following set of figures:
(i) 41, 26, 27, 64, 72, 65, 85, 20, 41
(ii) 32, 72, 99, 44, 57, 71
(iii) 80, 74, 74, 91, 66, 7, 51, 26, 59, 83
(iv) −13, 3, −2, 12, −11, −3, 4, 3.

6 Calculate (i) the mean, (ii) the median, (iii) the mode of the following figures:
(a) 3, 4, 4, 4, 5, 8, 8, 9, 9
(b) 10, 19, 17, 14, 12, 19, 18, 14, 16, 17, 19, 11.

7 The mean age of a class of 20 boys is 12 years 6 months. Four new boys join the class, their average age being 12 years. By how much is the average age of the class lowered?

8 Write down a set of five integers:
(i) whose mean is 7 and whose median is 5
(ii) whose median is 7 and whose mean is 5.

9 The number of trains per hour through Stenworth station for a period of 24 hours were as follows:

3 5 2 1 2 5 7 4
2 6 3 5 6 2 5 1
4 7 3 2 3 2 5 6.

Find (i) the mean, (ii) the mode, (iii) the median of the distribution.

10 A sample of 50 pea pods was taken and the peas counted:

number of peas:	1	2	3	4	5	6	7
frequency:	3	5	8	17	13	3	1

Calculate (i) the median, (ii) the mean number of peas per pod.

17.4 Histograms

A *histogram* is the display of data in the form of a block graph where the area of each rectangle is proportional to the frequency. When the rectangles are the same width, their heights too are proportional to the frequency and the histogram is synonymous to the bar chart.

An example of a histogram with unequal class intervals follows.

EXAMPLE 8

The following table gives the distribution of the number of employees in the 50 factories in Stenworth:

number of employees:	0–39	40–59	60–79	80–99	100–139
number of factories:	5	15	13	10	7

Construct a histogram to show the distribution.

The widths of the five class intervals are:

40 20 20 20 40

As the widths of the first and last intervals are twice the width of the other three, the heights of the rectangles are reduced in proportion, i.e. if the height of the second rectangle is 15 units, the height of the first rectangle is $\frac{5}{2} = 2\frac{1}{2}$ units and the height of the last rectangle is $\frac{7}{2} = 3\frac{1}{2}$ units.

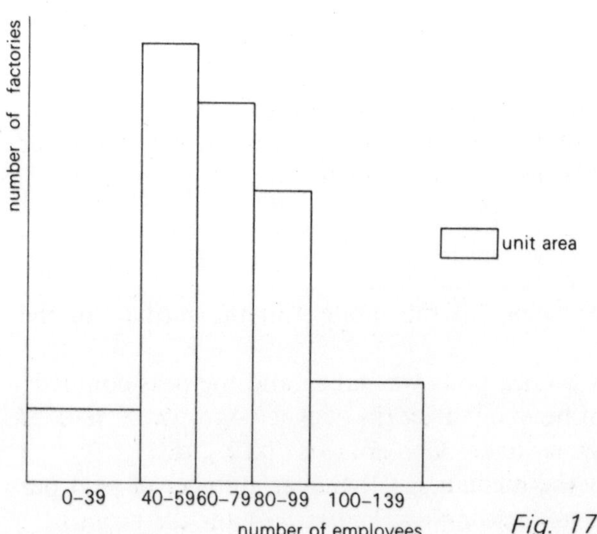

Fig. 17.3

Why do you think there is no vertical scale on the diagram?

17.5 Grouped frequencies

EXAMPLE 9

Class 3YZ were asked to estimate a time interval of 1 minute. The results in seconds are shown below:

37, 63, 47, 41, 59, 47, 79, 51, 43, 83,
75, 45, 53, 31, 52, 52, 41, 51, 44, 66,
73, 63, 65, 57, 58, 36, 51, 66, 49, 66.

Let us draw up a frequency table and place the data in groups 30–39, 40–49, 50–59, etc.

estimated time (seconds)	tally	frequency
30–39	111	3
40–49	~~1111~~ 111	8
50–59	~~1111~~ 1111	9
60–69	~~1111~~ 1	6
70–79	111	3
80–89	1	1

The histogram for this table is given in Fig. 17.4.

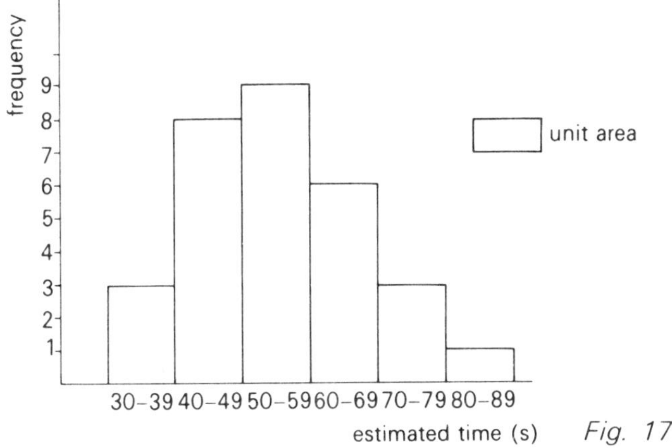

Fig. 17.4

Try the same problem but place the data in (i) groups of 2 seconds, (ii) groups of 20 seconds; then compare histograms.

It will be seen from the three diagrams that common sense should be used in deciding on the best size for the groups.

The width of the bars is known as the class interval and in the example above this is equal to 10 (not 9).

17.6 Calculation of the mean from a grouped frequency table

EXAMPLE 10

Suppose from the previous example we were only given the table, so that we did not know individual times.

estimated time (seconds)	frequency	half-way mark	half-way mark × frequency
30–39	3		
40–49	8		
50–59	9	54.5	490.5
60–69	6		
70–79	3		
80–89	1		
total:	30		

In order to calculate a value for the mean we make the estimation that the number in each class have all guessed the half-way value of that class, i.e. the 9 in the class interval 50–59 have all guessed 54.5. This would make a total of 490.5 in this class interval. Complete the table and make an estimate for the mean.

Now check this estimation with the true mean which you can calculate from the actual figures.

EXERCISE 17c

1 Over a 7-week period the output from Mathemania's only car factory was recorded:

week:	1	2	3	4	5	6	7
no. of cars:	220	190	235	175	120	90	30

Construct a histogram to show this information.

2 The following table gives the score of 50 throws at a dartboard:

38, 27, 32, 17, 45, 46, 31, 28, 78, 54,
47, 51, 39, 5, 12, 19, 63, 58, 47, 47,
56, 23, 31, 28, 25, 42, 3, 81, 56, 37,
32, 29, 27, 3, 67, 54, 48, 21, 18, 22,
17, 47, 35, 34, 34, 28, 57, 48, 36, 33.

Construct a histogram for the data using class intervals 0–9, 10–19, 20–29, etc. What is the modal class of this distribution?

3 Mathemania has 24 football teams in the First Division. The number of wins for these teams during a season were as follows:

5, 42, 27, 1, 14, 8, 38, 7,
2, 17, 13, 40, 6, 11, 15, 18,
37, 3, 4, 38, 12, 16, 2, 5.

Construct a histogram for the data using class intervals 1–5, 6–10, 11–15, etc.

4 The results of an examination taken by 100 candidates are given in the following table:

mark obtained:	0–19	20–39	40–59	60–79	80–89
number of candidates:	10	15	35	27	13

(i) Draw a histogram to illustrate these results.
(ii) Compute an estimation for the mean.

5 The number of deals on the Stock Exchange on 50 business days were as follows:

number of deals:	70–79	80–89	90–99	100–109	110–119	120–129
frequency:	4	7	13	15	8	3

(i) Draw a histogram to illustrate these results.
(ii) Compute an estimation for the mean.

17.7 Cumulative frequency curves

40 families were interviewed in Stenworth to find the number of children in the family. The results were as follows:

0, 2, 3, 1, 2, 4, 0, 1, 3, 3,
3, 3, 4, 2, 3, 1, 3, 2, 1, 2,
2, 3, 3, 2, 5, 2, 0, 4, 3, 1,
5, 3, 2, 3, 1, 3, 2, 2, 1 3.

We have already seen how to group these figures in a frequency table and how to construct a histogram. Now let us add an extra column to the frequency table headed 'running total' or *cumulative frequency*.

The figure in a given row of the cumulative frequency column is computed from all the figures in the frequency column up to and including the corresponding row, e.g. $21 = 11+7+3$.

number of children	tally	frequency	running total (cumulative freq.)
0	111	3	3
1	~~1111~~ 11	7	10
2	~~1111~~ ~~1111~~ 1	11	21
3	~~1111~~ ~~1111~~ 1111	14	35
4	111	3	38
5	11	2	40

If we plot the number of children against the running total, we have a cumulative frequency curve. (See Fig. 17.5.)

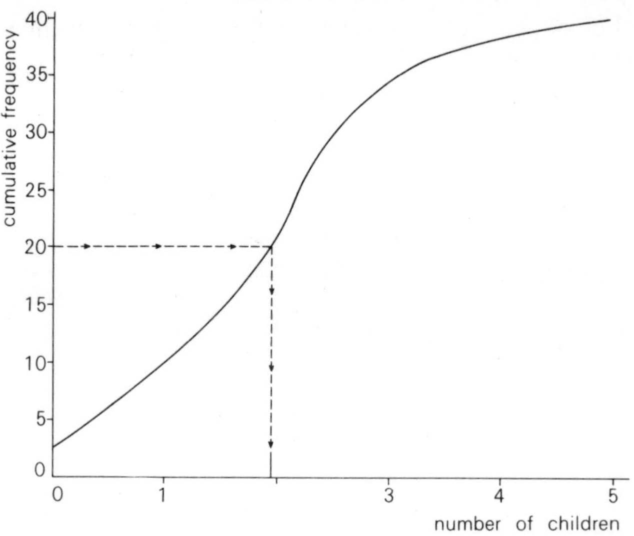

Fig. 17.5

Without considering the raw data we can estimate the median from the graph by reading off the half-way mark on the vertical scale. In the above graph we follow the dotted lines to see that the estimate for the median is 2.

EXAMPLE 11

100 commuters were interviewed to find the distance travelled to work daily to the nearest kilometre. From the results, the following table was constructed:

distance in kilometres	number of commuters	cumulative frequency
0–5	6	6
5–10	11	17
11–15	17	34
16–20	25	59
21–25	21	80
26–30	12	92
31–35	8	100

To be able to draw the cumulative frequency curve, we need to know the co-ordinates to plot. Taking the first class interval, 6 commuters travel a distance of between 0 and 5 kilometres. As the distance is measured to the nearest kilometre, the 6 commuters all travel less than 5.5 km. Therefore the point to plot is (5.5 km, 6). Similarly, the next point is (10.5 km, 17), that is, 17 commuters travel less than 10.5 km.

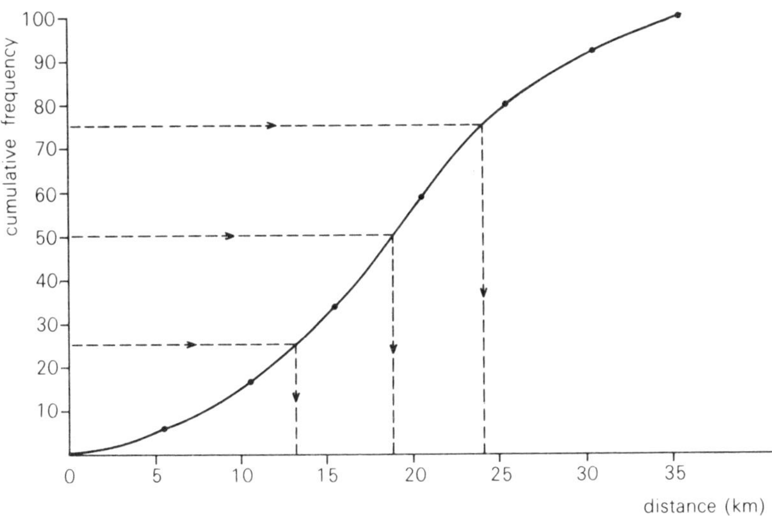

Fig. 17.6

If we divide the number of commuters into quarters, then the second quarter position gives the median. An estimate for the median is 19 kilometres. The distance travelled by the 25th commuter, 13 km, is known as the lower quartile. The distance travelled by the 75th commuter, 24.5 km, is known as the upper quartile.

17.8 The semi-interquartile range

$$\text{(in example 11)} = \frac{\text{upper quartile} - \text{lower quartile}}{2}$$
$$= \frac{24.5-13}{2}\text{ km}$$
$$= 5.75\text{ km.}$$

EXERCISE 17d

1 Five coins were tossed 500 times, and the numbers of heads were recorded as follows:

number of heads:	0	1	2	3	4	5
frequency:	40	60	140	170	70	20

(i) Construct a cumulative frequency curve.
(ii) Estimate the median number of heads.

2 The number of people waiting at bus stops was as follows:

number of people:	0	1	2	3	4	5	6
number of stops:	3	8	12	15	13	7	2

Construct a cumulative frequency table and draw the cumulative frequency curve. From the curve find the median number of people per stop.

3 The following results were obtained in an examination:

marks:	0–20	21–40	41–60	61–80	81–100
number of candidates:	18	37	86	40	19

Construct a cumulative frequency curve and obtain, to the nearest whole number, the median mark and the semi-interquartile range.

HARDER EXAMPLES 17

1 The following table shows the percentage of steel produced in the world in 1970:

Western Europe	USSR	Asia	N. America	others
27%	23%	20%	22%	8%

Represent this information by a pie chart.

2 Construct a frequency table of occurence of the digits 0 to 9 from the last digits in column 0 of the first page of logarithms from the top line 10 to line 49. Represent the table in two ways; by a bar chart and a pie chart. [L]

3 Draw a pie chart to represent the information:

age	frequency
less than 5 years	35
5–18	72
19–64	130
over 64	33

4 The mean mark of 15 candidates in an examination was 56. The 5 weakest candidates had a mean of 24. Find the mean mark of the other 10 candidates.

5 The mean reading of the barometer for Sunday, Monday and Tuesday was 753 mm. The mean reading for Monday and Tuesday was 756 mm. Find the reading on Sunday.

6 The mean of the numbers 7, 11, 5, x, 4, 2, 8 is 6. Find x.

7 In a game the score x occurs with the frequency shown in the table:

x:	1	2	4	6	12
frequency:	12	6	3	2	1

(i) Find the mean score.
(ii) If the different values of the frequencies are represented on a pie chart, calculate the number of degrees in the sector representing $x = 6$.

[C]

8 John Scragg batted 24 times for Stenworth Ramblers Cricket Club in a season. His scores were:

24	0	17	83	16	51	42	62
0	1	3	15	22	17	41	9
21	33	8	0	14	112	26	8

Calculate the median score.

9 The mean of 8 numbers is 7 and the mean of a further 2 numbers is 12. What is the mean of all 10 numbers?

10 Mr Dickens recorded the following sizes of shoe sold from his shop on one particular day:

6	3	$4\frac{1}{2}$	2	7	8	5	$7\frac{1}{2}$	6	$4\frac{1}{2}$
$6\frac{1}{2}$	7	7	8	5	7	3	8	8	7
6	7	$7\frac{1}{2}$	5	$6\frac{1}{2}$	$6\frac{1}{2}$	$6\frac{1}{2}$	7	8	8

Calculate (i) the median; (ii) the mode; (iii) the mean size shoe bought.

11 The numbers of letters per day received by the mail order company EMS of Stenworth in one working week were as follows:

day:	M	Tu	W	Th	F	Sat
number of letters:	220	150	70	90	50	20

Draw a histogram to represent these results.

12 The marks for an aptitude test were recorded as follows:

3, 7, 9, 8, 6, 5, 5, 5, 6, 7, 2, 1, 3, 5, 7,
8, 7, 6, 5, 6, 6, 4, 4, 3, 4, 6, 7, 5, 1, 2.

Represent this information in a histogram. Calculate the mean mark.

13 Two dice are rolled 100 times and the total score recorded each time:

total score of 2 dice:	2	3	4	5	6	7	8	9	10	11	12
frequency:	1	5	5	10	18	23	16	9	8	4	1

(i) Draw a histogram to illustrate the results.
(ii) Calculate the mean combined score.

14 An experiment was performed on 200 animals to estimate their surface area. The results obtained were as follows:

area (cm^2):	0–2	3–5	6–8	9–11	12–14	15–17
frequency:	12	21	50	76	30	11

Rewrite this information as a cumulative frequency table and from it draw the cumulative frequency curve. Use this to estimate:

(i) the median area;
(ii) the percentage of animals whose area was less than 10 cm^2;
(iii) the semi-interquartile range of the areas.

15 The table below shows the cumulative mark distribution of 560 candidates in an examination. Construct a cumulative frequency curve, labelling your axes clearly. How many candidates fail if the pass mark is 45? What must the pass mark be if 60 per cent of the candidates are to pass? Determine the median of the distribution.

mark:	10	20	30	40	50	60	70	80	90	100
cumulative frequency:	18	43	78	130	240	372	462	523	552	560

[SU]

18 Probability

18.1 Sample spaces

The set of possible outcomes to an experiment is known as the *sample space*. The most frequently used methods of representation are set notation or diagrams.

EXAMPLE 1

The rolling of a fair die:

sample space, S = {1, 2, 3, 4, 5, 6}.

EXAMPLE 2

A coin is tossed twice and the sequence of heads (H) and tails (T) observed:

sample space, S = {HH, HT, TH, TT}.

EXAMPLE 3

A blue die and a red die are rolled and the respective scores are observed:

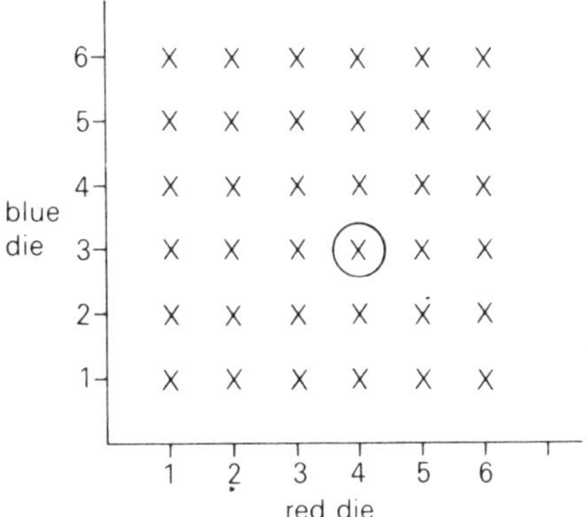

Fig. 18.1

The circled cross in Fig. 18.1 represents the rolling of a 3 on the blue die and a 4 on the red die.

(This is the sample space which contains all possible outcomes of the rolling of the two dice.)

EXAMPLE 4

A coin and die are tossed:

sample space, S = {H1, H2, H3, H4, H5, H6, T1, T2, T3, T4, T5, T6}.

18.2 Empirical definition of probability

For this definition, the measure of probability is based on experimental data. That is, if we attempt an experiment n times having m successes, then the rate m/n is called the empirical or experimental probability of the event. So, if we toss a coin 20 times out of which there are 9 heads, then the empirical probability is 9/20 or 0.45. As n increases, however, the ratio m/n becomes stable, i.e. approaches a limit. This limit coincides with the value derived from the second definition.

18.3 Classical definition of probability

Historically, probability developed from the study of games of chance such as cards or dice.

The chance or probability of any event happening was defined as:

$$\text{Probability} = \frac{\text{the number of ways the event can happen}}{\text{the total number of outcomes to the experiment}}$$

This is only true when each outcome is equally likely.
We can now make use of sample spaces and earlier work on sets to derive some probabilities.

EXAMPLE 5

A fair die is rolled. What is the probability of an even score?

Let the set A represent the ways the event can happen:

A = {2, 4, 6}.

The total number of outcomes is simply the sample space S = {1, 2, 3, 4, 5, 6,}. Then the probability of an even score, or

$$P(\text{A}) = \frac{\text{number of elements in A}}{\text{number of elements in S}}$$

$$= \frac{n\ (\text{A})}{n\ (\text{S})}$$

$$= \frac{3}{6}$$

$$= \tfrac{1}{2}.$$

EXAMPLE 6

Draw a card from a fair pack of cards. What is the probability of drawing a black picture card?

Let B represent the set containing the black picture cards:

$$\text{B} = \{\text{J}\clubsuit,\ \text{Q}\clubsuit,\ \text{K}\clubsuit,\ \text{J}\spadesuit,\ \text{Q}\spadesuit,\ \text{K}\spadesuit\}.$$

The sample space, S, is the number of cards in the pack.

$$P(\text{B}) = \frac{n\ (\text{B})}{n\ (\text{S})}$$

$$= \frac{6}{52}$$

$$= \tfrac{3}{26}.$$

This example can also be considered in the following way:

Let C be the event of drawing a black card.
Let D be the event of drawing a picture card.
Therefore

$$n(\text{C}) = \text{number of black cards}$$
$$n(\text{D}) = \text{number of picture cards}$$
$$n(\text{C} \cap \text{D}) = \text{number of black picture cards.}$$

The probability of drawing a black card *and* a picture card

$$= \frac{n(\text{C} \cap \text{D})}{n(\text{S})}.$$

See Fig. 18.2

EXAMPLE 7

Draw a card from a fair pack of cards. What is the probability of drawing a black card or a picture card?

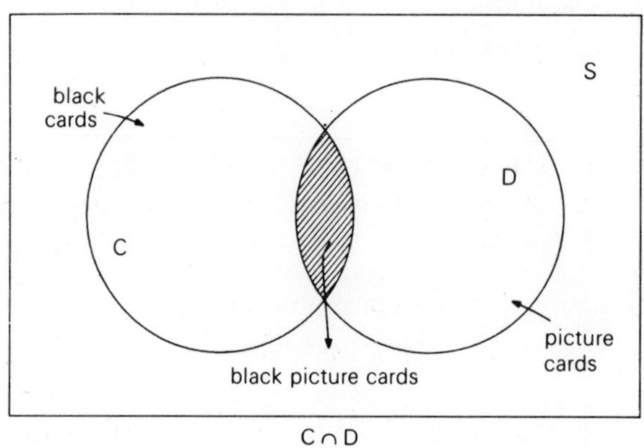

Fig. 18.2

Let E represent the set containing a black card or a picture card.

$n(\text{E}) = 32$ (13 clubs; 13 spades; J, Q, K diamonds; J, Q, K hearts)

$n(\text{S}) = 52$

$$P(\text{E}) = \frac{n\,(\text{E})}{n\,(\text{S})}$$

$$= \frac{\overset{8}{\cancel{32}}}{\underset{13}{\cancel{52}}} = \frac{8}{13}$$

Referring to Fig. 18.2, the number of black cards or picture cards (or both)

$= n(\text{C} \cup \text{D})$

So the probability of drawing a black card *or* a picture card

$$= \frac{n(\text{C} \cup \text{D})}{n(\text{S})}.$$

18.4 Complement

EXAMPLE 8

A bag contains 4 red marbles, 2 white marbles and 4 blue marbles. What is the probability of (i) drawing a red marble, (ii) not drawing a red marble?

Let R = {red marbles}. Then

$$P(\mathrm{R}) = \frac{4}{10} = \tfrac{2}{5}.$$

The probability of *not* drawing a red marble is the *complement* of the probability of drawing a red marble. Using set notation, the complementary probability is:

$$P(\mathrm{R}') = \frac{\text{number of blue and white marbles}}{\text{total number of marbles}}$$

$$= \frac{6}{10} = \tfrac{3}{5}.$$

Therefore $P(\mathrm{R}) + P(\mathrm{R}') = 1$. This is always the case with an event and its complement.

EXERCISE 18

1 List all the possible outcomes if a coin is tossed 4 times.

2 A box contains 1 red, 1 white and 1 blue ball. Draw up a sample space for the possible outcomes when two balls are drawn (the first being replaced before the second is drawn).

3 A die is numbered 1, 1, 2, 5, 6, 6. Construct a sample space for rolling the die twice.

4 Determine the probabilities of the following events:

(i) A King appears in drawing a single card from an ordinary pack of 52 cards.

(ii) A white marble appears in drawing a single marble from an urn containing 4 white, 3 red and 5 blue marbles.

(iii) Choosing a vowel from the word *probability*.

(iv) A prime number is drawn from a pack of 25 cards numbered 1 to 25.

(v) Not drawing a red ball from a box containing 6 red, 4 white and 5 blue balls.

5 $\mathscr{E} = \{1, 2, 3, \ldots 50\}$ A = {multiples of 7}
B = {multiples of 5} C = {multiples of 3}

If any element x is chosen at random from $\mathscr{E}$, find the probability that

(i) $x \in \mathrm{A} \cup \mathrm{B}$ (ii) $x \in \mathrm{B} \cap \mathrm{C}$ (iii) $x \in \mathrm{A} \cap \mathrm{C}'$.

18.5 Combined events

For certain problems we can construct an adequate sample space and approach the problem as before.

EXAMPLE 9

A coin and a die are tossed together. In example 1 we saw that:

$$S = \{H1, H2, H3, H4, H5, H6, T1, T2, T3, T4, T5, T6\}.$$

(a) The probability of both a head and a six.
Let F represent the ways this can happen.

$$F = \{H6\}$$
$$\therefore\ P(F) = \frac{n\,(F)}{n\,(S)} = \tfrac{1}{12}.$$

(b) The probability of either a tail or a six or both.
Let G represent the ways this can happen.

$$G = \{T1, T2, T3, T4, T5, T6, H6\}$$
$$\therefore\ P(G) = \frac{n\,(G)}{n\,(S)} = \tfrac{7}{12}.$$

EXAMPLE 10

A fair die is tossed twice. Find the probability of getting a 4, 5 or 6 on the first toss and a 1, 2, 3 or 4 on the second.

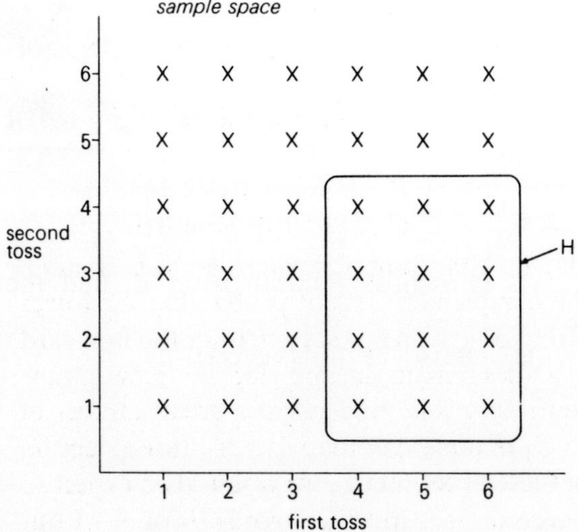

Fig. 18.3

Let H represent the ways this can happen.

$$P(\mathrm{H}) = \frac{n\,(\mathrm{H})}{n\,(\mathrm{S})}$$
$$= \frac{12}{36}$$
$$= \tfrac{1}{3}.$$

18.6 Tree diagrams

Suppose we wish to find the probability of drawing two aces from a fair pack of cards, replacing the first card before drawing the second. Clearly it would demand a lot of time and patience to produce a complete sample space, so instead we make use of a *tree diagram.*

EXAMPLE 11

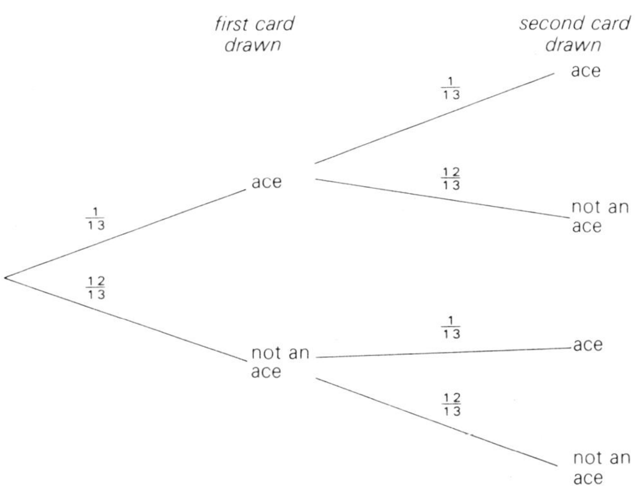

Fig. 18.4

The probability that the first card drawn is an ace is quite simply $\frac{1}{13}$, and by the complement law of probability the lower branch has a probability of $\frac{12}{13}$. The replacement of the first card means that the probabilities remain the same for the second draw.

If we considered repeating this experiment a great number of times we could expect $\frac{1}{13}$ of our experiments to produce an ace on the first draw. Out of these experiments, we would then expect to find $\frac{1}{13}$ producing a second ace. In other words $\frac{1}{13}$ of $\frac{1}{13}$ of our experiments would be expected to produce two aces. This leads us to the multiplication rule for combined events.

18.7 Multiplication rule

The probability of two combined events occurring together is the probability of one multiplied by the probability of the other.

In the last example the probability of 2 aces $= \frac{1}{13} \times \frac{1}{13} = \frac{1}{169}$.

EXAMPLE 12

A bag contains 4 red balls and 5 blue balls. A second bag contains 2 red balls and 5 blue balls. If a ball is taken at random from each bag, find the probability that both balls are red.

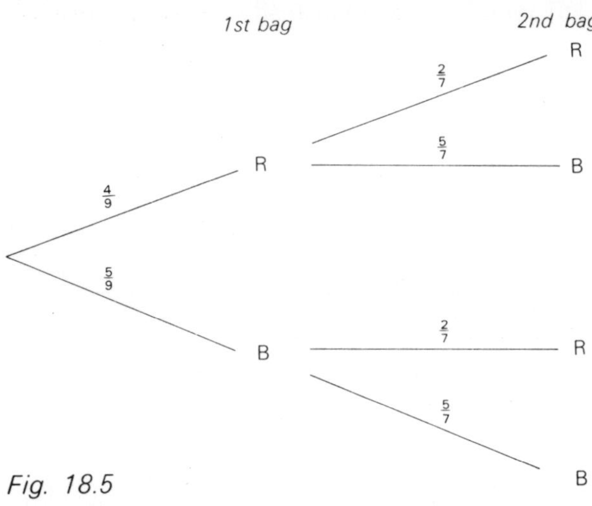

Fig. 18.5

The sample space is different for the two bags.

The probability that both balls are red $= \dfrac{4}{9} \times \dfrac{2}{7} = \dfrac{8}{63}$.

EXAMPLE 13

With the information of example 12, find the probability that one ball is red and the other blue.

If we look at the tree diagram, we see that *two* branches affect our result.

The probability of the first bag producing a red ball and the second a blue ball $= \dfrac{4}{9} \times \dfrac{5}{7} = \dfrac{20}{63}$.

The probability of the first bag producing a blue ball and the second a red ball $= \frac{5}{9} \times \frac{2}{7} = \frac{10}{63}$.

As either path is suitable, the probability of a red ball and a blue ball $= \frac{20}{63} + \frac{10}{63} = \frac{30}{63}$.

EXAMPLE 14

Five cards numbered 1, 2, 3, 4, 5 were placed in a bag. Two cards were drawn, the first not being replaced before the second was drawn. Find the probability that both cards drawn are odd.

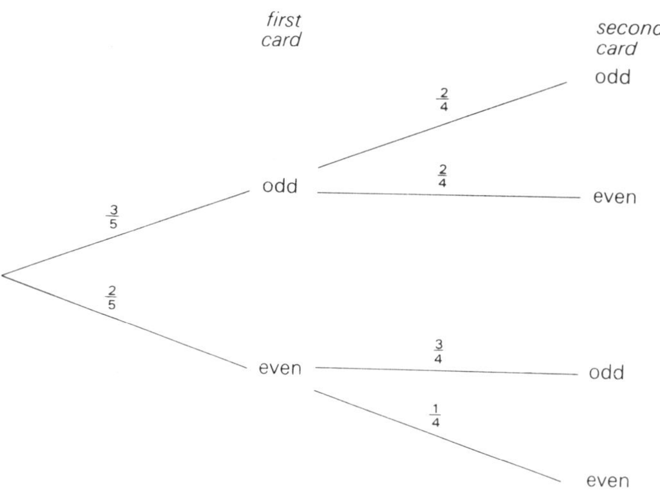

Fig. 18.6

It is important to note in this example that the first card drawn affects the respective probabilities of the second in two ways:

(i) The sample space for the second draw is reduced by one (from 5 to 4).

(ii) The first draw either reduces the number of odd numbers left or the number of even numbers.

In this example, the required probability is $\frac{3}{5} \times \frac{2}{4} = \frac{3}{10}$. (As the first card drawn is odd, the number of odd cards for the second draw is reduced to 2.)

EXAMPLE 15

A point is selected at random inside a circle. Find the probability that the point is closer to the centre of the circle than to its circumference.

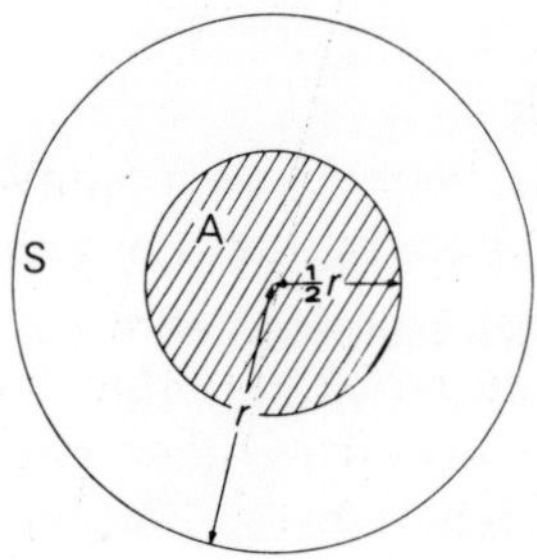

Fig. 18.7

The point must lie within the circle A of radius $\frac{1}{2}r$ (see Fig. 18.7). The required probability

$$= \frac{\text{area of A}}{\text{area of S}} = \frac{\pi(\frac{1}{2}r)^2}{\pi r^2} = \frac{1}{4}.$$

HARDER EXAMPLES 18

1 A fair six-sided die is thrown. Are the following statements true or false? The probability of throwing
(i) a 6 is less than that of throwing a 1;
(ii) an even number is $\frac{1}{2}$;
(iii) a number less than 5 is $\frac{2}{3}$;
(iv) an even number or a number less than 5 is $\frac{7}{6}$.

2 A football captain wins the toss for four consecutive matches. What is the probability that he will call correctly for the fifth match?

3 What is the probability that a letter drawn at random from the word *statistics* is a consonant?

4 Bill was asked, 'What is the probability that when a die is thrown the score is odd or at most 3?' He answered as follows:
'P (odd) $= \frac{3}{6}$, P (at most 3) $= \frac{3}{6}$. Hence P (odd or at most 3) $= \frac{3}{6} + \frac{3}{6} = 1$'.
Where is Bill's mistake?

5 Raffle tickets numbered 1 to 100 are placed in a drum and one is drawn at random. What is the probability that the number is divisible by (i) 2; (ii) 5; (iii) 2 and 5?

6 A circle is inscribed in a square. Taking the sample space as all the points inside the square, find the probability that a point chosen at random lies inside the square but outside the circle.

7 The probability of an event A happening is $\frac{1}{5}$ and the probability of an event B happening is $\frac{1}{4}$. Given that A and B are independent, calculate the probability that:
(i) neither event happens;
(ii) just one of the two events happens.

8 A die numbered 1, 1, 2, 3, 4, 4 is thrown twice. Draw a sample space for the results. What is the probability that the total score on the two throws is (i) odd; (ii) greater than 6?

9 Two fair dice are thrown. What is the probability that the difference in the two scores is (i) two; (ii) greater than two?

10 A bag contains 5 red balls and 4 white balls. Find the probability that two balls drawn at random are both red if the first ball drawn is (i) replaced; (ii) not replaced.

11 The probability that John wins the 100 metres is $\frac{1}{3}$ and the probability that Mike wins the pole vault is $\frac{2}{5}$. What is the probability that neither of them wins?

12 Three cards are drawn from a fair pack of cards. What is the probability that they are all aces?

13 A coin is biased in such a way that the probability of a head appearing in one toss of the coin is $\frac{2}{3}$. The coin is tossed three times. What is the probability that there are:
(i) three heads;
(ii) at least two heads?

14 Two regular tetrahedrons whose four faces were numbered 1, 2, 3 and 4 were thrown. The numbers on the faces resting on the ground were recorded. List the points of the sample space as ordered pairs and find the probability that:
(i) the difference in the two scores is one;
(ii) the total on the two dice is greater than 3.

15 In an experiment the probability of an event happening is p. What is the probability that the event does not happen? If the experiment is repeated, what is the probability that on both occasions the event does not occur?

16 Given that the probability of a male birth is 0.52 and that a woman has three children, calculate the probability that:
(i) all three are boys;
(ii) at least two are boys.

17 Two cards are drawn at random from a pack of playing cards without replacement. What is the probability that:
(i) they are not diamonds;

(ii) there is at least one diamond?

18 (i) Two unbiased coins are tossed together. State the probability that they will both (a) be heads, (b) be the same (i.e. either both heads or both tails).

(ii) One unbiased coin is tossed together with a 'loaded' coin which has a probability of $\frac{2}{5}$ of landing heads, $\frac{3}{5}$ of landing tails. State the probability that they will both (a) be heads, (b) be the same.

19

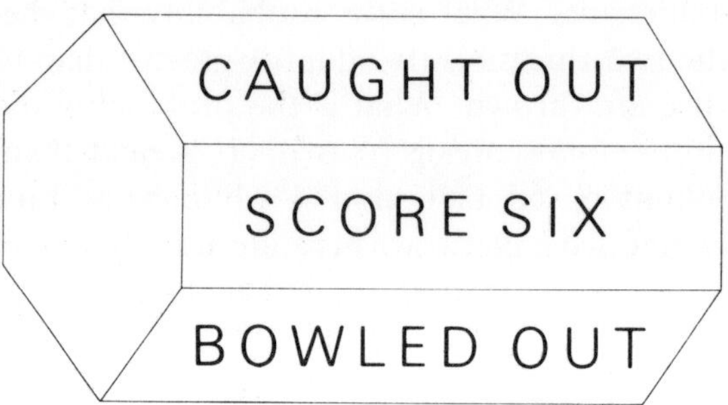

Fig. 18.8

In a simulated game of cricket, a player rolls a prism (see Fig. 18.8). The sides are inscribed:

(a) bowled out; (b) score two; (c) score four;
(d) score six (e) score zero (f) caught out.

Calculate the probabilities of the following results:

(i) The player is out first time.

(ii) He scores two, four, six in that order, with three throws.

(iii) He scores two, four and six, in any order, with three throws.

(iv) After three throws, his score is ten not out.

[JMB]

20 (i) Write down the probability that a three-digit number chosen at random is divisible by 4.

(ii) Find how many different three-digit numbers can be formed using the digits 1, 2, and 3 if repetitions are allowed.

21 A bag contains x red balls and y blue balls. If a ball is taken at random from the bag, the probability that it is red is $\frac{3}{7}$. Write down an equation connecting x and y.

If there had been 5 more red balls in the bag, the probability would have been $\frac{1}{2}$. Find x and y.

19 The solution of the general triangle

19.1 The sine rule

So far, we have only considered the numerical solution of right-angled triangles. We now consider two methods for the solution of non-right-angled triangles.

Sine rule: we will show that for any triangle ABC

$$\frac{a}{\sin A} = \frac{b}{\sin B} = \frac{c}{\sin C}.$$

In the acute-angled triangle we first draw the perpendicular from B to AC. (See Fig. 19.1.)

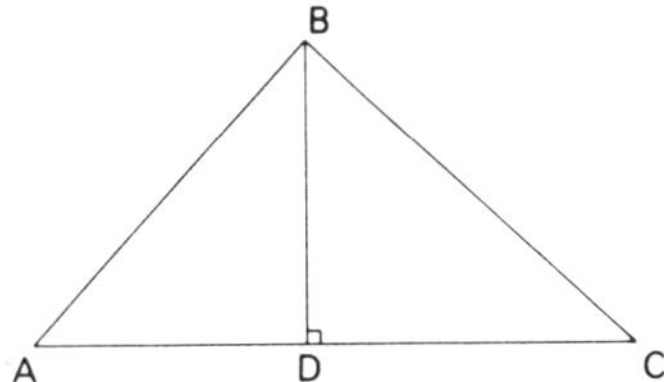

Fig. 19.1

In triangle ABD, BD $= c \sin A$.
In triangle BDC, BD $= a \sin C$.
We have thus obtained two expressions for the same quantity BD; hence $c \sin A = a \sin C$.

$$\therefore \frac{a}{\sin A} = \frac{c}{\sin C}.$$

If we had drawn the perpendicular from C to AB we would have obtained

$$\frac{a}{\sin A} = \frac{b}{\sin B}.$$

Hence

$$\frac{a}{\sin A} = \frac{b}{\sin B} = \frac{c}{\sin C}.$$

[*Note:* this proof is for an acute-angled triangle.]

For an obtuse-angled triangle, the perpendicular from A to CB produced must be drawn. (See Fig. 19.2.)

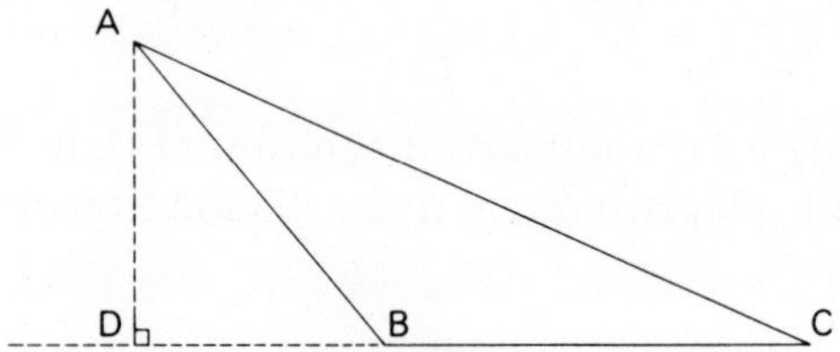

Fig. 19.2

In triangle ADB, AD $= c \sin(180° - B) = c \sin B$
$\therefore$ AD $= c \sin B$.
In triangle ADC, AD $= b \sin C$.
Hence $c \sin B = b \sin C$, and the result follows as before.

19.2 Use of the sine rule

In order to use the sine rule in a triangle you must know (i) *two sides and a non-included angle*; or (ii) *two angles and a side.*

EXAMPLE 1

In the triangle BAC, AB = 6 cm, $\angle A = 74°$, $\angle B = 37°$. Find BC.

In order to use the sine rule, we must first calculate $\angle C = 69°$. Since AB $= c$,

$$\frac{6}{\sin 69°} = \frac{a}{\sin 74°}$$

$$\therefore\ a = \frac{6 \sin 74°}{\sin 69°} = 6.18$$

i.e. BC = 6.18 cm.

EXAMPLE 2

In the triangle RST, $\angle R = 20°$, RS = 100 m, ST = 80 m. Find $\angle T$.

The appropriate form of the sine rule this time gives us:

$$\frac{80}{\sin 20^\circ} = \frac{100}{\sin T}$$

$$\sin T = \frac{100 \sin 20^\circ}{80} = 0.4275$$

(*Note:* there are two angles between 0° and 180° which satisfy this.)

Hence $\angle T = 25^\circ 18'$ or $180^\circ - 25^\circ 18' = 154^\circ 42'$.

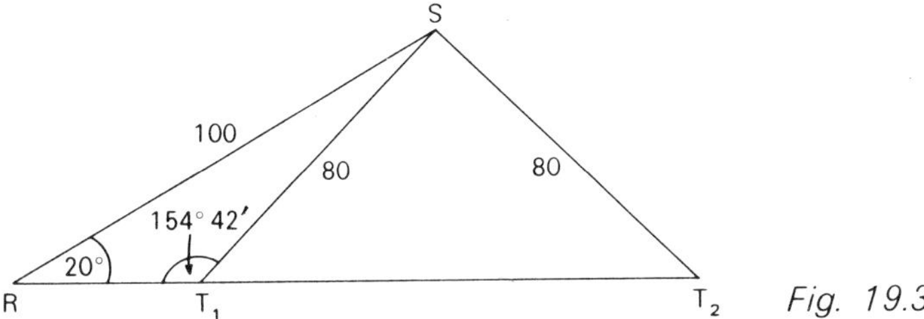

Fig. 19.3

The two possible solutions are shown in Fig. 19.3. It is clear that more information is needed to make the correct choice.

EXAMPLE 3

A boat sets a course of 065° and sails for 3 km until it reaches buoy A. It then changes direction and sails on a course of 200° until it reaches buoy B. At B, the bearing of the boat from its initial position is 160°. Find the distance between the buoys.

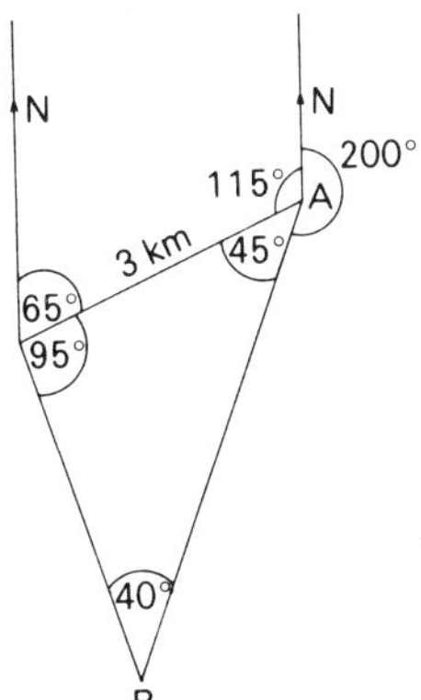

Fig. 19.4

In this type of problem, you will notice that there is quite a bit of working necessary to find B before the sine rule can be used.

$$\frac{AB}{\sin 95^\circ} = \frac{3}{\sin 40^\circ}$$

$$\therefore AB = \frac{3 \sin 95^\circ}{\sin 40^\circ} = \frac{3 \sin 85^\circ}{\sin 40^\circ} = 4.65$$

i.e. distance between the buoys = 4.65 km.

19.3 The cosine rule

The cosine rule enables us to cope with the situation where no angles are given. We will show that:

$$a^2 = b^2 + c^2 - 2bc \cos A$$

with the similar results

$$b^2 = c^2 + a^2 - 2ac \cos B$$
$$c^2 = a^2 + b^2 - 2ab \cos C.$$

Consider an acute-angled triangle.

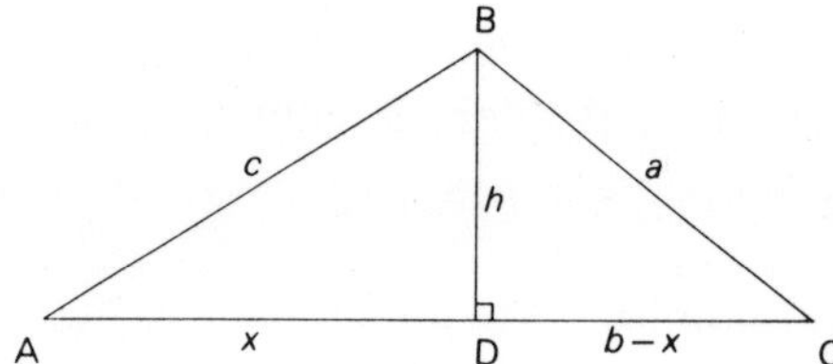

Fig. 19.5

First draw the perpendicular from B to AC. Let BD = h and AD = x; then DC = $b - x$. (See Fig. 19.5.) Pythagoras' theorem in triangle ABD gives

$$c^2 = x^2 + h^2$$

and in triangle BDC gives

$$a^2 = h^2 + (b - x)^2.$$

If we subtract the second equation from the first, h^2 cancels and we get

$$c^2 - a^2 = x^2 - (b - x)^2 = x^2 - (b^2 + x^2 - 2bx)$$
$$= x^2 - b^2 - x^2 + 2bx$$
$$\therefore c^2 - a^2 = -b^2 + 2bx$$
$$\therefore a^2 = b^2 + c^2 - 2bx.$$

But in triangle ABD, $x = c\cos A$ by projection on AC. Hence

$$a^2 = b^2 + c^2 - 2bc\cos A.$$

Once again we need to prove the formula for an obtuse-angled triangle.

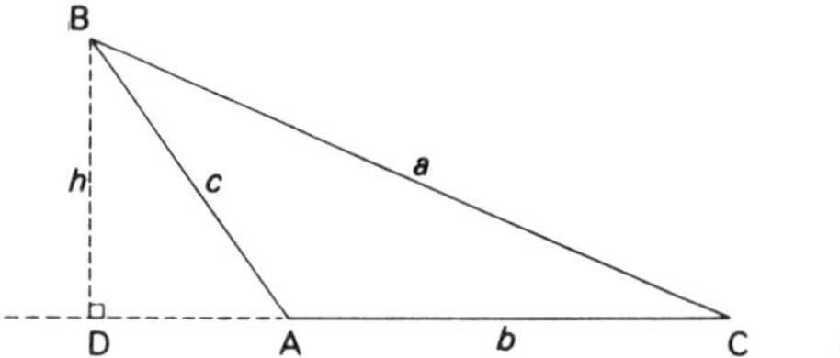

Fig. 19.6

This time the perpendicular is drawn from B to CA produced. (See Fig. 19.6.) In triangle DBC, Pythagoras' theorem gives

$$a^2 = h^2 + (b+x)^2$$

and in triangle BDA,

$$c^2 = h^2 + x^2.$$

Subtracting as before,

$$\begin{aligned} a^2 - c^2 &= (b+x)^2 - x^2 \\ &= b^2 + 2bx + x^2 - x^2 \\ \therefore\ a^2 - c^2 &= b^2 + 2bx \Rightarrow a^2 = b^2 + c^2 + 2bx. \end{aligned}$$

In triangle DBA as before, $x = c\cos(180^\circ - A)$.
Now we have shown that $\cos(180^\circ - A) = -\cos A$.

$$\begin{aligned} \therefore\ x &= -c\cos A \\ \therefore\ a^2 &= b^2 + c^2 - 2bc\cos A. \end{aligned}$$

19.4 Use of the cosine rule

We can see in all the different forms of the cosine formula there are three sides and one angle. To use the formula you need to know either *three* sides or two sides and an angle, usually the *included* angle between the two sides.

EXAMPLE 4

In triangle XYZ, XY = 8 km, YZ = 5 km, $\angle Y = 70^\circ$. Find XZ.

$$\begin{aligned} y^2 &= x^2 + z^2 - 2xz\cos Y \\ \therefore\ y^2 &= (5^2 + 8^2) - (2 \times 5 \times 8\cos 70^\circ). \end{aligned}$$

(*Note:* brackets have been included to show that there are two separate terms, a common mistake being to write $y^2 = ((5^2+8^2 - 2\times 5\times 8)\cos 70°.)$

$$\therefore\ y^2 = 25+64-80\times 0.342$$
$$= 89-27.36 = 61.64.$$

(Do not forget to take the square root.)

$$\therefore\ y = 7.85.$$

EXAMPLE 5

As example 4, with $\angle Y = 110°$.

This time $\cos Y = \cos 110° = -\cos 70°$. We get a sign change which gives:

$$y^2 = 5^2+8^2+80\times 0.342$$
$$= 89+27.36 = 116.36$$
$$\therefore\ y = 10.8.$$

The change in sign should not be surprising as by increasing $\angle Y$ we would expect to increase XZ.

EXAMPLE 6

In triangle BAC, $a = 6.2$, $b = 4.1$, $c = 3.7$. Find A.

Using $a^2 = b^2+c^2-2bc\cos A$, we get

$$(6.2)^2 = (4.1)^2+(3.7)^2-2\times 4.1\times 3.7\cos A$$
$$\therefore\ 38.44 = 16.81+13.69-30.34\cos A$$
$$\therefore\ 30.34\cos A = 16.81+13.69-38.44 = -7.94$$
$$\therefore\ \cos A = \frac{-7.94}{30.34} = -0.2617.$$

Because the result is negative, we must subtract the angle whose cosine is 0.2617 from 180°.

$$\therefore\ A = 180° - 74°\,50'$$
$$= 105°\,10'.$$

EXERCISE 19a

1 In the following examples, use the sine rule to find the sides marked with a letter:

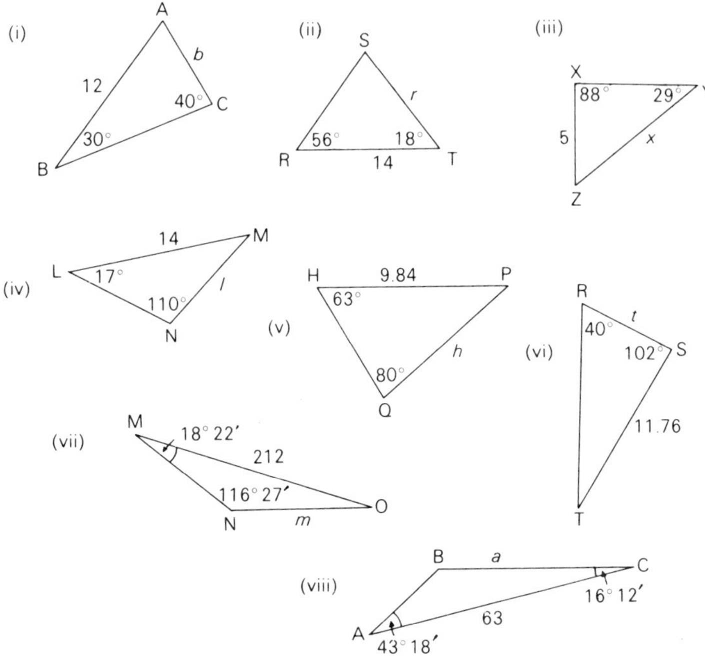

2 In the following examples, use the sine rule to find the marked angles:

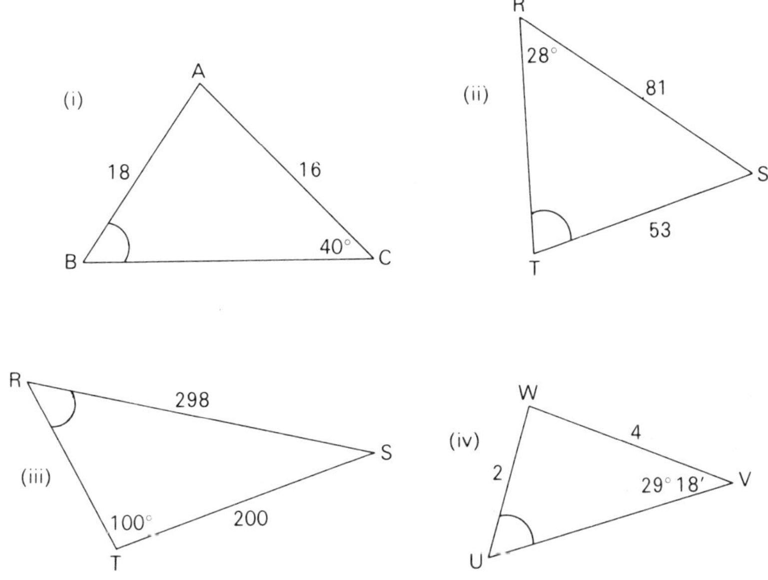

3 In the following examples, use the cosine formula to find the sides marked with a letter:

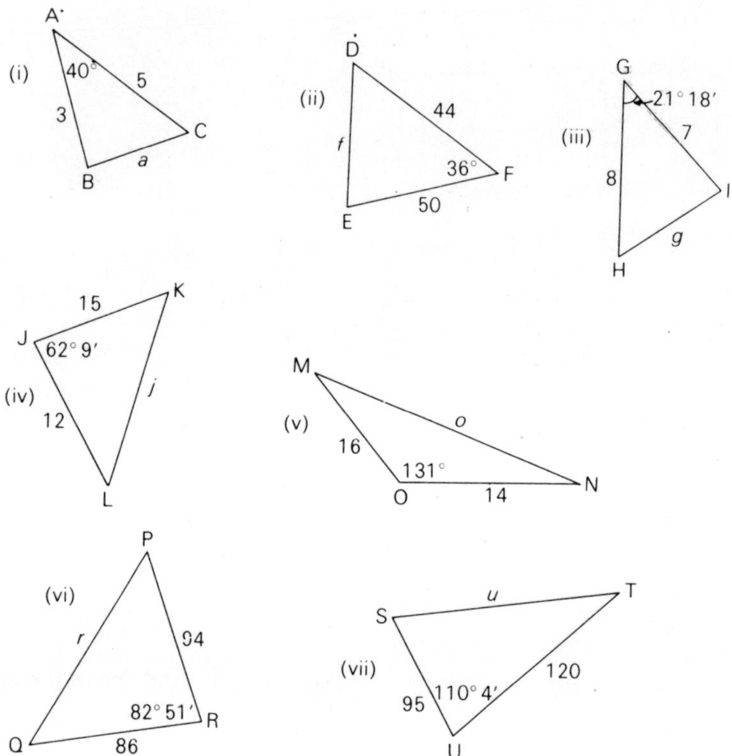

4 In the following example, use the cosine formula to find the marked angles:

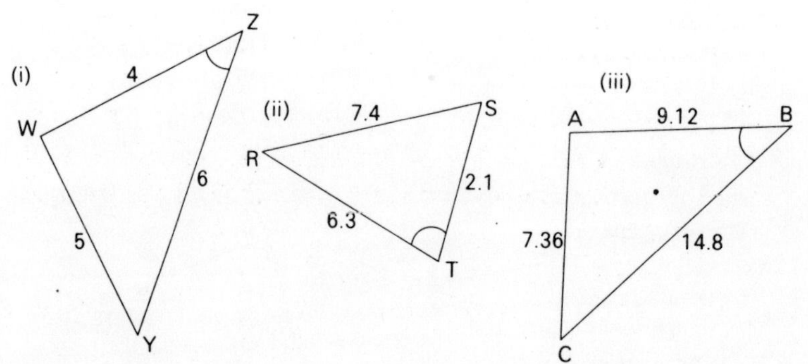

19.5 Area of a triangle

In proving the sine rule we used the fact that BD $= a \sin C$ (see Fig. 19.1). But the area of a triangle

$$= \tfrac{1}{2} \times \text{base} \times \text{perpendicular height}$$
$$= \tfrac{1}{2} \times b \times \text{BD}$$
$$= \tfrac{1}{2} ab \sin \text{C}.$$

Similarly, the area is also equal to $\tfrac{1}{2}bc \sin \text{A}$ and $\tfrac{1}{2}ac \sin \text{B}$.

EXAMPLE 7

Find the area of the triangle ABC where AB = 4 cm, BC = 6 cm, $\angle$B = 150°.

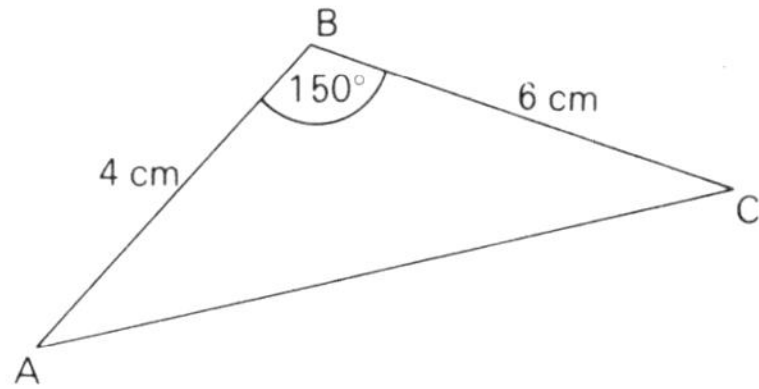

$$\begin{aligned}\text{Area} &= \tfrac{1}{2} \times 4 \times 6 \times \sin 150^\circ \\ &= \tfrac{1}{2} \times 4 \times 6 \times \sin 30^\circ \\ &= 12 \sin 30^\circ \\ &= 12 \times 0.5 \\ &= 6 \text{ cm}^2.\end{aligned}$$

EXERCISE 19b

1 Find the area of the following triangles:
(i) triangle ABC: AC = 10.2 mm, BC = 4.6 mm, $\angle$C = 42°
(ii) triangle DEF: EF = 5.35 km, FD = 0.028 km, $\angle$F = 72°
(iii) triangle JKL: JK = 6.85 cm, KL = 0.49 cm, $\angle$K = 37° 45′
(iv) triangle PQR: PR = 0.25 cm, PQ = 35 mm, $\angle$P = 14°
(v) triangle XYZ: YZ = 8.12 cm, XZ = 11.4 cm, $\angle$C = 110°.

2 Find the marked angles and sides given that all the triangles are acute-angled:

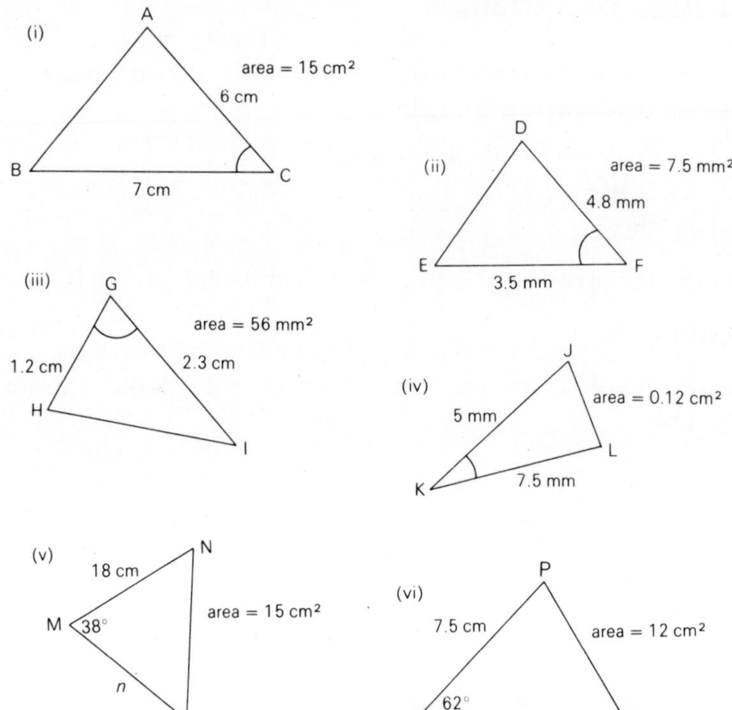

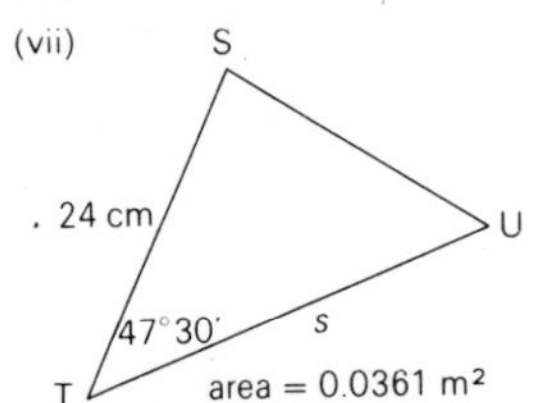

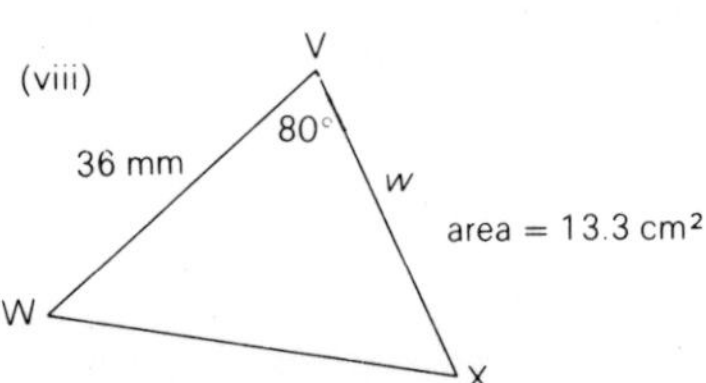

19.6 Bearings

The *bearing* of B from A is angle x in Fig. 19.7. (*Note:* the angle is measured clockwise.) We write: the bearing of B from A is 040°.

In the same diagram, to find the bearing of A from B, we find angle y (measured clockwise from north).

The bearing of A from B is $180° + 40° = 220°$.

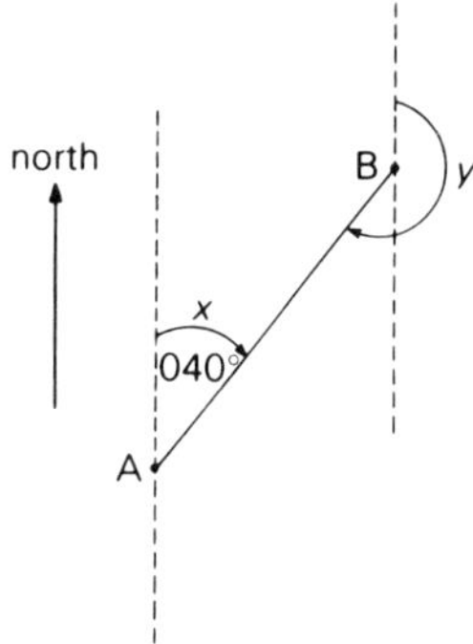

Fig. 19.7

19.7 Triangle of velocities

Ferryboats.

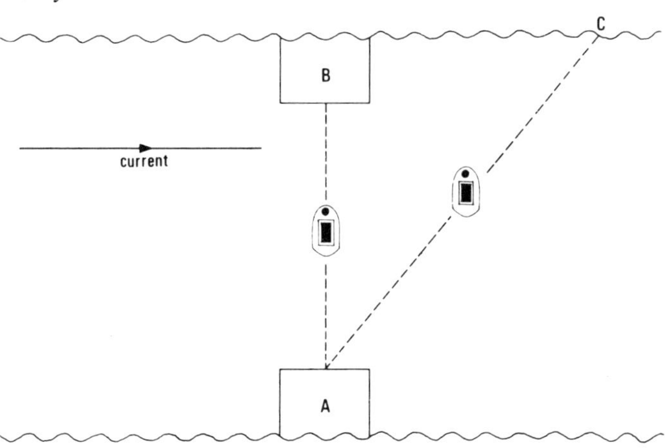

Fig. 19.8

A ferryboat leaves A and heads straight across the river to the landing stage B. Due to the current, however, the boat follows the path AC. The difference in angle between the course set by the pilot (AB) and the actual track taken (AC) is known as the *angle of drift*.

(*Note:* although the boat is actually moving in the direction of the path AC, it is still facing along a path parallel to AB.)

If we represent the boat's velocity in still water by the vector **b** and the current velocity by the vector **c**, then the resultant velocity of the boat along the track is the vector **b** + **c**.

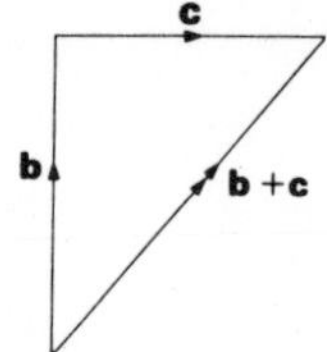

Fig. 19.9

This is known as a *velocity triangle*, the magnitude and direction of the three velocities being represented by the three sides of the triangle (see Fig. 19.9).

(*Note:* the component velocities follow each other around the triangle in a clockwise direction, whereas the resultant velocity is in the opposite direction.)

There are two methods for calculating the magnitude or direction of a velocity:

(a) by calculation,
(b) by drawing.

EXAMPLE 8

A ferry plies between two landing stages which are directly opposite each other on opposite banks of a river 300 metres apart. The ferry's velocity is 10 km/h and the current is 7 km/h. Find

(i) the angle of drift;
(ii) the boat's actual velocity;
(iii) the time taken to cross the river.

(a) By calculation. As the direction of the current is perpendicular to the actual direction of the boat we can use the trigonometry of a right-angled triangle.

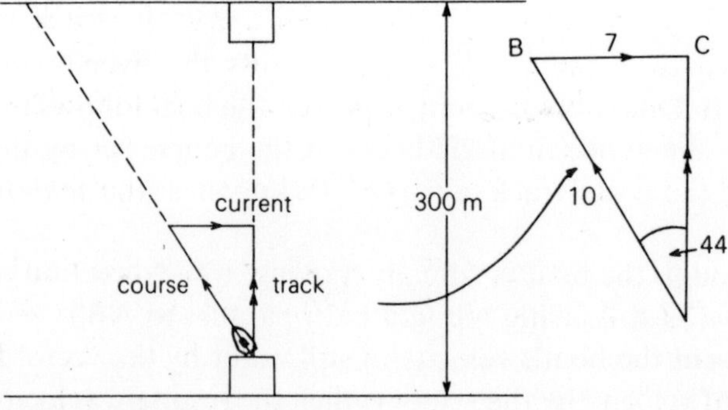

Fig. 19.10

(i) The angle of drift.

$$\sin \text{BAC} = \frac{\text{BC}}{\text{AB}} = \frac{7}{10}$$
$$\Rightarrow \angle\text{BAC} = 44^\circ\ 25'.$$

(ii) The boat's actual velocity.

$$\text{AC}^2 = 10^2 - 7^2$$
$$\Rightarrow \text{AC} = \sqrt{51}$$
$$= 7.14.$$
$$\text{Actual velocity} = 7.14\,\text{km/h}.$$

(iii) The time taken

$$= \frac{0.3\ \text{km}}{7.14\ \text{km/h}}$$
$$= 0.042\ \text{h}$$
$$= 2 \text{ minutes } 31 \text{ seconds.}$$

(b) By drawing.

1 Choose an appropriate scale, e.g. 1 cm represents 2 km/h.
2 Draw BC $3\frac{1}{2}$ cm in length.
3 Draw a line through C perpendicular to BC.
4 With centre B and radius 5 cm, draw an arc to cut the line representing the track, at A.
5 Measure AC and $\angle$BAC.

Solutions to problems similar to the example above involve the use of the trigonometry of the right-angled triangle. Using the sine and cosine rule we can extend solutions to cover any triangle.

Aircraft. As the course of the ferryboat was affected by the current, so the course of an aircraft is affected by the wind. The speed along the course set by the pilot is the *air speed.* The resultant speed of the aircraft or the speed along the track is known as the *ground speed.*

(*Note:* the wind is always given in the direction from which it blows.)

EXAMPLE 9

A aircraft is flying on a course 070° at an air speed of 500 km/h. The wind is blowing at 90 km/h from 350°. Find the track and ground speed.

(a) By calculation.

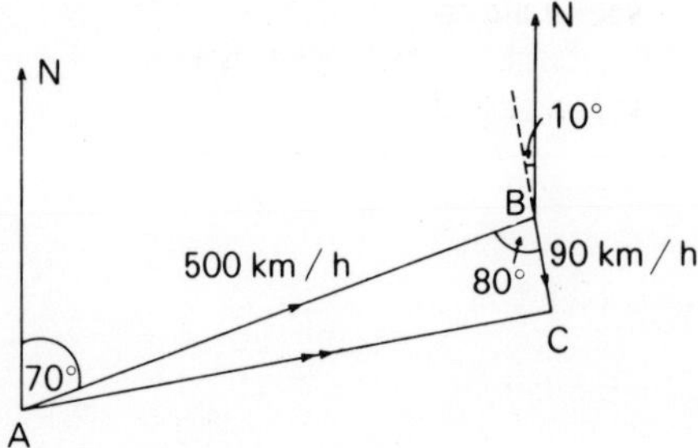

Fig. 19.11

By cosine rule:

$$\begin{aligned} AC^2 &= 500^2 + 90^2 - 2(500)(90)\cos 80^\circ \\ &= 250\,000 + 8100 - 90\,000 \cos 80^\circ \\ &= 258\,100 - 15\,640 \\ &= 242\,460 \\ \Rightarrow AC &= 492 \\ \therefore \text{ ground speed} &= 492 \text{ km/h.} \end{aligned}$$

By sine rule.

$$\frac{90}{\sin \angle BAC} = \frac{492}{\sin 80^\circ}$$

$$\Rightarrow \frac{90 \times \sin 80^\circ}{492} = \sin \angle BAC$$

$$\Rightarrow \angle BAC = 10^\circ 23'.$$

Direction of the track is therefore 080° 23′.

(b) By drawing.
1 Choose an appropriate scale, e.g. 1 cm represents 50 km/h.
2 Draw AB 10 cm in length at an angle of 70° to a north line.
3 At B, draw a line 1.8 cm in length to represent the wind ($\angle ABC = 80^\circ$).
4 Measure AC and $\angle BAC$.

Sometimes velocities are given in knots, where 1 knot = 1.85 km/h, instead of km/h.

EXERCISE 19c

1 The following data refer to ferryboat crossings. Assuming that the current flows at right angles to the track, determine by calculation or drawing the missing data.

	speed in still water	current speed	speed along track (actual speed)	angle drift
(i)	6.5 km/h	3.5 km/h		
(ii)	7 km/h		5 km/h	
(iii)		15 km/h	13 km/h	
(iv)	9 km/h			20°
(v)			5.5 km/h	15°
(vi)		7 km/h		17°30′

2 By calculation or drawing, find the missing data for the following aircraft.

	course	air speed	wind	speed	track	ground speed
(i)	090°	400 km/h	180°	88 km/h		
(ii)	120°	425 km/h	270°	75 km/h		
(iii)	290°	300 knots	000°	50 knots		
(iv)	072°	550 km/h			075°	530 km/h
(v)	200°	310 knots			180°	320 knots
(vi)	320°	300 km/h			325°	280 km/h
(vii)			325°	60 km/h	210°	420 km/h
(viii)			020°	45 knots	175°	290 knots
(ix)			270°	75 km/h	170°	180/km/h
(x)		375 km/h	055°	55 km/h	170°	
(xi)	190°		225°	35 knots		290 knots
(xii)		360 km/h	160°	60 km/h	226°	

HARDER EXAMPLES 19

1 In the triangle PQR, $p = 7$, $r = 5$ and $\angle Q = 120°$. Find q.

2 Find the smallest angle of the triangle whose measurements are 5 cm, 6 cm and 7 cm.

3 The ratio of the angles of a triangle to one another is 2:3:4. If the longest side of the triangle is 8 cm, calculate the length of the other two sides.

4 The distance from a tee to a hole on a golf course is 350 m. If a golfer's drive of 200 m is 180 m short of the hole, find how many degrees his drive was off the direct line.

5 The hands of a clock are 5 cm and 7 cm long. Find the distance between the ends of the hands at 17.00 hours.

6

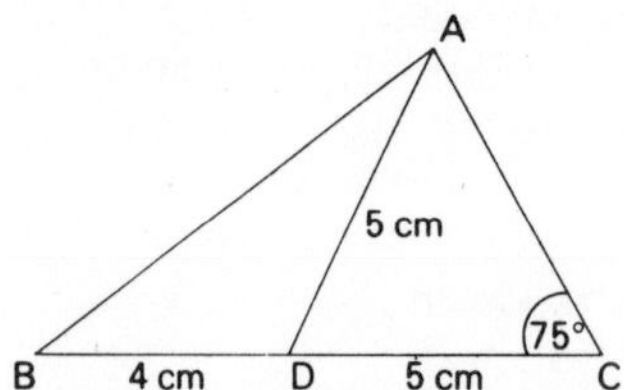

In the triangle ABC, the angle ACB = 75°. D is the point on BC such that BD = 4 cm and DC = 5 cm. If AD = 5 cm, find

(i) the length of AC,
(ii) the length of AB.

7 A boat sails from port A on a bearing of 050° to a port B, a distance of 75 km. It then sails from B on a bearing 160° a distance of 150 km to a port C. Find the distance of C from A.

8 Three beacons A, B and C are situated on a moorland. Beacon A is 1500 metres due north of B and beacon C is 3000 metres on a bearing 320° from B. What is the distance and bearing of C from A?

9 In the square ABCD of sides 6 cm, AF and AE are such that E and F bisect sides CD and BC respectively. Calculate

(i) AF,
(ii) ∠FAE,
(iii) area of triangle AFE.

10 In the triangle ABC, if AB = 4 cm, AC = 6 cm and ∠BAC = $x°$ ($120 < x < 180$), calculate the limits y and z between which the length BC must lie.

11 A triangular playing area is bounded by two walls which meet at an angle of 47°. Find the total playing area if the walls are 20 m and 35 m long.

12

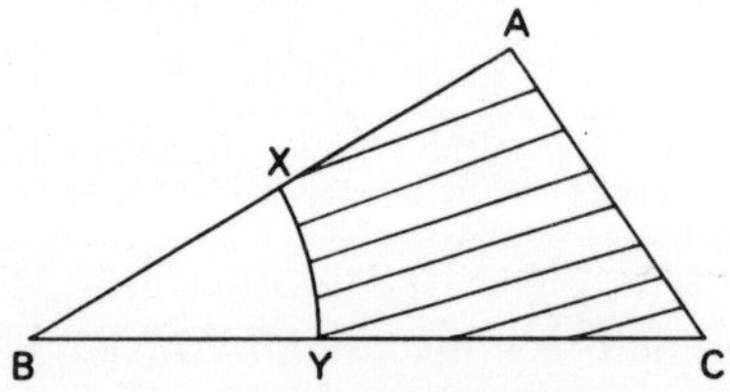

ABC is a triangle in which AB = 40 cm, BC = 80 cm and ∠ABC = 30°. XY is an arc of a circle centre B and radius 10 cm. Calculate

(i) the area of triangle ABC,
(ii) the area of the shaded portion.

[C]

13

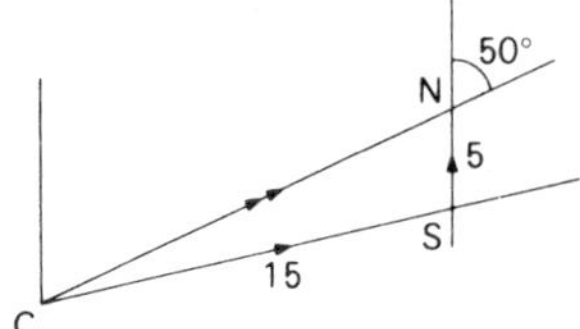

A captain wishes to set a course in order to travel on a bearing 050°. His ship can travel at 15 knots in still water and a current is flowing at 5 knots from due south. Calculate the ∠NCS in the given velocity triangle and hence find to the nearest degree the course the captain should set. [C]

14 An aircraft flies at 300 knots in still air. The pilot sets a course of 270°. The wind speed is steady at 35 knots but its direction varies between 045° and 225°. Find the maximum and minimum values for the aircraft's ground speed.

15 Town A is 1500 km from B on a bearing 320°. An aircraft which can travel at 400 km/h in still air leaves B at 11.00 hours for A. The prevailing wind is 50 km/h from the north. Calculate (i) the track and ground speed of the aircraft, (ii) the time of arrival at A.

16 A river is flowing due south at 3 km/h. The point X is on the west bank and Y, on the east bank, is such that the bearing of Y from X is 150° (S 30° E). A boat, which travels at 2 km/h in still water, goes from X to Y. Using an accurate drawing, to a scale of 4 cm to 1 km/h, show that there are two courses on which the boat can be steered. State the bearings of these courses. If, however, the speed of the boat in still water is V km/h, state

(i) the value of V above which there is only one course on which the boat can be steered,

(ii) the value of V below which the boat cannot possibly reach Y. [C]

17 A man who can row at 5 km/h in still water intends to travel due east. A current of 8 km/h is flowing from a direction 300° (N 60° W). Sketch a velocity diagram to show the two possible directions in which the man can row. Find by calculation

(i) the angle between these two directions,

(ii) the two resultant velocities of the rowing boat. [C]

18 An aircraft C is observed simultaneously at radar stations A and B which are 30 km apart. If the angle of elevation at A is 6° and at B is 9°, calculate the distance AC. Hence find the height of the aircraft above the ground.

20 Techniques of drawing and construction

Although a thorough treatment of the methods of accurate drawing is beyond the scope of this book, this chapter sets out a number of the simpler constructions and an introduction to the use of simple plans and elevations.

20.1 Constructions using ruler and compass

(a) To bisect a line XY at right angles.

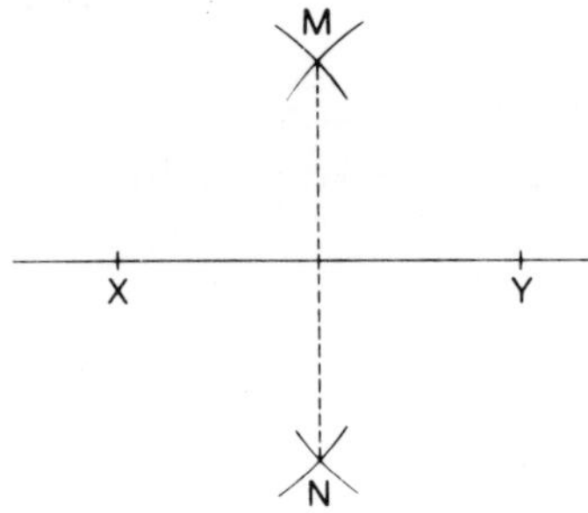

Centre the compass on point X and with a radius greater than one half of XY make two arcs, one above and one below XY. Repeat this, using the same radius, centre Y, to make two crosses M and N. The line MN bisects XY at right angles.

(b) To bisect a given angle ABC.

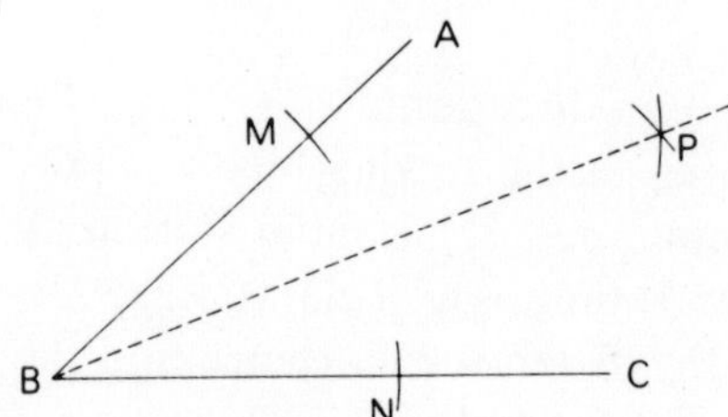

Centre the compass on B and make two arcs at M and N. Use the same radius, centre M and then N, to make a small cross at P. The line BP bisects angle ABC.

(c) To construct a line from a point P perpendicular to a line XY.

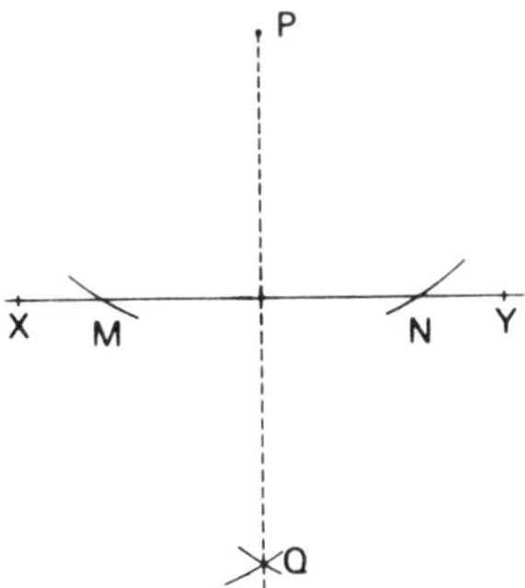

Centre the compass at P and make two arcs at M and N on XY. Use the same radius centre at M and then N to obtain a cross at Q. The line PQ will be perpendicular to XY.

(d) To divide a line XY into four (or any number of) equal parts.

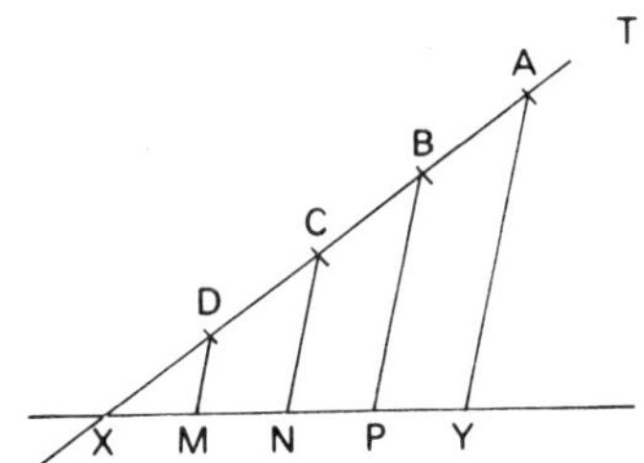

Through X (or Y), draw any line XT. Starting at X, mark four equal lengths on XT, i.e. A, B, C, D. Join AY. Use two set squares to draw three lines BP, CN, DM, parallel to AY. These lines divide XY into four equal parts.

(e) To construct an angle of 60° on AB at A.

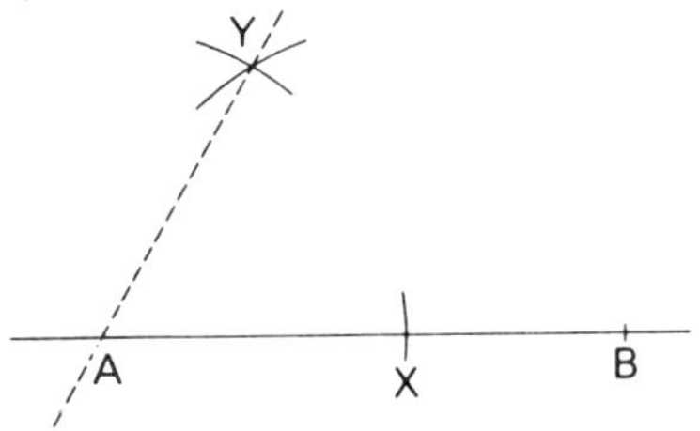

Use a radius less than AB, centre the compass at A and make an arc X (on AB) and Y. With the same radius, centre X, complete the cross at Y. The angle YAB is 60°.

(f) To construct angles of 30°, 15°, $7\frac{1}{2}$°, etc. These angles can be obtained by bisecting 60°, 30°, 15°, etc.

(g) To construct an angle of 90° on AB at B.

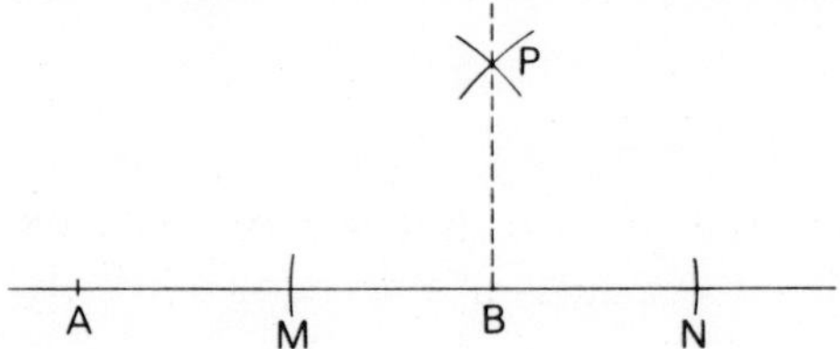

Produce AB. Centre the compass at B and make two arcs at M and N. Increase the radius and using M and N as the centre, make a cross at P. PBA is now 90°.

(h) To construct an angle of 45°. This can be obtained by bisecting an angle of 90°.

EXAMPLE 1

Construct the triangle RST, where ST = 4 cm, ∠TSR = 75° and TR = 5 cm.

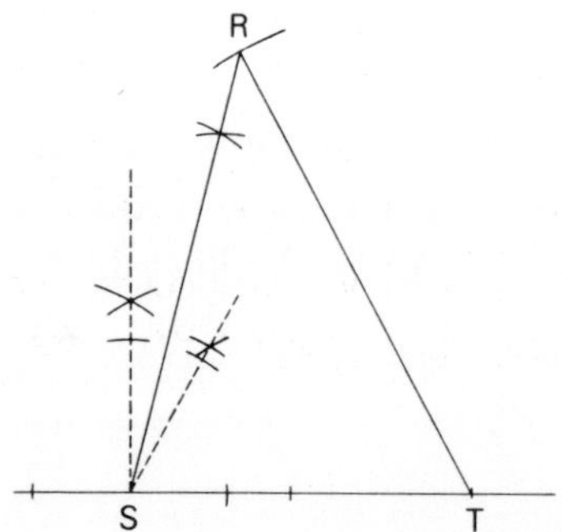

The angle 75° has been constructed by using 60° and 30° bisected. TR is measured using a compass, centre T, and radius 5 cm.

[*Note:* all construction lines (shown dotted in these diagrams) should be continuous lines, and as faint as possible.]

(i) To construct the circumscribing circle of a triangle ABC.

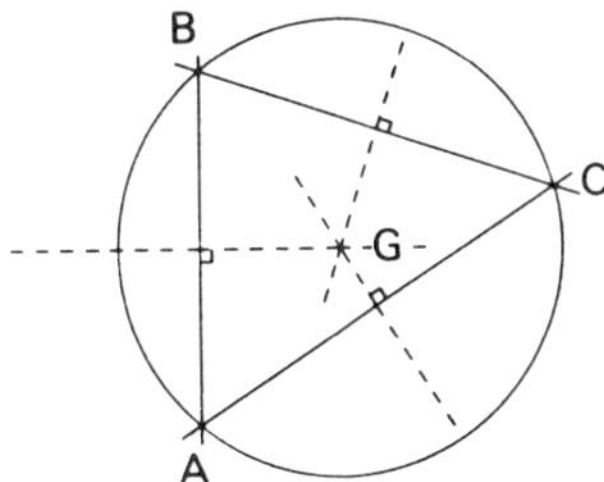

The centre of the circle is at the point of intersection G of the perpendicular bisectors of the three sides of the triangle (usually only two are necessary).

(j) To construct the inscribed circle of a triangle ABC.

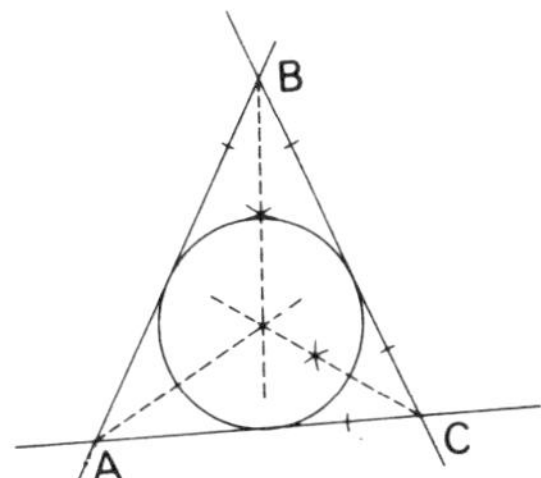

The centre of the circle is at the point of intersection of the bisectors of the three angles of the triangle (usually only two are necessary).

EXAMPLE 2

Construct the locus of a point P which moves with respect to two fixed points A and B distance 5 cm apart, such that $\angle APB = 60^\circ$.

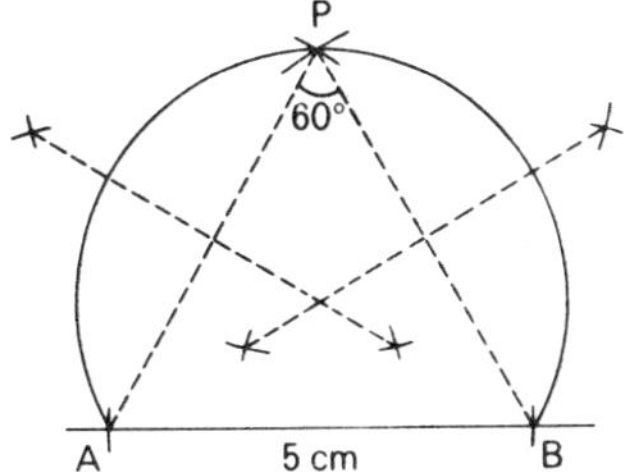

Fig. 20.1

Since $\angle APB$ is constant, P must move on the arc of a circle (angles in the same segment). We have, therefore, to find any one position of P and construct the circumscribing circle. The simplest position is when APB is an equilateral triangle. (See Fig. 20.1.)

EXERCISE 20a

1 Using ruler and compass, construct the following triangles, measuring the sides and angles not given:
(i) $AB = 4$ cm, $\angle BAC = 60°$, $\angle ABC = 90°$
(ii) $AB = 6$ cm, $\angle BAC = 15°$, $BC = 2$ cm
(iii) $AB = 5$ cm, $\angle ABC = 75°$, $AC = 7$ cm
(iv) $BC = 3$ cm, $AB = 5$ cm, $AC = 4\frac{1}{2}$ cm
(v) $AC = 7$ cm, $\angle ACB = 45°$, $\angle CAB = 37\frac{1}{2}°$
(vi) $AB = 5$ cm, $\angle ABC = 150°$, $\angle BAC = 15°$.

2 The following ordered pairs represent the length of the base (in cm) and the vertical angle (in degrees) respectively of an isosceles triangle. In each case, construct the triangle and measure one of the two equal sides.
(i) (4, 30) (ii) (3, 90) (iii) (7, 45).

3 Construct the triangle with sides of length 3 cm, 4 cm and 5 cm. Construct the circumscribing circle of the triangle. What is the radius of this circle?

4 Construct the triangle ABC, where $AB = 12$ cm, $BC = 7$ cm and $AC = 8$ cm. Construct the circumscribing circle of this triangle and the point P on the circle, such that the area of triangle APC equals the area of triangle ABC.

5 AB is a fixed line of length 5 cm. Construct the locus of a point P, which moves so that $\angle APB = 45°$.

6 ABCD is a rectangle with $AB = 15$ cm, $BC = 8$ cm. Construct this rectangle, and the position of a point P on AB such that

$$\frac{\text{area of triangle APD}}{\text{area of rectangle}} = \frac{1}{6}.$$

7 Find the radius of the largest sphere which will fit inside a cone of base radius 3 cm and vertical height 7 cm.

8 PQRS is a square of side 8 cm. X is a point on PQ such that $PX = 2XQ$. l is the line which is at right angles to XR, and passes through the mid-point of XR. If $l \cap PS = Y$, construct the position of Y, and measure YS.

20.2 Plans and elevations

The solid shown in Fig. 20.2 is similar to the shape of a house. Hidden edges are shown dotted. Let us consider what it looks like if viewed from different directions.

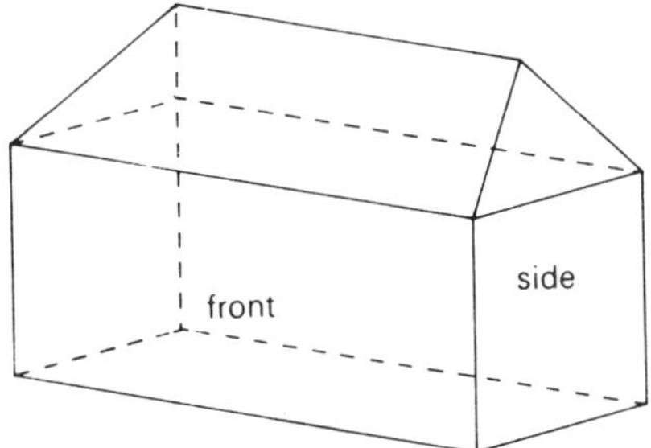

Fig. 20.2

It does not need too great a stretch of the imagination to visualise the views would be as in Fig. 20.3. The bird's eye view is known as the *plan*. The side views are called *elevations*.

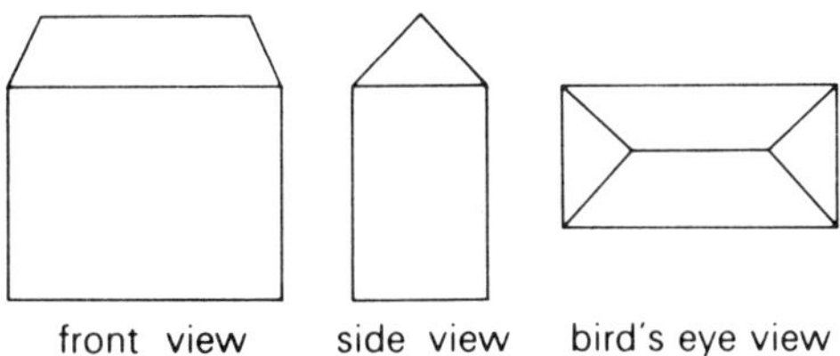

Fig. 20.3

These three diagrams can be combined into one, as shown in Fig. 20.4.

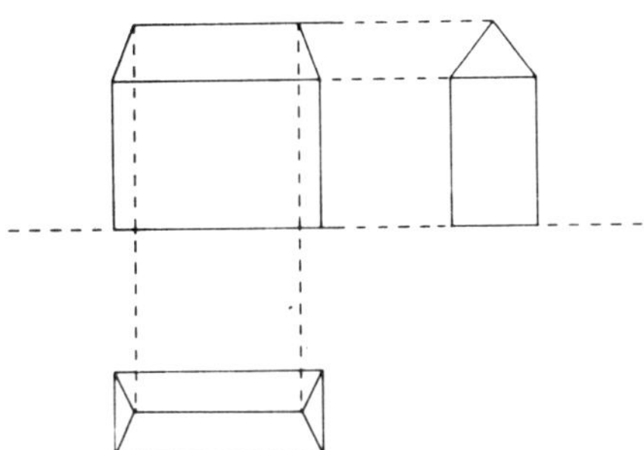

Fig. 20.4

Note: in many books, the plan is put above the front elevation.

EXAMPLE 3

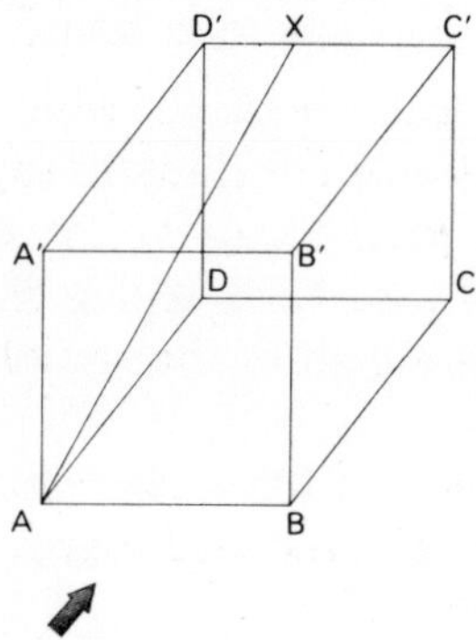

Fig. 20.5

Figure 20.5 shows a cube constructed from 12 equal lengths of wire of length 10 cm. A further piece joins A and X where D′X = 3 cm. Draw the plan and elevation in the direction shown. By a further construction, find the length of AX.

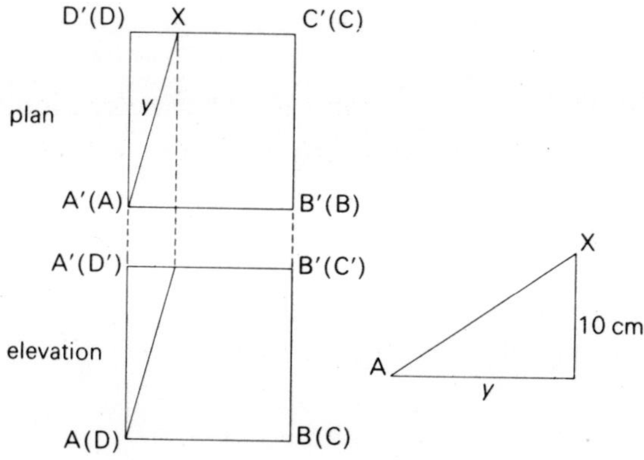

Fig. 20.6

In order to find the length of AX, we measure the horizontal distance between A and X from the plan (y) and draw the vertical cross-section. We see that AX = 14.5 cm.

EXERCISE 20b

1 Three equal spheres stand on a plane so that they touch each other. An identical sphere is placed on top. Draw the plan of this configuration.

2 A pyramid with a square base stands on the top of a cube. The sides of the base of the pyramid are 50 % greater than the edges of the cube. Draw the plan view.

3 A lamp shade is made by joining 3 circular hoops of wire of radius 20 cm, 25 cm and 30 cm, the narrowest being at the top, and all are held equally spaced in horizontal planes, a distance 10 cm apart, by three straight pieces of wire. Draw a plan and suitable elevation in order to find the length of the straight pieces of wire.

4 A rectangular block is cut in half so as to form two equal triangular prisms. The two prisms are then fastened together to form a new block, as shown in Fig. 20.7.

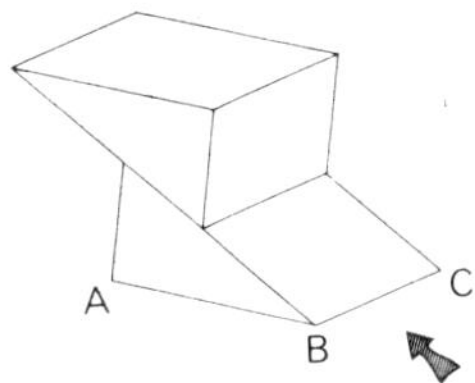

Fig. 20.7

If the edges AB and BC are horizontal, sketch the elevation of the new block on a vertical plane parallel to BC as viewed
(i) in the direction of the arrow,
(ii) in the opposite direction.

[C]

5 Figure 20.8 shows the front and side elevations of a solid object. Sketch a possible plan of this object.

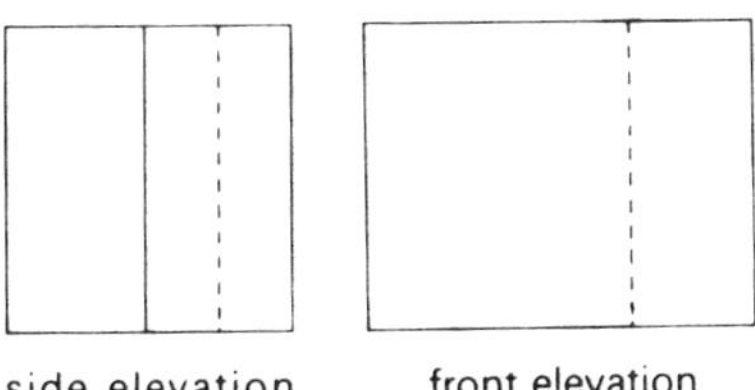

Fig. 20.8

HARDER EXAMPLES 20

1 Draw a line XY of length 7 cm. Construct a line PQ parallel to XY, and distant 4 cm from XY. Construct the circle which

passes through X and Y, touching PQ. Measure the radius.

2 Draw any acute-angled triangle BAC. Produce BC to D where CD = CA. Produce CB to E where BE = BA. Construct the bisectors of angles ABE and ACD. Label the point of intersection of these two bisectors O. By drawing a circle through A with centre O, deduce that OB is the line of symmetry of the quadrilateral OABE. Construct also the bisector of angle BAC. What do you notice?

3 Construct a hexagon of side 3 cm.

4 The shape of a field is a quadrilateral ABCD in which AB = 120 m, angle BAC = 90° and BC = 175 m. Angle BCD = 120° and CD = 100 m. A straight fence is erected from D to AB, being at right angles to AB. Find by calculation, or accurate drawing, the length of this fence and also find the total area of the field.

[SU]

5

Fig. 20.9

Figure 20.9 shows a tetrahedron constructed from 6 equal rods of length 10 cm. X is the mid-point of AC and Y is the mid-point of VB. Draw the plan and suitable elevations, to find the length of XY.

6

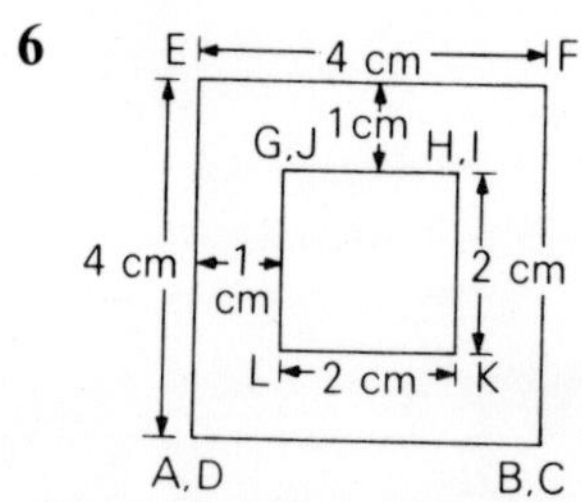

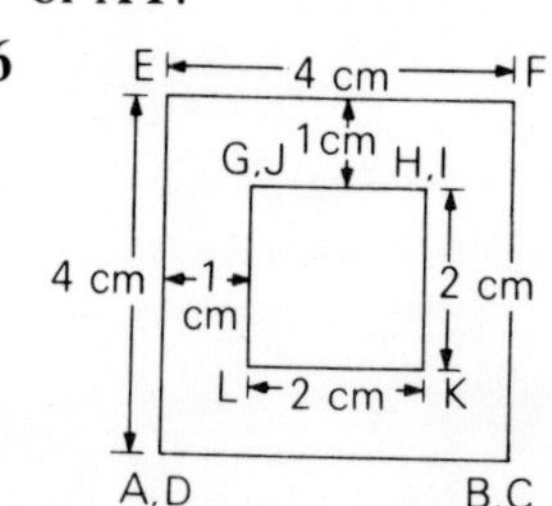

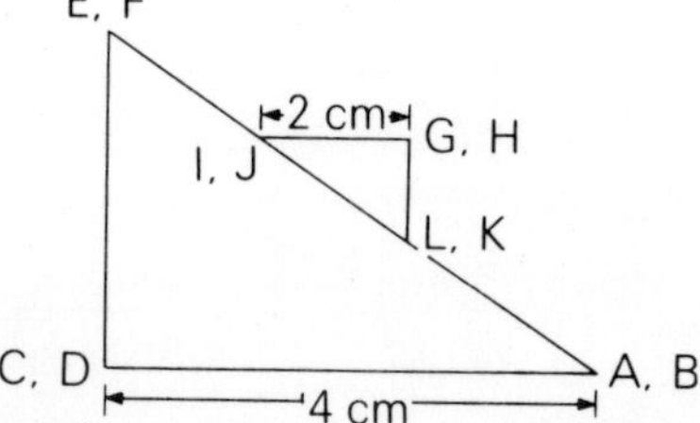

Fig. 20.10

Figure 20.10 shows the front and side elevations of a solid block of wood.

(i) Draw a neat diagram of the plan, showing all the points A to L correctly positioned.

(ii) Draw a three-dimensional sketch of the block in which G and H are visible, and letter correctly all the other visible corners.
(iii) The block is cut into two pieces along the plane ADF. Sketch the cross-section and letter its corners. (An accurate drawing is *not* required.)

[C]

7

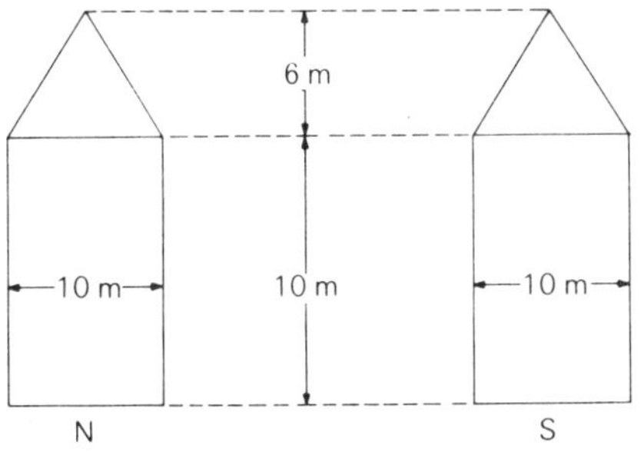

Fig. 20.11

Figure 20.11 shows the north elevation N (i.e. the elevation as viewed from the north) and the south elevation S of a house, details such as windows and chimneys being omitted. The overall length of the house is 20 m and the length of the top roof ridge is 15 m.

Sketch the plan and the west elevation W. An accurate drawing is not required, but all 'visible' lines must be shown.

Calculate the areas of the smallest and largest cross-sections by planes which are vertical and are in a north–south direction. [L]

8 

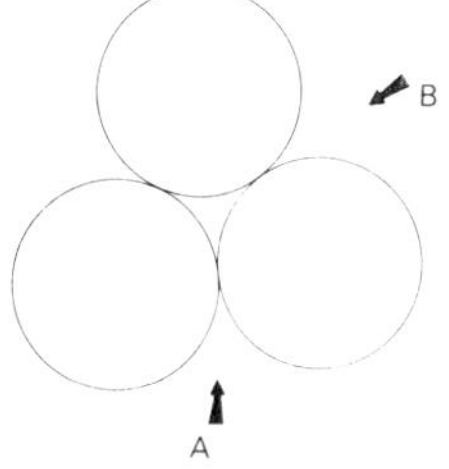

Fig. 20.12

Figure 20.12 shows the plan of three identical spheres standing on a horizontal table with each touching the other two. Draw the elevation looking in the direction of (i) the arrow A; (ii) the arrow B. [C]

21 Latitude and longitude

21.1 Length of a circular arc

In Fig. 21.1, AB is an arc of a circle, centre O. If we construct the radii AO and OB, the angle θ° formed is the angle which the arc AB subtends at the centre of the circle. As there are 360° in a complete turn, θ° corresponds to $\theta/360$ of a complete turn.

$$\therefore \text{ the length of AB is } \frac{\theta}{360} \text{ of the circumference}$$

$$\text{i.e. AB} = \frac{\theta}{360} \times 2\pi r.$$

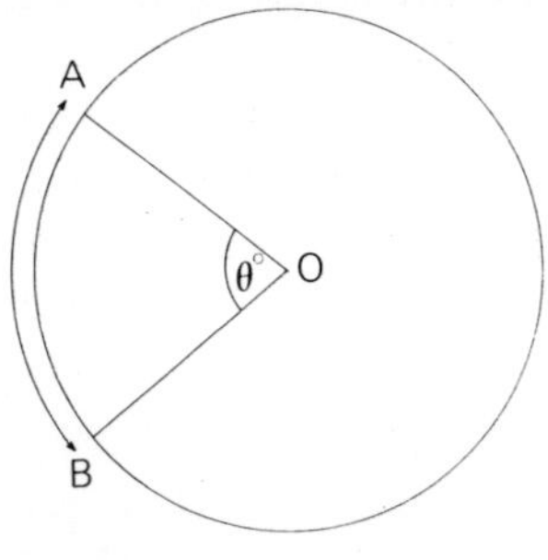

Fig. 21.1

EXAMPLE 1

Find the length of an arc subtending an angle of 80° at the centre of a circle radius 14 cm.

$$\begin{aligned}\text{Circumference of the circle} &= 2\pi r\\ &= 2 \times \frac{22}{\not{7}_1} \times \not{14}^2 \text{ cm}\\ &= 88 \text{ cm}\end{aligned}$$

$$\therefore \text{ length of } \mathbf{AB} = \frac{8\not{0}}{36\not{0}} \times 88 \text{ cm}$$
$$= \frac{176}{9} \text{ cm}$$
$$= 19\tfrac{5}{9} \text{ cm.}$$

EXAMPLE 2

Find the angle subtended by an arc of length 180 km at the centre of a circle of radius 6400 km.

$$\text{Length of arc} = 180 \text{ km} = \frac{x}{\cancel{360}_9} \times 2 \times \pi \times \overset{160}{\cancel{6400}} \text{ km}$$

$$\Rightarrow 180 = x \times \frac{320\pi}{9}$$

$$\Rightarrow x = \frac{9 \times \cancel{180}^9}{\cancel{320}_{16}\pi}$$
$$= \frac{81}{16\pi}$$
$$= 16.1$$

$\therefore$ angle subtended at the centre $= 16^\circ$ (to nearest degree).

21.2 Radian measure

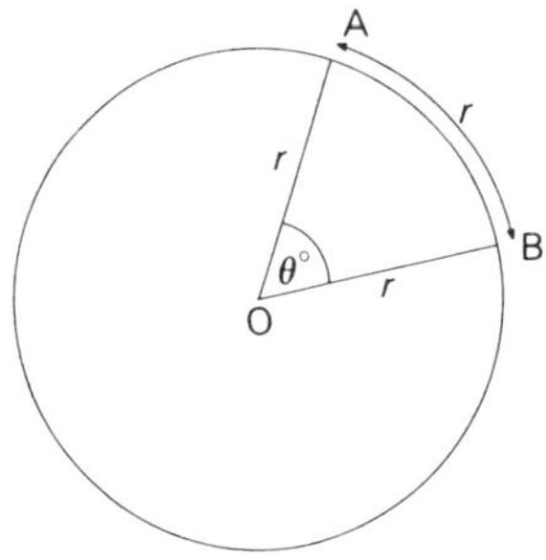

Fig. 21.2

When the length of the arc is equal to the radius of the circle, the angle subtended at the centre, θ, is defined as 1 *radian*. The ratio of the angle to the circle $= \dfrac{\theta}{360^\circ}$.

The ratio of AB to the circumference $= \dfrac{r}{2\pi r}$.

$$\therefore \frac{\theta}{360^\circ} = \frac{1}{2\pi}$$

$$\therefore \theta = \frac{180^\circ}{\pi}$$

$$\therefore 1 \text{ radian} = \left(\frac{180}{\pi}\right)^\circ$$

From this we obtain:

$$\pi \text{ radians} = 180 \text{ degrees}$$

$$1 \text{ degree} = \frac{\pi}{180} \text{ radians.}$$

EXAMPLE 3

Change 45° to radians.

$$45^\circ = \frac{\pi}{180} \times 45 \text{ radians}$$

$$= \frac{\pi}{4} \text{ radians}$$

$$= \frac{3.142}{4}$$

$$= 0.785 \text{ radians.}$$

EXERCISE 21a

Take $\pi = 3.14$.

1 What fraction of 360° are the following angles?

(i) 30° (ii) 15° (iii) 12° (iv) 7° (v) 4° 30′ (vi) 1° 30′ (vii) 56° 15′ (viii) 11° 36′ (ix) 32° 45′ (x) 157° 30′.

2 Calculate the lengths of the arcs for the following:

	radii	angle subtended at centre
(i)	7 cm	70°
(ii)	6370 km	130°
(iii)	5.7 m	4° 30′
(iv)	6400 km	22° 30′

3 Calculate the angles subtended by the following arcs to the nearest degree:

	radii	length of arc
(i)	8 cm	12 cm
(ii)	6370 km	6370 km
(iii)	528 m	49 m
(iv)	6370 km	150 km

4 Calculate the radii of the circles for the following:

	arc length	angle subtended at centre
(i)	5 m	20°
(ii)	7.3 cm	13°
(iii)	825 km	75°
(iv)	400 km	112° 30′

5 Change the following to radians:

(i) 30° (ii) 72° (iii) 180°
(iv) 135° (v) 24° (vi) $7\frac{1}{2}°$.

6 Change the following angles in radians to degrees:

(i) 0.2 (ii) 2 (iii) 0.7
(iv) $\frac{\pi}{3}$ (v) $\frac{2\pi}{3}$ (vi) 3.14.

21.3 Latitude and longitude

Our home planet Earth is a sphere with an approximate radius of 6370 km and rotates about its axis once every 24 hours.

If we take a plane through the centre of the earth, then the circumference of the circle formed is called a great circle, and its radius is 6370 km.

A *meridian of longitude* is such a circle which passes through the north and south poles.

A *parallel of latitude* is a circle formed by a plane through the earth perpendicular to the line joining the north and south poles.

The equator is a parallel of latitude which divides the earth into two hemispheres. (See Fig. 21.3.)

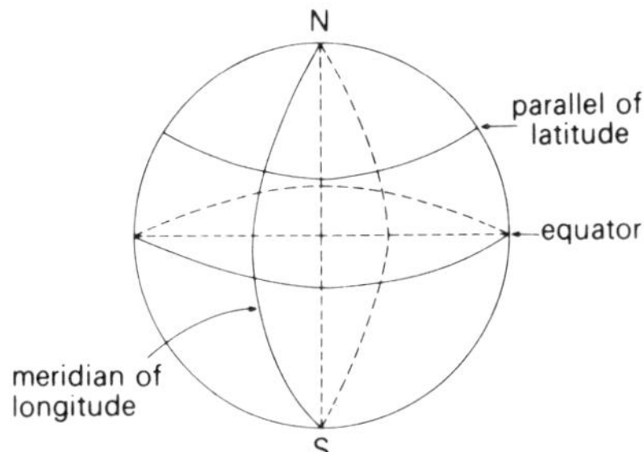

Fig. 21.3

21.4 Nautical mile

A *nautical mile* is the length of the arc of the equator which subtends an angle of 1′ or $\frac{1}{60}$ of a degree.

1 nautical mile = 1.85 km

1 knot is 1 nautical mile per hour.

21.5 Greenwich

Lines of longitude are labelled east and west of a reference line which passes through Greenwich (0°). These labels are determined by the angle subtended by the arc of the equator which joins the Greenwich meridian to the required meridian. Calcutta, for example, lies on the 90° east line of longitude.

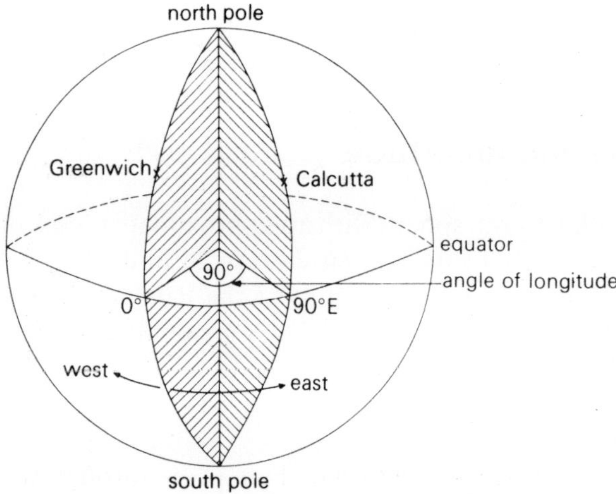

Fig. 21.4

The lines of longitude are labelled from 0° to 180° east and west at which point they coincide.

Lines of latitude are labelled in a similar manner, starting at 0° at the equator to 90° north and 90° south at the poles. To determine the line of latitude of a required place, let us take an example.

EXAMPLE 4

Determine the line of latitude of Greenwich.

We must first take a plane through the two poles and Greenwich. The circle formed is a great circle. (See Fig. 21.5.)

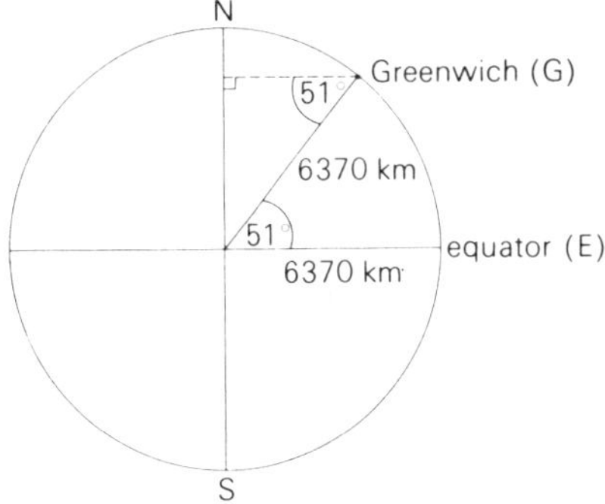

Fig. 21.5

The line of latitude is determined by the angle subtended at the centre of the earth by the arc GE, which in this case is 51°. So Greenwich lies on the line of latitude 51° north.

EXAMPLE 5

Greenwich and Accra lie on the 0° line of longitude. Find the distance between them if Greenwich is 51°N and Accra is 5°N.

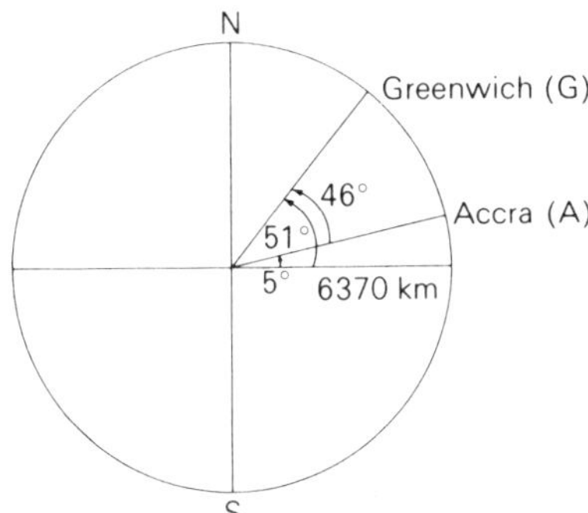

Fig. 21.6

The angle subtended by the arc of the great circle joining Greenwich and Accra is 46°.

$$\therefore \text{ length of GA} = \frac{46}{36\not{0}} \times 2 \times \frac{22}{\not{7}} \times \not{6}\not{3}\not{7}\not{0}^{91} \text{ km}$$

$$= \frac{46 \times \not{4}\not{4}^{11} \times 91 \text{ km}}{\not{3}\not{6}_9}$$

$$= 5120 \text{ km.}$$

EXAMPLE 6

Two places P and Q lie on the line of latitude 55°N. Find the distance between them measured along the line of latitude if the difference in longitudes is 42°.

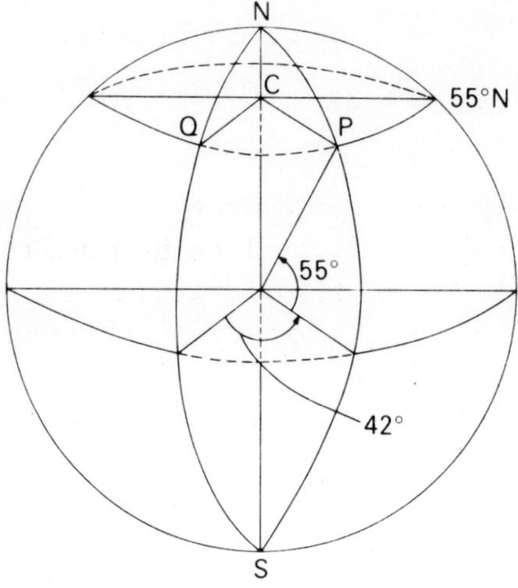

Fig. 21.7

To find the length of the arc PQ we need to find the radius of the line of latitude 55°N. To do this we take a plane through the two poles.

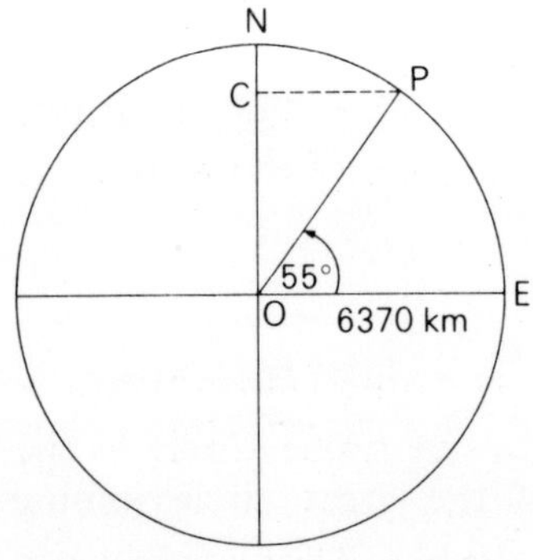

Fig. 21.8

$$\frac{\text{CP}}{6370 \text{ km}} = \cos 55^\circ$$

$$\Rightarrow \text{CP} = 6370 \text{ km} \times \cos 55^\circ$$

$$= 3650 \text{ km}.$$

The length of the arc PQ can now be determined from the angle 42° which it subtends at C and the radius of the circle, 3650 km.

$$PQ = \frac{42}{360} \times 2 \times \pi \times 3650 \text{ km}$$
$$= 2680 \text{ km}.$$

EXERCISE 21b

Radius of the earth = 6370 km; $\pi = 3.14$.

1 Calculate the distance around the equator.
2 Calculate a degree of longitude at the equator.
3 Quito and Kampala lie on the equator and the difference in longitudes is 111°. Calculate the distance between them.
4 The distance between two places on the equator is 500 km. Find the difference in their longitudes.
5 A ship sails 300 km due west along the equator. Calculate the change in longitude.
6 Two cities on the same line of longitude differ in latitude by 12°. Find the distance between them.
7 Find the distance moved by the end of a minute hand of a clock in 3 minutes if the length of the hand is 4 cm.
8 Find the radius of the circle of latitude 45° N.
9 Find the radius of the circle of latitude 14° S.
10 A ship sails due north at a speed of 30 knots. How long does it take for the ship to change its latitude by 3°?

HARDER EXAMPLES 21

Radius of the earth = 6370 km.

1 Find (a) the radius, and (b) the length of the line of latitude which passes through the following cities of the world:
(i) Moscow, 56° N;
(ii) Cape Town, 34° S;
(iii) Bangkok, 15° N.
2 Find the length of the Arctic Circle, latitude 66° $32\frac{1}{2}'$N.
3 Find the distance from a point on the Arctic Circle to the north pole.
4 Find the distance measured along the line of latitude between New Orleans (90° W, 30° N) and Cairo (30° E, 30° N).
5 Calculate a degree of longitude at 36° N.
6 Find the distance along the line of longitude between John O'Groats (3° W, 58° 30′N) and Lyme Regis (3° W, 50° 30′N).
7 Calculate the distance between two towns A and B which are on the same meridian but whose latitudes differ by 110°.

8 Calculate the distance between two places S and T which are on the same line of latitude of 30° N but whose longitudes differ by 20°.

9 The longest stretch of straight railway track in the world is from Nuringa (126° E, 31° S) to Ooldea (132° E, 31° S) in Australia. Calculate its distance.

10 State the latitude and longitude of the place which is at the opposite end of a diameter of the earth from Greenwich (0° W, 51° N).

11 A ship sails 300 nautical miles due west and finds that her longitude has altered by 5°. What is her latitude?

12 Two places P and Q are both on the line of latitude 50° N. The longitudes of P and Q are 15° E and 70° W respectively. A plane leaves P and flies to B at a speed of 300 knots. Calculate how long the journey takes.

13 A satellite circles the earth in such a way that it remains stationary above a point A on the equator. B is a place due north of A in latitude 65° N. If B is the furthest point north at which the satellite can be seen with a telescope, what is the height of the satellite above the earth's surface?

14 P, Q and R are points on the earth's surface all in the same longitude. P is in latitude x°N, Q is in latitude 21°N and R is on the equator.

(i) If the distance from P to R is three times the distance from Q to R, find x.

(ii) If, however, the circumferences of the circles of latitude through P and Q are in the ratio 1 : 3, find the new value of x correct to the nearest whole number. [C]

15 (i) The circumference of the circle of latitude x° N is 7200 nautical miles. Find the value of x.

(ii) The distance measured along the meridian from a point in latitude y°N to the North Pole is 3600 nautical miles. Find the value of y. [C]

16 X is on the surface of the earth in latitude 71° N, longitude 30° W, and Y is in latitude 71° N, longitude 150° E. Find, to the nearest ten km, the difference between the distances from X to Y measured (i) along the meridian (through the North Pole), and (ii) along the circle of latitude eastwards from X to Y.

An aeroplane followed the North Pole route from X to Y flying at an average speed of 300 km per hour. Find the time taken for the journey, correct to the nearest minute.

[O]

22 Group theory

22.1 The symmetry group of a rectangle

Make a card similar to the one in Fig. 22.1, and label the corners. Notice that there are two lines of symmetry which are marked on the diagram. If we rotate the card through 180° about the vertical line of symmetry, only the letters change position. (See Fig. 22.2.) We will call this operation *a*.

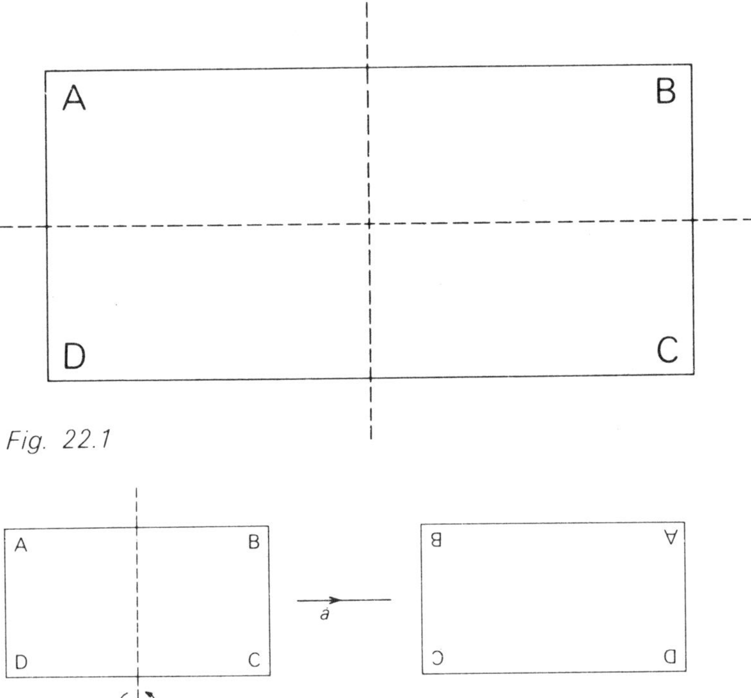

Fig. 22.1

Fig. 22.2

We have a similar situation with the rotation of 180° about the horizontal line of symmetry; again the letters are in a different position. We will call this operation *b*. (See Fig. 22.3.)

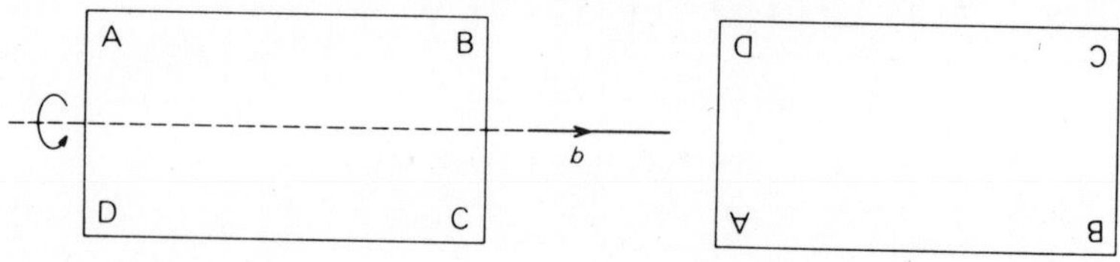

Fig. 22.3

If we rotate the rectangle through 180° about its centre we again have no apparent movement of the card but the letters have again changed position. (See Fig. 22.4.) We will call this operation c.

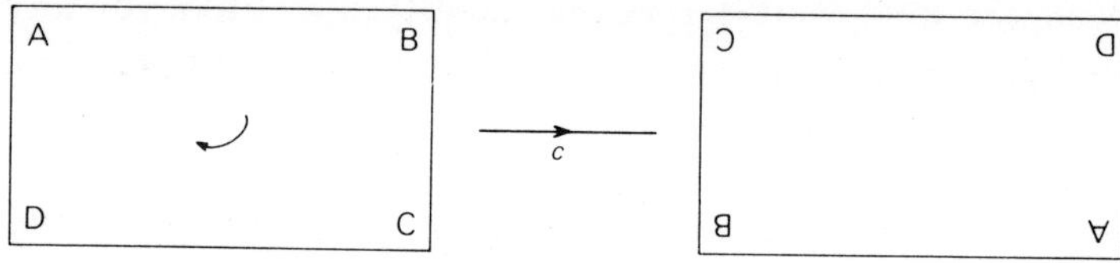

Fig. 22.4

Finally, the operation which leaves the card alone we will call e.

Let us now define the combination of two operations by the sign $\otimes$ as follows:

'$a \otimes b$' means 'do the operation b followed by the operation a'.

Figure 22.5 shows the result of this.

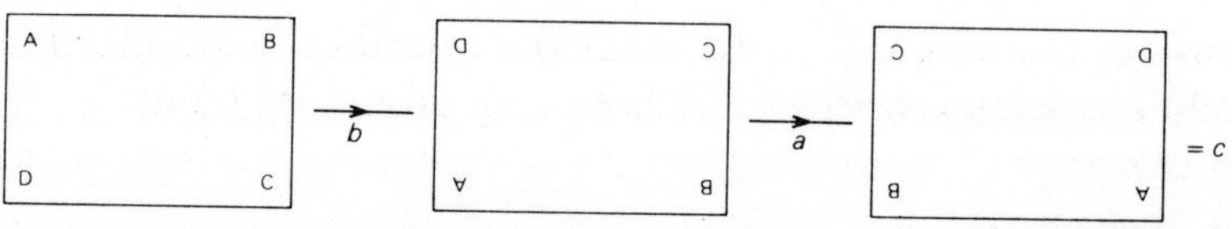

Fig. 22.5

We see that $a \otimes b = c$.

Check that $b \otimes c = a$ and $c \otimes a = b$. What do $a \otimes a$, $b \otimes b$, $c \otimes c$ equal?

Now complete the table in Fig. 22.6.

$\otimes$	e	a	b	c	
e					
a			c		$\leftarrow$ (i.e. $c = a \otimes b$; etc.)
b				a	
c		b			

Fig. 22.6

From the body of the table you should see that the combination of any two operations from the set $\{e, a, b, c\}$ also belongs to this set. This is the closure property that we met in chapter 5. In chapter 5, we also discussed associativity, identities and inverses. Is $\otimes$ associative for all the elements in the set?

i.e. does $a \otimes (b \otimes c) = (a \otimes b) \otimes c$?

Clearly in the above table e is the identity operation, for any operation preceded or followed by e remains the same operation,

i.e. $a \otimes e = a = e \otimes a$.

Looking at one of the diagonals we see that it contains the identity element,

i.e. $$\begin{aligned} e \otimes e &= e \\ a \otimes a &= e \\ b \otimes b &= e \\ c \otimes c &= e. \end{aligned}$$

So for each element in the set there exists an element (namely itself in this case) such that one operation followed by the other produces the identity. This second element is the inverse of the first element.

EXAMPLE 1

Consider the set $\{1, 2, 3, 4\}$ under the operation multiplication modulo 5. Check whether the following properties hold:

(i) closure;
(ii) associativity;
(iii) existence of identity;
(iv) existence of inverses.

In modulo arithmetic we are concerned only with the remainder on division by the modulo number; i.e. in modulo 5,

$$3 \times 4 = 2$$
$$2 \times 3 = 1, \text{ etc.}$$

The operation table is as follows:

×	1	2	3	4
1	1	2	3	4
2	2	4	1	3
3	3	1	4	2
4	4	3	2	1

(i) Clearly the operation is closed, as all the resultant elements are in the original set $\{1, 2, 3, 4\}$.
(ii) By exhaustive check, the operation is associative, i.e.

$$3 \times (4 \times 2) = 3 \times 3 = 4$$

and

$$(3 \times 4) \times 2 = 2 \times 2 = 4.$$

(iii) The identity element is 1.
(iv) The inverses of 1, 2, 3, 4 are 1, 3, 2, 4 respectively. Here, all four properties hold.

This example and the table for the symmetries of the rectangle exhibit similar properties, namely (a) closure; (b) associativity; (c) existence of an identity; and (d) existence of inverses. We define any operation on a set with the above properties to be a group.

A more formal way of defining these properties is as follows:

The set $G = \{a, b, c, \ldots\}$ with an operation $*$ is a *group* if the following are true:

1 Closure. For every element $a, b \in G$, $a * b \in G$. That is, the resultant of $a * b$ also belongs to the set.

2 Associativity. For all a, b and $c \in G$,

$$a*(b*c) = (a*b)*c.$$

3 Identity. For all $a \in G$ there is an element called the identity element e such that

$$a * e = a = e * a.$$

4 Inverse. For each element $a \in G$ there is an element $a^{-1} \in G$ such that

$$a*a^{-1} = e = a^{-1}*a.$$

EXAMPLE 2

Does the set $\{1, 2, 3, 4\}$ under the operation $\times$ form a group?

The operation table is:

$\times$	1	2	3	4
1	1	2	3	4
2	2	4	6	8
3	3	6	9	12
4	4	8	12	16

← these elements $\notin \{1, 2, 3, 4\}$

This does not form a group as the operation is not closed: e.g. $3 \times 4 = 12$ and $12 \notin \{1, 2, 3, 4\}$.

EXAMPLE 3

Does the set $\{1, -1, i, -i\}$, where $i = \sqrt{-1}$ under the operation multiplication, form a group?

The operation table is:

$\times$	1	-1	i	$-i$
1	1	-1	i	$-i$
-1	-1	1	$-i$	i
i	i	$-i$	-1	1
$-i$	$-i$	i	1	-1

$i \times i = \sqrt{-1} \times \sqrt{-1} = -1$

$i \times -i = \sqrt{-1} \times -\sqrt{-1} = 1$

Every entry in the body of the table belongs to the set. Therefore, the operation is closed. By exhaustion, the operation is associative, e.g.

$$1 \times (-1 \times i) = 1 \times -i = -i$$

and

$$(1 \times -1) \times i = -1 \times i = -i$$

The identity element is 1.

The inverses of $1, -1, i, -i$ are $1, -1, -i, i$ respectively; so the set $\{1, -1, i, -i\}$ under the operation multiplication forms a group.

In the last example, what do you notice about $1 \times -i$ and $-i \times 1$? Do all such pairs have the property that they are equal? If a group with group operation $*$ has the additional property that for all a and $b \in G$, $a*b = b*a$, then we say that it is a commutative group.

Check to see if example 1 is commutative.

EXERCISE 22a

Which of the following forms a group under the given operation?

1 $\{0, 1, 2, 3\}$ under the operation + modulo 4.

2 $\{1, 2, 3\}$ under the operation $\times$ modulo 4.

3 $\{1, 3, 5, 7, 9\}$ under the operation $\times$ modulo 10.

4 $\{1, 3, 5, 7\}$ under the operation $\times$ modulo 8.

5 $\{1, -1\}$ (i) under addition; (ii) under multiplication.

6 $\{0, 1, 2, 3, 4, 5\}$ under the operation $\times$ modulo 6.

7 $\{1, 5\}$ under the operation $\times$ modulo 10.

8 $\{1, 5, 7, 11\}$ under the operation $\times$ modulo 12.

9 $\{\emptyset, \{a\}, \{b\}, \{a, b\}\}$ under the set operation $\cap$.

10 $\{\emptyset, \{a\}, \{b\}, \{a, b\}\}$ under the set operation $\cup$.

11 $\{0, 1, 2, 3\}$ under the operation $\sim$, defined as the difference between two elements – i.e. $1 \sim 2 = 1$, $3 \sim 1 = 2$.

12 $\{E, O\}$ under addition, where E is the set of even integers and O the set of odd integers. (E + E means take any two even numbers from E and add, and find to which set the answer belongs.)

13 As question 12, but under multiplication.

14

*	*a*	*b*	*c*	*d*
a	*c*	*a*	*d*	*b*
b	*a*	*b*	*c*	*d*
c	*d*	*c*	*b*	*a*
d	*b*	*d*	*a*	*c*

15 $\left\{\begin{pmatrix} 0 & 0 \\ 0 & 0 \end{pmatrix}, \begin{pmatrix} 1 & 0 \\ 0 & 1 \end{pmatrix}\right\}$

(i) under matrix addition, (ii) under matrix multiplication.

16 $\{0, 1, 2, 3\}$ under the operation $*$ modulo 4, where

$$a * b = a^2 + b^2$$

i.e.

$$2 * 3 = 2^2 + 3^2 = 1 \text{ (mod 4)}.$$

17 Show that

$$\left\{\begin{pmatrix} 1 & 0 \\ 0 & 1 \end{pmatrix}, \begin{pmatrix} -2 & -3 \\ 1 & 1 \end{pmatrix}, \begin{pmatrix} 1 & 3 \\ -1 & -2 \end{pmatrix}\right\}$$

under matrix multiplication forms a group.

18 Does the set

$$\left\{\begin{pmatrix}1 & 0\\0 & 1\end{pmatrix}, \begin{pmatrix}-1 & 0\\0 & -1\end{pmatrix}, \begin{pmatrix}0 & -1\\1 & 0\end{pmatrix}, \begin{pmatrix}0 & 1\\-1 & 0\end{pmatrix}\right\}$$

form a group under matrix multiplication?

19

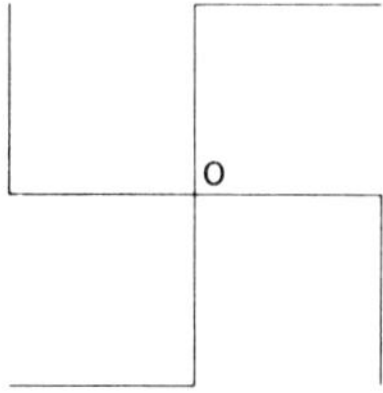

Fig. 22.7

For the swastika, denote a positive turn about O of 90° by p. Then pp or p^2 is a rotation of 180°, p^3 a rotation of 270° and $p^4 = e$ the identity. (See Fig. 22.7.)

Does the set $\{e, p, p^2, p^3\}$ under the combination of operations form a group?

20

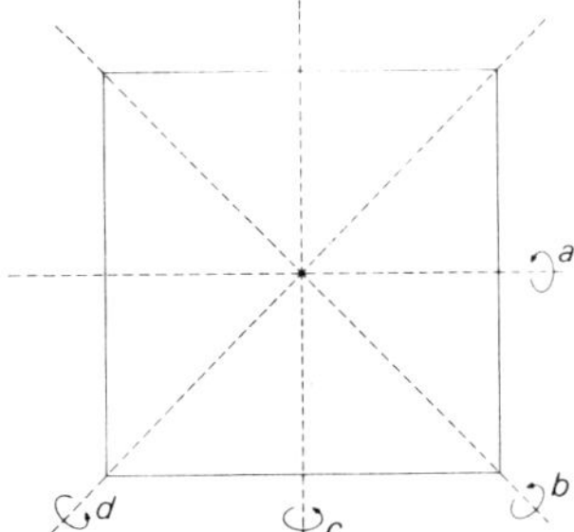

Fig. 22.8

The symmetries of the square are rotations of 90°, 180°, 270° and 360° about the centre (denoted by p, p^2, p^3 and e) and rotations of 180° about the four lines of symmetry (denoted by a, b, c, d).

Does $\{e, p, p^2, p^3, a, b, c, d\}$ form a group under the combination of operations?

21

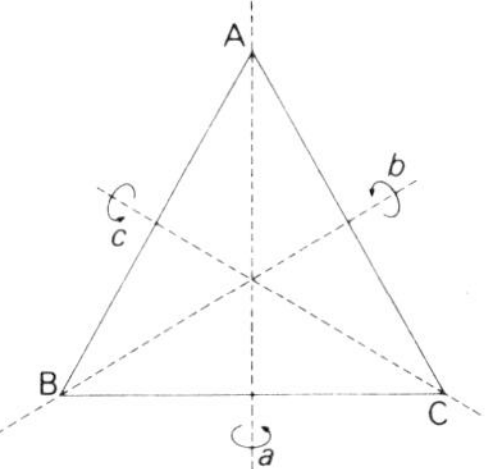

Fig. 22.9

The symmetries of the equilateral triangle are rotations of 120°, 240° and 360° about the centre (denoted by t, t^2, e), and rotations of 180° about the three lines of symmetry (denoted by a, b, c).

Show that $G = \{e, t, t^2, a, b, c\}$ forms a group under the combination of operations.

22 Which groups in the above questions are not commutative?

22.2 Subgroups

The operation for question 21 in the previous exercise is as follows:

$\circ$	e	t	t^2	a	b	c
e	e	t	t^2	a	b	c
t	t	t^2	e	c	a	b
t^2	t^2	e	t	b	c	a
a	a	b	c	e	t	t^2
b	b	c	a	t^2	e	t
c	c	a	b	t	t^2	e

What do you notice about the set $\{e, t, t^2\}$ under the operation? Does it form a group under the given operation?

This group is called a *subgroup* of G. The two conditions for a subgroup to exist are

(i) the set defined is a subset of G;

(ii) under the operation defined on G, the subset forms a group.

Is $\{a, b, c\}$ under the operation defined above a subgroup of G?

EXAMPLE 4

In example 3, is the set $\{1, -1\}$ under the operation $\times$ a subgroup?

(i) $\{1, -1\}$ is a subset of $\{1, -1, \mathrm{i}, -\mathrm{i}\}$.

(ii) Operation table:

$\times$	1	-1
1	1	-1
-1	-1	1

It is easily seen that the set $\{1, -1\}$ under the operation $\times$ forms a group. Therefore, $\{1, -1\}$ under multiplication is a subgroup of the original group.

EXERCISE 22b

1 In question 21 of the last exercise, are there any more subgroups?

2 For the symmetry group of the swastika, exercise 22a, question 19, find a subgroup.

3 Find how many subgroups there are of the group in exercise 22a, question 18.

4 Are there any subgroups of the group in exercise 22a, question 4?

5 Find all the subgroups of the group $\{0, 1, 2, 3, 4, 5\}$ under addition modulo 6.

6 For the symmetry group of the square, exercise 22a, question 20, find any subgroups.

22.3 Isomorphism

Let us look at the group tables for the following:
(i) $\{1, 3, 7, 9\}$ under multiplication mod 10.
(ii) Symmetry group for the swastika.

$\times$	1	3	7	9
1	1	3	7	9
3	3	9	1	7
7	7	1	9	3
9	9	7	3	1

(i)

$\circ$	e	p	p^2	p^3
e	e	p	p^2	p^3
p	p	p^2	p^3	e
p^2	p^2	p^3	e	p
p^3	p^3	e	p	p^2

(ii)

Do you see any resemblance between the two tables? If not, look at a rearrangement of the second table:

$\circ$	e	p	p^3 ⟷	p^2
e	e	p	p^3	p^2
p	p	p^2	e	p^3
↑ p^3	p^3	e	p^2	p
↓ p^2	p^2	p^3	p	e

(iii)

All that has been done is to exchange p^3 and p^2. Now you should see some resemblance between this table and table (i). In fact, all we have done is replaced each number by a corresponding letter in the set $\{e, p, p^2, p^3\}$; that is, there is a one-to-one correspondence between the set $\{1, 3, 7, 9\}$ and the set $\{e, p, p^2, p^3\}$, i.e.

$1 \leftrightarrow e,\ 3 \leftrightarrow p,\ 7 \leftrightarrow p^3,\ 9 \leftrightarrow p^2.$

As the two groups have the same structure they are said to be *isomorphic to* each other.

HARDER EXAMPLES 22

1 Find how many groups from the questions in exercise 22a are isomorphic to the above two groups.

2 Show that the symmetry group for the rectangle is isomorphic to the group $\{1, 5, 7, 11\}$ under multiplication mod 12.

3 The combination table for a set of 4 elements $\{a, b, c, d\}$ under the operation $*$ is given by:

$*$	a	b	c	d
a	b	c	a	d
b	c	d	b	a
c	a	b	c	d
d	d	a	d	c

(i) State the identity element.
(ii) State the inverses of a and of d.
(iii) Is the operation commutative?
(iv) By considering $a * b * d$ show that the set is not associative under $*$.

[O & C (SMP)]

4 The operations $\circ$ and $*$ are defined on the sets $\{a, b, c, d\}$ and $\{P, Q, R, S\}$ by the following tables:

$\circ$	a	b	c	d
a	b	a	d	c
b	a	b	c	d
c	d	c	a	b
d	c	d	b	a

$*$	P	Q	R	S
P	S	R	Q	P
Q	R	P	S	Q
R	Q	S	P	R
S	P	Q	R	S

(i) Show that these two systems are isomorphic. Listing a suitable one – one correspondence between the elements of the two sets will be sufficient.
(ii) Show that they are both groups by stating a geometrical group to which they are each isomorphic.
(iii) In $\{P, Q, R, S\}$, state
(a) the identity;
(b) the inverse of each element.

[O & C (SMP)]

5 The operation $*$ is defined such that $a * b = c$ where c is the remainder when the product ab is divided by 5, e.g. $2 * 4 = 3$. Copy and complete the following table for the set $S = \{1, 2, 3, 4\}$ where $2 * 4 = 3$ is shown:

$*$	1	2	3	4
1				
2				3
3				
4				

(i) Explain how it is known that S is closed under the operation $*$.
(ii) Find the inverses of 3 and 4.
(iii) If x, y are members of S, solve the equations
$3 * x = 3$
$y * y = 4$. [W]

6 The operation $*$ is defined on the set $\{2, 4, 6, 8\}$ by the relation $a * b =$ the units digit in the product $a \times b$. Construct a table which shows the values of $a * b$ for all pairs of elements in the set. Give reasons why this set, together with the given operation, is a group. (You may assume that the operation is associative.)

The above group is isomorphic to the group $\{I, A, B, C\}$ under the operation defined by the following table:

$\otimes$	I	A	B	C
I	I	A	B	C
A	A	B	C	I
B	B	C	I	A
C	C	I	A	B

Find a correspondence between the elements in $\{2, 4, 6, 8\}$ under $*$ and $\{I, A, B, C\}$ under $\otimes$. Give another table defining an operation on $\{I, A, B, C\}$ which gives a group not isomorphic to the one above. [JMB]

7 The operation $*$ is defined on the set of integers by the relation $a * b =$ the units digit in the value of $a + b + ab$.
(i) Construct a table to show the value of $a * b$ for all pairs of elements a, b belonging to the set $\{0, 2, 4, 6, 8\}$.
(ii) State how from your table it is possible to determine whether or not the operation $*$ is commutative.
(iii) Give an example to illustrate that $*$ is associative.
(iv) Give a reason why the set $\{0, 2, 4, 6, 8\}$ under $*$ is not a group. Write down a subset of $\{0, 2, 4, 6, 8\}$ containing four elements which under $*$ is a group. [JMB]

23 The calculus

23.1 Differentiation

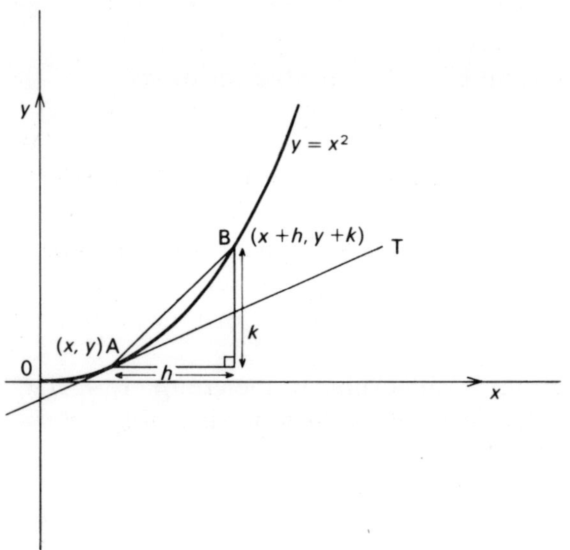

Fig. 23.1

In Fig. 23.1, A(x, y) and B$(x + h, y + k)$ are two points on the curve $y = x^2$, and AT is the tangent to the curve at A.
Since A is on the curve,

$$y = x^2. \tag{1}$$

B is also on the curve,

$$\therefore\ y + k = (x + h)^2$$
$$\text{i.e. } y + k = x^2 + 2hx + h^2. \tag{2}$$

Subtract equation (1) from equation (2):

$$k = 2hx + h^2. \tag{3}$$

Divide equation (3) by h:

$$\therefore \frac{k}{h} = 2x + h.$$

Now k/h is the gradient of AB, hence we obtain the result that the gradient of chord AB is $2x + h$.

If B is close to A, then although the length of AB becomes small, its gradient gets nearer and nearer to the gradient of the tangent at A, and if eventually, $h = 0$, then we get the result that the gradient of the tangent (i.e. the gradient of the curve), at A is given by

$$\text{gradient} = 2x.$$

The gradient of a curve is denoted by $\frac{\mathrm{d}y}{\mathrm{d}x}$ (pronounced d y by d x). Hence we have the result that if $y = x^2$,

$$\frac{\mathrm{d}y}{\mathrm{d}x} = 2x.$$

If we adopt similar techniques for $y = x^n$, we can show that

$$\frac{\mathrm{d}y}{\mathrm{d}x} = nx^{n-1}.$$

The process of finding $\frac{\mathrm{d}y}{\mathrm{d}x}$ is called *differentiation*, and nx^{n-1} is called the derivative of y with respect to x.

The letters used for the variables in this process are of course irrelevant, and if $z = t^4$, then $\frac{\mathrm{d}z}{\mathrm{d}t} = 4t^3$, and we have found the derivative of z with respect to t.

Note: (i) If $y = kx^n$, where k is a constant, then

$$\frac{\mathrm{d}y}{\mathrm{d}x} = knx^{n-1}$$

e.g. if $y = 6x^3$, then $\frac{\mathrm{d}y}{\mathrm{d}x} = 18x^2$.

(ii) If $y = kx$, where k is a constant, then since $x = x^1$,

$$\frac{\mathrm{d}y}{\mathrm{d}x} = k \times 1 \times x^0 = k.$$

This result is to be expected, since $y = kx$ is a straight line passing through the origin with gradient k.

(iii) If $y = k$, where k is a constant, then since $x^0 = 1$, we can write $y = k \times x^0$.

$$\therefore \frac{dy}{dx} = k \times 0 \times x^{-1} = 0.$$

Again, this result is to be expected, since $y = k$ is a straight line parallel to the x-axis, with zero gradient.

23.2 Polynomial functions

A polynomial function of x consists of a sum of terms, each of which is a different power of x.

e.g. $y = 3x^4 + 7x^2 - 5x + 1.$

In order to differentiate such a function, you simply differentiate each term and add the results. Hence in the above example,

$$\frac{dy}{dx} = 12x^3 + 14x - 5.$$

23.3 Derivative of x^n when n is not a positive integer

It can be proved that the rule for differentiating $y = x^n$ applies if n is not a positive integer.

EXAMPLE 1

Differentiate with respect to x:

(i) $x^{2/3}$ (ii) $\frac{2}{x^2}$ (iii) $3x^{0.15}$

(i) If $y = x^{2/3}$, $\frac{dy}{dx} = \frac{2}{3}x^{-1/3}$.

(ii) If $y = \frac{2}{x^2}$, then $y = 2x^{-2}$. Hence $\frac{dy}{dx} = -4x^{-3}$.

(iii) If $y = 3x^{0.15}$, $\frac{dy}{dx} = 0.45x^{-0.85}$.

Here is a more difficult example, which still uses the one simple rule.

EXAMPLE 2

Differentiate with respect to x:

(i) $2x^4 - x^2 + \frac{3}{x} + \sqrt{x}$ (ii) $(x+5)^3$ (iii) $\frac{x^2+1}{x^2}$

(i) If $y = 2x^4 - x^2 + \frac{3}{x} + \sqrt{x}$, first of all rewrite this as

$$y = 2x^4 - x^2 + 3x^{-1} + x^{1/2}$$

then

$$\frac{dy}{dx} = 8x^3 - 2x - 3x^{-2} + \tfrac{1}{2}x^{-1/2}.$$

(ii) If $y = (x+5)^3$, expand this first:

$$\begin{aligned} y &= (x+5)(x+5)^2 = (x+5)(x^2+10x+25) \\ &= x^3 + 10x^2 + 25x + 5x^2 + 50x + 125 \\ &= x^3 + 15x^2 + 75x + 125. \end{aligned}$$

$$\begin{aligned} \frac{dy}{dx} &= 3x^2 + 30x + 75 = 3(x^2 + 10x + 25) \\ &= 3(x+5)^2. \end{aligned}$$

(iii) $y = \frac{x^2+1}{x^2}$ can be rewritten as

$$y = \frac{x^2}{x^2} + \frac{1}{x^2} = 1 + x^{-2}.$$

$$\therefore \quad \frac{dy}{dx} = -2x^{-3}$$

EXERCISE 23a

Differentiate the following functions with respect to x:

1 x^4

2 $3x^5$

3 $\frac{4}{x^2}$

4 $x^2 + 9$

5 $x^3 - 7x + 1$

6 $x^2 + 3x + 5$

7 $2x^4 - x^2 + 1$

8 $8x^3 - 3x$

9 $7x^6 + 5x^2 - 4$

10 $6x - \frac{2}{x^2}$

11 $\sqrt{x}$

12 $6x^{0.4}$

13 $x^{2/3} + x^{1/3}$

14 $\frac{1}{\sqrt{x}}$

15 $x(x+2)$

16 $(x+1)(x-1)$

17 $(x^2+1)^2$

18 $\dfrac{x^2+1}{x}$

19 $\dfrac{x+\sqrt{x}}{\sqrt{x}}$

20 $\dfrac{(x+1)^2}{x}$

21 $(x^2-1)(x^2+1)$

22 $\dfrac{(x-2)(x+1)}{x^2}$

23 $\dfrac{(x^2-1)^2}{x^2}$

24 $\left(x^2+\dfrac{1}{x}\right)(x^2+x)$

25 $(3x^2+4)^2$

23.4 Equation of a tangent to a curve

EXAMPLE 3

Find the equation of the tangent to the curve $y = x^3 - 5x^2 + 8x$ at the point (1, 3) and also the points on the curve where the tangent is parallel to the line $y = 5x + 2$.

(i) $\dfrac{dy}{dx} = 3x^2 - 10x + 8.$

If $x = 1$,

$$\frac{dy}{dx} = 3 - 10 + 8 = 1.$$

Using the equation of a straight line as $y = mx + c$, then

$$m = 1$$
$$\therefore \quad y = x + c.$$

But this tangent passes through (1, 3).

$\therefore \quad 3 = 1 + c \Rightarrow c = 2$

$\therefore$ the equation of the tangent is $y = x + 2$

(ii) the gradient of $y = 5x + 2$ is 5, hence we want $\dfrac{dy}{dx} = 5$.

$$3x^2 - 10x + 8 = 5$$
$$\therefore \quad 3x^2 - 10x + 3 = 0$$
$$\text{i.e. } (3x-1)(x-3) = 0.$$
$$\therefore \quad x = 3 \Rightarrow y = 3^3 - 5 \times 3^2 + 8 \times 3 = 6$$

(substituting into the original equation)

and $x = \frac{1}{3} \Rightarrow y = (\frac{1}{3})^3 - 5 \times (\frac{1}{3})^2 + 8 \times \frac{1}{3}$

$$= \frac{58}{27}.$$

$\therefore$ the two points are $(3, 6)$ and $\left(\frac{1}{3}, \frac{58}{27}\right)$.

23.5 Maximum and minimum values

Turning points on a curve are given the names maximum and minimum values as shown in Fig. 23.2.

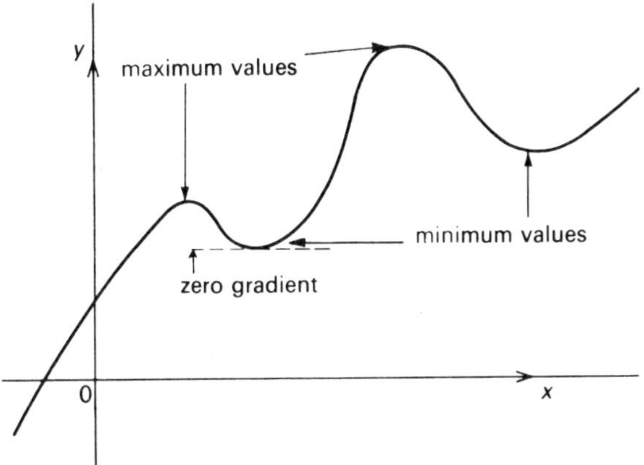

Fig. 23.2

Note that it is possible for a minimum value to be greater than a maximum value. At these points, the curve has zero gradient.

$$\text{i.e. } \frac{dy}{dx} = 0$$

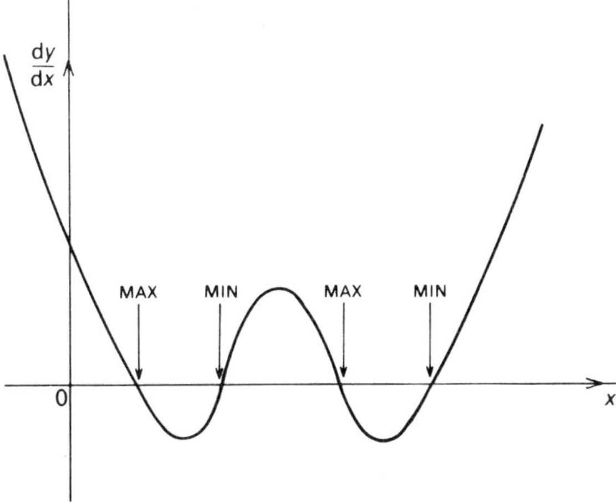

Fig. 23.3

Figure 23.3 shows the graph of $\frac{dy}{dx}$ against x, taken from the values of Fig. 23.2. The gradient of this graph is denoted by $\frac{d^2y}{dx^2}$ (read: d 2 y by d x squared).

It is the second derivative of y with respect to x, and is obtained by differentiating $\frac{dy}{dx}$. It can be seen that at a maximum,

$$\frac{d^2y}{dx^2} < 0,$$

and at a minimum

$$\frac{d^2y}{dx^2} > 0.$$

EXAMPLE 4

Find the maximum and minimum values of $y = 2x^3 - 15x^2 + 36x + 6$, and sketch the graph.

$$\frac{dy}{dx} = 6x^2 - 30x + 36 = 6(x^2 - 5x + 6)$$
$$= 6(x-3)(x-2) = 0 \text{ if } x = 2, 3$$
$$\frac{d^2y}{dx^2} = 12x - 30 = 24 - 30 = -6 \text{ at } x = 2 \text{ MAXIMUM}$$
$$= 36 - 30 = +6 \text{ at } x = 3 \text{ MINIMUM}$$

If $x = 2$, $y = 2 \times (2)^3 - 15 \times (2)^2 + 36 \times (2) + 6 = 34$.
If $x = 3$, $y = 2 \times (3)^3 - 15 \times (3)^2 + 36 \times (3) + 6 = 33$.
$\therefore$ the maximum point is (2, 34),
the minimum point is (3, 33).

A sketch of the graph is shown in Fig. 23.4.

EXERCISE 23b

Find all the maximum and minimum values of the following functions and distinguish between them. Sketch the curve in each case.

1 x^2
2 $x^2 + 5$
3 $2x^2 - 7$
4 $x^2 + 5x + 6$
5 $2 - x^2$
6 $x - x^3$
7 $x^3 + x$
8 $x + \frac{1}{x}$
9 $(2x-3)(x+4)$
10 $x^4 - 2x^2 + 1$

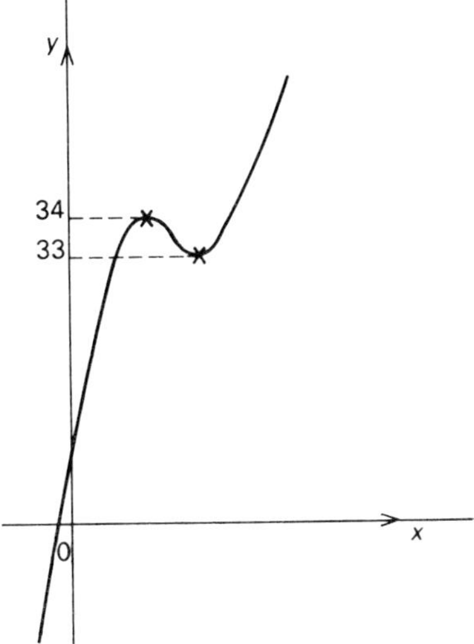

Fig. 23.4

23.6 Velocity and acceleration

In sections 3.8 and 3.9, we dealt with distance–time and speed–time graphs, composed of straight lines. Suppose now, the graphs are curved.

Since average speed $= \dfrac{\text{distance } (s)}{\text{time } (t)}$, this is just the slope of the line

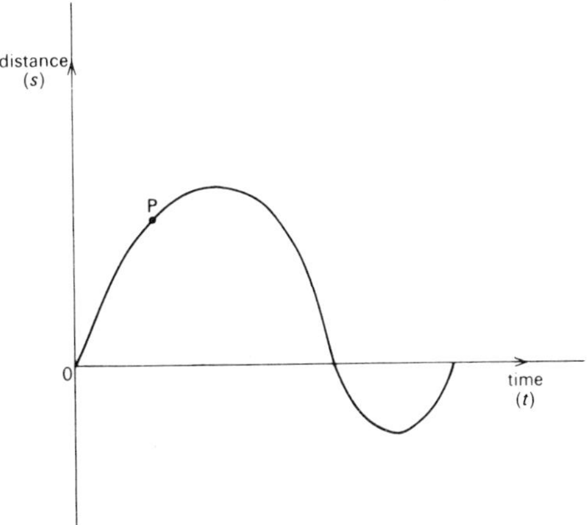

Fig. 23.5

if it is straight, i.e. constant speed. If the speed is variable as in Fig. 23.5, then the gradient of the curve at P, i.e. $\frac{ds}{dt}$, gives us the speed at that point. If $\frac{ds}{dt}$ is negative, then the body is moving in the opposite direction, hence $\frac{ds}{dt}$ gives the velocity (speed and direction).

We now define

$$\text{average acceleration} = \frac{\text{increase in velocity}}{\text{time}}.$$

If the graph is not a straight line, as in Fig. 23.6, we can find the acceleration at Q by finding the gradient:

i.e. $\frac{dv}{dt} = \frac{d^2s}{dt^2}$.

If $\frac{d^2s}{dt^2}$ is negative, then the speed must be decreasing, hence the body is decelerating. The units of acceleration are m/s^2.

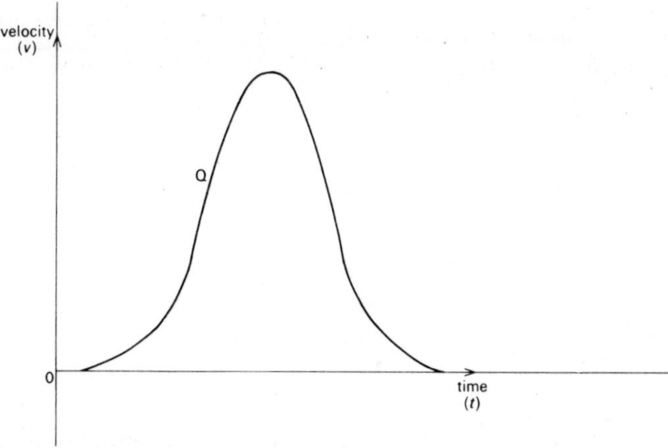

Fig. 23.6

EXAMPLE 4

A body leaves a point O, moving in a straight line so that t seconds later its displacement s m from O is given by $s = 2t^3 - 3t^2 - 36t$. *Note*: Since s measures the distance from O, it could easily be negative, and strictly speaking, s gives the displacement measured from O, rather than just the distance. See Fig. 23.5.

Calculate:

(i) The values of t when the velocity is zero.

$$\text{velocity} = \frac{\mathrm{d}s}{\mathrm{d}t} = 6t^2 - 6t - 36.$$

$$\therefore \text{ velocity zero} \Rightarrow 6t^2 - 6t - 36 = 0$$
$$\text{i.e. } 6(t^2 - t - 6) = 0$$
$$\therefore 6(t+2)(t-3) = 0.$$
$$\therefore t = 3 \text{ or } -2.$$

The value $t = -2$ has no meaning in this problem, hence the velocity is zero after 3 seconds.

(ii) The initial velocity of the body.

$$\text{If } t = 0, \frac{\mathrm{d}s}{\mathrm{d}t} = -36.$$

Hence the body starts by moving backwards at a speed of 36 m/s.

(iii) The acceleration when $t = 1$.

$$\frac{\mathrm{d}^2s}{\mathrm{d}t^2} = 12t - 6 = 6 \text{ at } t = 1.$$

$\therefore$ at $t = 1$ the acceleration $= 6\ \mathrm{m/s^2}$.

(iv) The total distance travelled in the first 4 seconds. Care has to be taken here, as the body comes to rest after 3 seconds.
When $t = 3$, $s = 2 \times 3^3 - 3 \times 3^2 - 36 \times 3 = -81$.
When $t = 4$, $s = 2 \times 4^3 - 3 \times 4^2 - 36 \times 4 = -64$.

$\therefore$ in the third second, distance travelled $= 81 - 64 = 17$ m.
$\therefore$ the total distance travelled in the first 4 seconds is $81 + 17 = 98$ m.

Finally, Fig. 23.7 shows the distance–time and velocity–time graphs for the body in the first six seconds.

EXAMPLE 5

The velocity v m/s of a body moving in a straight line through O, t seconds after passing through O is given by $v = 6t - 9t^2$. Find the distance from O after $\frac{1}{2}$ second.

Since $v = \frac{\mathrm{d}s}{\mathrm{d}t}$, we have

$$\frac{\mathrm{d}s}{\mathrm{d}t} = 6t - 9t^2.$$

Integrating (see section 23.7) gives

$$s = 3t^2 - 3t^3 + C,$$

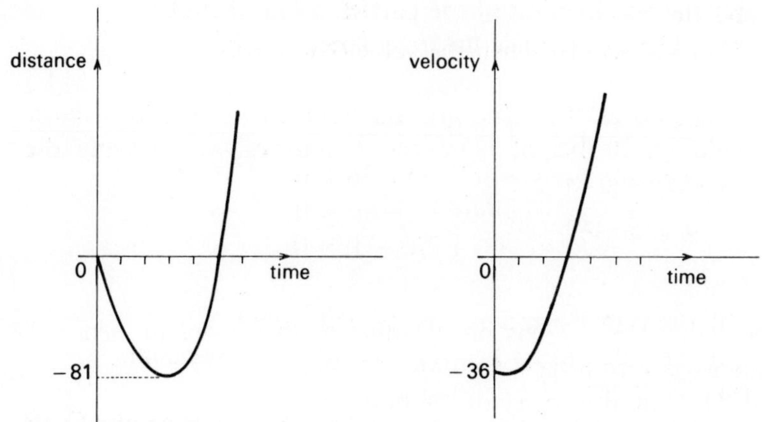

Fig. 23.7

but when $t = 0$, $s = 0$ $\therefore C = 0$.
Hence

$$s = 3t^2 - 3t^3.$$

When $t = \frac{1}{2}$, $s = 3 \times \frac{1}{4} - 3 \times \frac{1}{8} = \frac{3}{8}$.
$\therefore$ the distance from O is $\frac{3}{8}$ m.

EXERCISE 23c

1 The distance, s metres, from a given point, of a car after travelling for a time t seconds is given by the equation $s = 3t^2 - t + 1$. Calculate the value of t when the speed is 11 m/s, and find its position at that time.

2 A stone is thrown into the air, and its height h metres above the ground after t seconds is given by $h = 10t - 5t^2$.
Find:
(i) the velocity after t seconds
(ii) the initial velocity
(iii) the greatest height reached
(iv) the time that elapses before the stone returns to the ground.

3 O and P are points on a line. A particle moves along the line in such a way that, t seconds after it is at O, its velocity is v cm/s, where $v = kt - t^2$ and k is a constant. At the time when $t = 6$, the particle is momentarily at rest at P.
Find:
(i) the value of k
(ii) the distance OP
(iii) the average speed of the particle between O and P

(iv) the acceleration of the particle when it is at P
(v) when the particle has zero acceleration.

[L]

4 A lift moves from the ground floor to the top of a large building. Its height h metres, t minutes after leaving the ground, is given by:

$$h = 40(t^2 - \tfrac{1}{6}t^3).$$

Find:
(i) the velocity and acceleration at time t.
(ii) the value of t when the velocity is a maximum.
(iii) the height of the building.

5 A particle is moving along a straight line through a point O. Its displacement s cm from O after t seconds is given by:

$$s = 4t^3 - 12t^2 + 9t.$$

Find:
(i) the values of t when the velocity is zero
(ii) the initial velocity
(iii) the acceleration after 2 seconds
(iv) the distance travelled during the second second after passing O.
(v) the time that elapses before the particle returns to O.
Sketch the velocity–time graph for the first six seconds of the motion.

23.7 Indefinite integration

If $\dfrac{dy}{dx} = x^n$, what function y when differentiated gives this answer?

By inspection

$$y = \frac{x^{n+1}}{n+1} + C,$$

where C is an unknown constant. The result can be checked by differentiating y. We say that we have integrated x^n with respect to x.

We write this as: $$\underbrace{\int x^n dx}_{} = \frac{x^{n+1}}{n+1} + C$$

'the integral of x^n with respect to x is'

Because C is unknown, this is called an *indefinite integral*.

Note: This rule works for all values of n except $n = -1$: $\int \frac{1}{x}\,dx$ cannot be done at this stage.

EXAMPLE 5

Integrate the following functions with respect to x:

(i) $(x^2+1)^2$ (ii) $\dfrac{2x+x^3}{\sqrt{x}}$

(i) $$\int (x^2+1)^2 dx = \int x^4+2x^2+1\,dx = \frac{x^5}{5}+\frac{2x^3}{3}+x+C.$$

(ii) $$\int \frac{2x+x^3}{\sqrt{x}}\,dx = \int 2x^{1/2}+x^{5/2}\,dx = \frac{4x^{3/2}}{3}+\frac{2x^{7/2}}{7}+C.$$

23.8 Definite integration

$\int_1^2 x^2+3\,dx$ is called a *definite integral.*

The numbers at the top and bottom of the integral are called the limits. The upper limit need not be numerically greater than the lower limit.

$$\int_1^2 x^2+3\,dx = \left[\frac{x^3}{3}+3x\right]_1^2$$

substitute upper limit for x substitute lower limit for x

$$= \left[\frac{2^3}{3}+3\times 2\right] - \left[\frac{1^3}{3}+3\times 1\right]$$

$$= 8\tfrac{2}{3} - 3\tfrac{1}{3} = 5\tfrac{1}{3}.$$

Note that $+C$ does not appear in the answer. The process is self-explanatory.

23.9 Area under a curve

In order to find the area under a curve, the definite integral can be used.

In Fig. 23.8, in order to find the area shaded, we proceed as follows:

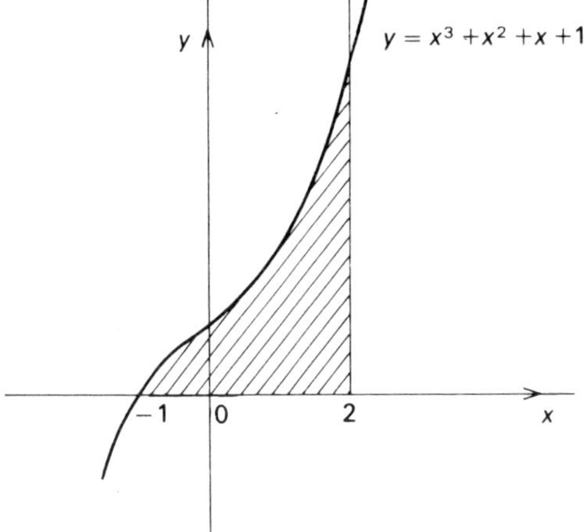

Fig. 23.8

$$\text{shaded area} = \int_{-1}^{2} x^3 + x^2 + x + 1 \, dx = 11\tfrac{1}{4} \text{ sq. units.}$$

If the area is negative, then the curve is below the x-axis.

In order to find the area in Fig. 23.9, two integrals must be found.

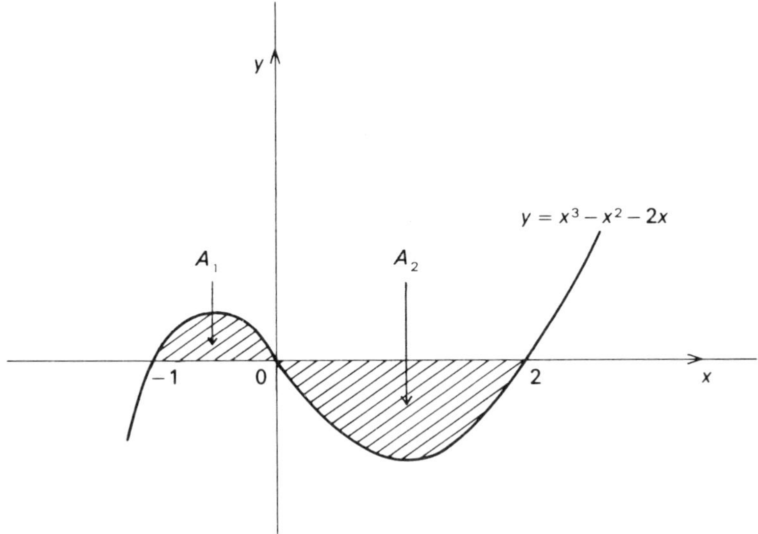

Fig. 23.9

$$A_1 = \int_{-1}^{0} x^3 - x^2 - 2x \, dx = \frac{5}{12} \text{ sq. units}$$

$$A_2 = \int_0^2 x^3 - x^2 - 2x \, dx = -2\tfrac{2}{3} \text{ sq. units.}$$

Hence,

$$\text{the total shaded area} = \frac{5}{12} + 2\tfrac{2}{3} = 3\tfrac{1}{12} \text{ sq. units.}$$

We can summarise this section by saying that the area under the curve $y = f(x)$ between $x = a$ and $x = b$, is given by

$$\text{area} = \int_a^b y \, dx, \text{ where } y \text{ is replaced by } f(x).$$

23.10 Volume of revolution

If the areas in the previous section had been rotated by 360° about the x-axis, a solid would have been generated. We can find the volume of that solid, by using the formula

$$\text{volume} = \pi \int_a^b y^2 \, dx.$$

It is not necessary to worry about whether the area is above or below the x-axis. Hence the volume of the solid generated in Fig. 23.9, is given by:

$$\text{volume} = \pi \int_{-1}^2 (x^3 - x^2 - 2x)^2 dx.$$

Now

$$\begin{aligned}&(x^3 - x^2 - 2x)(x^3 - x^2 - 2x)\\ &= x^6 - x^5 - 2x^4 - x^5 + x^4 + 2x^3 - 2x^4 + 2x^3 + 4x^2\\ &= x^6 - 2x^5 - 3x^4 + 4x^3 + 4x^2.\end{aligned}$$

Hence,

$$\begin{aligned}\text{volume} &= \pi \int_{-1}^2 x^6 - 2x^5 - 3x^4 + 4x^3 + 4x^2 \, dx\\ &= 4\tfrac{22}{35}\pi \text{ cubic units.}\end{aligned}$$

23.11 Area and volume with the *y*-axis

The area between a curve and the y-axis is given by:

$$\text{area} = \int_{y=a}^{y=b} x\,\mathrm{d}y$$

The volume when this area is rotated by 360° about the y-axis is given by:

$$\text{volume} = \pi \int_{y=a}^{y=b} x^2\,\mathrm{d}y.$$

Questions will not be set in this book which specifically require these formulae, and they are stated here for reference only.

EXERCISE 23d

1 Integrate with respect to x, the functions in questions 1–17 of exercise 23a.

2 Evaluate the following definite integrals:

(i) $\int_1^2 x^2 + 2\,\mathrm{d}x$

(ii) $\int_0^1 x^3 - 4\,\mathrm{d}x$

(iii) $\int_{-1}^1 x^2 + x + 1\,\mathrm{d}x$

(iv) $\int_{-1}^2 x + \frac{1}{x^2}\,\mathrm{d}x$

(v) $\int_1^2 x(x+2)\,\mathrm{d}x$

(vi) $\int_{-1}^0 x(x-1)\,\mathrm{d}x$

(vii) $\int_2^3 (x^2+1)^2\,\mathrm{d}x$

(viii) $\int_{-1/2}^{1/2} \frac{x^2+1}{x^2}\,\mathrm{d}x$

3 Find for the following curves, the area between the given values of x, the curve and the x-axis.

(i) x^2, between $x = -2$ and $x = 1$
(ii) $8 - x^2$, between $x = 0$ and $x = 2$
(iii) $x(x+4)$, between $x = -4$ and $x = 0$
(iv) $\frac{1}{x^2}$, between $x = \frac{1}{2}$ and $x = 10$
(v) $x^2 - 1$ between $x = 0$ and $x = 2$.

4 Find the volumes obtained when the areas in question 3 are rotated by 360° about the x-axis.

HARDER EXAMPLES 23

1 (a) Calculate the gradient of the tangent at P (2, −4), on the curve $y = 3x^2 - 8x$, and hence find the equation of the tangent at P.
(b) On the curve $y = x^3 - 6x^2 + 13$, calculate the maximum and minimum points, carefully distinguishing between them.
[W]

2 (a) (i) If $y = x^2 - x - 5$, write down the expression for $\frac{dy}{dx}$ and hence find the values of x for which $y = \frac{dy}{dx}$.
(ii) Evaluate $\int_{-1}^{2} (4 - x^2)dx$ and show clearly in a neat sketch the area represented by this integral.
(b) A lift moves from rest at ground level to rest at the top of a building. When it has been moving for t minutes its height, h metres, above the ground is given by $h = 20(t^2 - \frac{1}{3}t^3)$.
(i) Obtain expressions for its velocity and acceleration in terms of t.
(ii) Calculate the value of t when the velocity is a maximum.
(iii) Find the value of t when the lift comes to rest at the top of the building and hence calculate the height of the building.
[NI]

3 A particle moves in a straight line so that its distance, s cm, from a fixed point O is given after time t seconds by the equation $s = t^3 + t^2 - 8t + 7$.
Calculate:
(i) the distance from O after 2 seconds;
(ii) the velocity after 1 second;
(iii) the initial acceleration;
(iv) the time when the particle is at rest instantaneously.
[W]

4 The distance, s metres, from a given point, of a car after travelling for a time t seconds is given by the equation $s = 5t^2 - 30t + 2$. Calculate the value of t when the speed of the car is 10 metres per second.
[L]

5 A particle leaves a point O and moves in a straight line so that t seconds later its displacement from O is $t^3 - 6t^2 + 9t$ cm.

(i) Calculate the values of t when the velocity is zero;
(ii) Calculate the initial velocity of the particle;
(iii) Sketch the velocity–time graph for $0 \leq t \leq 4$;
(iv) Calculate the acceleration when $t = 2$.
(v) Calculate the distance travelled by the particle during the third second after leaving O;
(vi) State the value of t at which the particle again passes through O.

[JMB]

6 A particle moves in a straight line such that its velocity v cm/s is given after time t seconds by the equation $v = 10 + 3t - t^2$. Calculate:
(i) the initial velocity;
(ii) the acceleration after 4 seconds;
(iii) the time when the particle is at rest instantaneously;
(iv) the distance travelled in the third second. Explain the sign of your answer in part (ii).

[JMB]

7 (a) Integrate $9.6x^7 - 6x^3 + 5$ with respect to x.
(b) If $\int_{-1}^{a} (3x-2)^2 \, dx = 4a + 13$, find the two possible values of a.

[NI]

8 Calculate the coordinates of the points on the graph of $y = 12x - x^3$ at which the tangents are parallel to the x-axis.
(a) Sketch the graph of $y = 12x - x^3$.
(b) Calculate the area bounded by the curve, the x-axis and the ordinates $x = 1$ and $x = 3$.

[L]

9 Find the coordinates of the points of intersection of the straight line whose Cartesian equation is $y = x$ and the curve whose Cartesian equation is $y = 3x - x^2$.
The function f is defined for all real numbers x satisfying $0 \leq x \leq 3$ by

$$f : x \to \begin{cases} x & \text{for } 0 \leq x \leq 2 \\ 3x - x^2 & \text{for } 2 < x \leq 3. \end{cases}$$

(i) Sketch the Cartesian graph of the function;
(ii) Calculate the area of the region enclosed by this graph and the x-axis;
(iii) Given that the line $x = c$ divides this region into two regions of equal area, find the value of c correct to two decimal places.

10 The curve whose equation is $y = 2x - x^2$ intersects the x-axis at the origin and again at P.
(a) Calculate the x-coordinate of P;
(b) Sketch the graph of $y = 2x - x^2$;
(c) Calculate the coordinates of the point on the curve at which the gradient equals 1;
(d) The area bounded by the curve and the x-axis is rotated completely about the x-axis. Calculate the volume of the solid of revolution generated.

[L]

24 Test papers

I

1 Evaluate 6.09×1.07 exactly.
2 Evaluate $31.86 \div 4.9$ correct to two decimal places.
3 If $n(X) = 18$, and $n(X \cup Y) = 25$, what is the smallest value of $n(Y)$? Illustrate with a Venn diagram.
4 Find the matrix product $\begin{pmatrix} 3 & 1 & 0 \\ 2 & -1 & 6 \\ 0 & 1 & 0 \end{pmatrix} \begin{pmatrix} 1 \\ 1 \\ 2 \end{pmatrix}$.
5 What is the probability of drawing a picture card from a pack of 52 cards?
6 If $p * q = p^2 + q^2$, find $(1 * 2) * 3$.
7 Arrange in order of increasing magnitude $\frac{3}{7}, \frac{5}{9}, \frac{4}{11}$.
8 Solve, for x, $3x - 8 = x + 16$.
9 If $f: x \rightarrow 3x^2 - 1$, find (i) $f(0)$; (ii) $f(-2)$.
10 If $A = \{(x, y): x > -1, y > -2, x + y < 2\}$, illustrate this region on a diagram, and find its area.

II

1 Evaluate $1\frac{3}{5} + 2\frac{1}{2}$.
2 What is the probability of obtaining a score of 4 or more with the throw of a die?
3 If $a * b = a - 3b$, find $2 * 1$.
4 If $A = \begin{pmatrix} 1 & 0 \\ 2 & 1 \end{pmatrix}$ and $B = \begin{pmatrix} 3 & -1 \\ 1 & 0 \end{pmatrix}$, find $2A + 3B$.
5 Arrange in order of size, smallest first, the fractions $\frac{3}{4}, \frac{7}{9}, \frac{3}{5}$.
6 Find $2.86 + 1.34 - 2.66$.
7 If $A = \{2, 3, 5, 6\}$ and $B = \{4, 6, 9, 3\}$, find $A \cap B$.
8 Find 15 % of £24.

9 Find x if $x+8 = 3x-4$.
10 Simplify $2a+3b-4a+b$.

III

1 Find $1\frac{2}{5} \times 3\frac{1}{7}$.
2 Three coins are tossed at the same time; what is the probability of obtaining three heads?
3 If $a * b = a^2-2b$, find $1 * -1$.
4 If $X = \begin{pmatrix} 1 & 3 \\ 4 & 2 \end{pmatrix}$ and $Y = \begin{pmatrix} 1 & -1 \\ 2 & 0 \end{pmatrix}$, find XY.
5 Find the average of the numbers $2\frac{1}{2}$, $3\frac{3}{4}$, $7\frac{1}{2}$, $9\frac{1}{4}$.
6 How many prime numbers are there between 9 and 32?
7 How many axes of symmetry has a square?
8 What is the smallest whole number which is exactly divisible by 3, 5 and 12?
9 If $y = at+bt^2$, find y when $a = 1$, $b = -1$ and $t = 4$
10 Express $\frac{7}{8}$ as a percentage.

IV

1 Find 38_9+16_9, giving your answer in the octal scale.
2 Is the average of two odd numbers always an odd number?
3 Subtract $3x-y$ from $x+2y$.
4 ABC is a triangle, and $\angle ABC = 90°$. If BC = 3 cm and AB = 5 cm, find $\angle CAB$.
5 Draw a Venn diagram illustrating the relationship between three sets A, B and C if (i) $A \subset B$, (ii) $A \cap C = \emptyset$, (iii) B and C have some elements in common.
6 Draw a quadrilateral which has one axis of symmetry.
7 If $x = \dfrac{a}{b+c}$, find x when $a = b = c$.
8 If $A = \{x: x \text{ is an integer and } 1\frac{1}{2} \leq x < 4\frac{3}{4}\}$, find $n(A)$.
9 Divide £20 into two parts which are in the ratio 2:3.
10 Find x if $2+3(x+1) = 2(x+5)$.

V

1 What is the surface area of a cube of side 8 m?
2 If $f: x \rightarrow x^2-1$, find f(3).

3 If $a * b$ is the largest of a and b, find $-1 * (2 * 3)$.

4 A shopkeeper sells a fridge for £78. If he prices his goods so as to make a profit of 30 %, find how much he paid for the fridge.

5 Use tables to find 38.7×19.4, giving your answer correct to two significant figures.

6 If $C = 2\pi r$, express r in terms of C and π.

7 In triangle RST, $\angle S = 90°$. If RS = 12 cm and ST = 5 cm, find RT.

8 If $\frac{x^2}{2} = \frac{4}{x}$, find x.

9 Use logarithms to find $86.5 \div 312$.

10 If $2(x-1) < 3x+5$, which of the following statements is true?
(i) $x < -7$ (ii) $-x < 7$ (iii) $x > 7$.

VI

1 Simplify $\frac{7}{8} \div \frac{3}{4}$.

2 Simplify $2x^2 + 2x + x^2 - 5x + x(2-x) + 7$.

3 Express 1111_2 in the octal scale.

4 What can you say about x, if $2x+5 > 5x+17$?

5 What is the mean of the numbers 906, 905, 957, 935, 913, 904?

6 Solve, for x, $\frac{x}{2} = \frac{8}{x}$.

7 Write $2 \times 10^3 + 3 \times 10^4$ as a single number, using standard form for the answer.

8 A box measuring 60 cm × 32 cm × 25 cm is filled with small metal bars, whose measurements are 5 cm × 6 cm × 8 cm. What is the maximum number of bars that can be placed in the box?

9 If $x * y = x^y + y^x$, evaluate $2 * 3$.

10 Use logarithms to find $(0.00623)^2 \div 0.0009$.

VII

1 Which number must be removed from the following to make the mean of those remaining 9?
11, 15, 2, 18, 14, 3, 6.

2 If $A = \begin{pmatrix} 2 & 0 \\ -3 & -1 \end{pmatrix}$ and $B = \begin{pmatrix} 1 & 2 \\ 1 & 3 \end{pmatrix}$, find BA.

3 If $f: x \to 1 + x + x^2$, find (i) $f(0)$; (ii) $f(-1)$.

4 If $a * b = \frac{a+b}{2}$, find $1 * (2 * 3)$.

5 Solve, for x, $x+2(x+1) = 11$.

6 The longest side of a rectangle is 6 cm. The length of the longest side of a similar rectangle is 72 cm. If the area of the larger rectangle is 288 cm^2, what is the area of the smaller?

7 Find $1011_2 + 111_2 + 101_2$ in base 2.

8 What is the probability when throwing two dice, of an even number appearing on either die?

9 Simplify $x^2+3x+x(x-1)-5$.

10 If $A = \{x : x \text{ is an odd number and } 18 < x \leq 31\}$, what is $n(A)$?

VIII

1 What number must be added to 8, 7, 3, 9, 12, 15 to make the new average 10?

2 If $A = \begin{pmatrix} 3 & 1 \\ 2 & 6 \end{pmatrix}$ and $B = \begin{pmatrix} -1 & 2 \\ -3 & 0 \end{pmatrix}$, find AB.

3 If $f : x \to 2-3x^2$, find (i) f(0); (ii) f(−2).

4 If $a * b = \sqrt{a^2+b^2}$, find $(3 * 4) * 12$.

5 Solve, for x, $2x+3(x-1) = 7x+3$.

6 The height of a box is 2 cm and its volume is 10 cm^3. The height of a similar box is 3 cm. What is its volume?

7 Find $10111_2 \times 101_2$ in base 2.

8 What is the probability of scoring more than 9 with two dice?

9 Simplify $x^2+2x(x-2)+3x-5$.

10 If $A = \{x : x \text{ is a prime number and } 30 < x < 40\}$, list the elements of A.

IX

1 Find the mean of 328, 328, 309, 384, 361.

2 Evaluate $0.017 \div 0.0031$ correct to two decimal places.

3 Use logarithms to find 816.5×3.16 correct to 3 significant figures.

4 If $A = \begin{pmatrix} 3 & 1 \\ 2 & 6 \end{pmatrix}$ and $B = \begin{pmatrix} 1 & -1 \\ 5 & -2 \end{pmatrix}$, find AB − 2BA.

5 If $f : x \to 1-x^2$, find (i) f(−1); (ii) f(3).

6 If $n(A) = 15$, $n(B) = 8$, and $n(A \cap B) = 7$, what is $n(A \cup B)$?
7 If $x * y = 2x + y$, find $1 * (2 * 3)$.
8 Sketch on a diagram the region
$\{x, y): x + y > 2, x \geq 1 \text{ and } y \geq 0\}$.
9 Solve, for x, $3(x + 1) + 1 = 2x - 5$.
10 Simplify $a(b + 2a) - b(a - b)$.

X

1 Evaluate 3.03×2.53 exactly.
2 Evaluate $2.16 \div 1.62$ correct to two decimal places.
3 If $n(X \cup Y) = 25$, $n(X \cap Y) = 5$ and $n(Y) = 14$, draw a Venn diagram to illustrate these data and find $n(X)$.
4 Evaluate the matrix product $\begin{pmatrix} 2 & 1 \\ 3 & -1 \end{pmatrix}\begin{pmatrix} 1 \\ 3 \end{pmatrix}$.
5 What is the probability of scoring 3 or more with a throw of a die?
6 If $p * q = \sqrt{pq}$, find $24 * (4 * 9)$.
7 State a fraction between $\frac{1}{2}$ and $\frac{4}{9}$.
8 Solve, for m, $2m - 7 = 3$.
9 $(2, 1)$, $(2, -1)$, $(-2, -1)$, $(-2, 1)$ are the vertices of a quadrilateral. What is its area?
10 Prove that, if $0 < x < 90°$ and $\sin x = \frac{5}{11}$, then $\cos x < 2\sin x$.

XI

1 What is the interest on £250 at $7\frac{1}{2}\%$ for 1 year 3 months?
2 Express 1050 as a product of prime factors.
3 Express $\dfrac{5.83 \times 10^2}{4.78 \times 10^{-4}}$ in standard form.
4 Given that $A = 2\pi r(r + h)$, express h in terms of A, π and r.
5 Factorise $x^2 - 7x + 10$. Find the range of values of x for which $x^2 - 7x + 10 > 0$.
6 Given that y varies directly as x, and $y = 3\frac{1}{2}$ when $x = 2\frac{1}{2}$, find x when $y = 5\frac{1}{2}$.
7 ABC is a right-angled triangle with $\angle B = 90°$ and $\sin A = \frac{5}{13}$. Find cos A and tan B.
8 The points $(-1, 1)$, $(1, 0)$ and $(4, k)$ lie on a straight line. Find k.
9 Solve the equation $(x - 3)^2 - 11 = 0$.
10 If $5 : y = 9 : 15$, find y.

XII

1 Find the largest integral value of x such that $5-2x > x+8$.

2 Given that the mean of the set $\{5, 7, x, 3, 6, 11\}$ lies between 4.8 and 7.8, find the limits for the value of x.

3 If $f: x \rightarrow 3x - \frac{2}{x} + 1.5$, evaluate (i) f(5); (ii) f(−2).

4 The volume of a cone is $\frac{1}{3}\pi r^2 h$. If the radius is increased by 20%, what is the percentage increase in volume?

5 Given that x belongs to the set of positive integers, F(x) denotes the sum of the odd integers less than or equal to x, e.g.

F(8) = 1 + 3 + 5 + 7 = 16.

(i) Find F(11).

(ii) Give possible values for x such that F(x) = 81.

6 Find a and b such that $a\begin{pmatrix}3\\2\end{pmatrix}+b\begin{pmatrix}-1\\4\end{pmatrix}=\begin{pmatrix}7\\0\end{pmatrix}$.

7 A chord of a circle of radius 5 cm is 8 cm long. Find the angle subtended by this chord at the centre of the circle.

8 Make d the subject of the formula $c = 1 + \frac{1}{d}$.

9 If $\tan\theta = \frac{5}{12}$, find, without using tables, (i) $\sin\theta$; (ii) $\cos\theta$.

10 Write down the median of the following set of numbers: $\{44, 57, 18, 37, 58, 71\}$.

XIII

1 What is $12\frac{1}{2}$% as a fraction?

2 Solve, for x, $\frac{1}{3}(2x-1)+\frac{2}{3} = x$.

3 Evaluate $\sqrt[3]{(0.053)}$.

4 Find the mean of the following set of numbers:

$\{5, 5, 5, 6, 7, 8, 9, 9, 9, 10, 11, 11\}$.

5 What is the number-scale of the correct multiplication $332 \times 14 = 11303$?

6 Find the value of a and of b if $\begin{pmatrix}1 & a\\3 & 2\end{pmatrix}\begin{pmatrix}b\\-1\end{pmatrix}=\begin{pmatrix}-3\\4\end{pmatrix}$.

7 If $6 \leq M \leq 9$ and $4.3 \leq N \leq 4.6$, calculate the limits between which $M \times N$ must lie.

8 Factorise $x^2+8x-20$.

9 The line joining the points A(3, a), B($5-a$, 8) has a gradient of $\frac{1}{2}$. Find a.

10 Simplify $\dfrac{1}{x^2+3x+2}-\dfrac{1}{x+1}$.

XIV

1 What is $\dfrac{1.3}{2}$ as a percentage?

2 What is the number-scale of the correct division $414 \div 15 = 22$?

3 If $x \times \frac{3}{4} \times \frac{5}{8} \times 6 \times \frac{8}{9} \times 2 \times \frac{1}{3} = \frac{3}{2}$, what is x?

4 Use factors to find the exact value of

$$\frac{15.9^2 - 2.1^2}{13.8 \times 4.6 + 13.8 \times 5.4}.$$

5 Make x the subject of the formula $y = p(t - zx)$.

6 The interior angles of a pentagon are $(x+30)^\circ$, $(2x+70)^\circ$, $(x+45)^\circ$, $(170-2x)^\circ$ and $(3x+40)^\circ$. Find x.

7 The mean of the following set of numbers is $7: \{6, y, 7, x, 9\}$. Express y in terms of x.

8 If $f: x \to x^2 - \dfrac{1}{x^2}$, find f(2) and f(−4).

9 Given that the points (2, 5) and (−1, 8) lie on the line $y = mx + c$, find m and c.

10 If $P = \{x : x$ is a prime factor of $210\}$, and $Q = \{x : 10 - x$ is a prime number$\}$, write down the value of (i) $n(P)$; (ii) $n(P \cap Q)$.

XV

1 Find the cost of 35 litres of petrol at 8.2p per litre.

2 Calculate exactly $\dfrac{0.84}{0.36} \times 1.08$.

3 Express $(5^3)^2$ as a power of 5.

4 If

$$P = \begin{pmatrix} 0 & -1 \\ -1 & 0 \end{pmatrix} \text{ and } Q = \begin{pmatrix} 1 & 0 \\ 2 & 1 \end{pmatrix},$$

calculate the matrix $\mathbf{M} = PQP$. State what geometrical transformation $\mathbf{M}$ represents.

5 Use tables to evaluate 0.0235×0.587. Write your answer in standard form.

6 The set $A = \{1, 2, 3, 4\}$. Write down all possible subsets of A.

7 $f: x \to x^2 + x + 1$. Find (i) f(2); (ii) f(−1); (iii) f(0).

8 Solve the simultaneous equations $x+2y=7$, $x-3y=12$.

9 If two fair coins are tossed together, what is the probability of obtaining (i) two tails; (ii) no tails?

10 Sketch the region $y > x+4$.

XVI

1 Factorise a^3-4ab^2.

2 What is 15 % of £6.50?

3

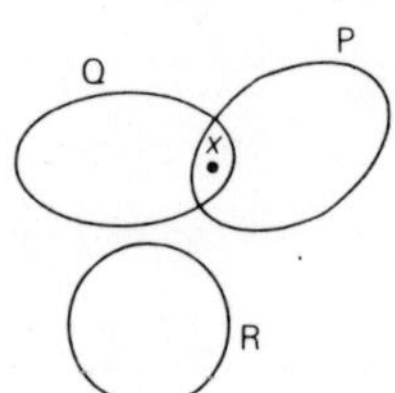

Fig. 24.1

Which of the following statements about the sets in Fig. 24.1 are true?

(i) $x \in R$;

(ii) $x \subset P$;

(iii) $x \in Q$;

(iv) $P \cap Q \subset P$;

(v) $P \cap R = \emptyset$.

4 Write the following as a power of 2:

(i) 32×512;

(ii) $2^{12} \times 2^7 \div 2^{14}$.

5 Factorise $3a^2+2a$.

6 Find $\begin{pmatrix} 1 & 6 \\ 2 & 5 \end{pmatrix}\begin{pmatrix} 3 \\ -4 \end{pmatrix}$.

7 How many $5\frac{1}{2}$p stamps can be bought for £1, and how much change will be given?

8 Solve, for x, $\frac{1}{2}(x+5) = 17$.

9 In triangle ABC, angle A $= 45°$, angle B $= 90°$ and AC $= 4$ cm. Find AB.

10 What is the median of the following set of numbers? $\{2, 7, 6, 5, 3, 2, 1, 6\}$.

XVII

1 Use tables to evaluate $(0.562)^{1/3}$.

2 Simplify $81^{3/4}$

3

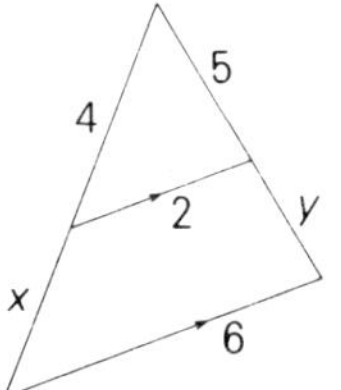

Fig. 24.2

Find x and y using the information in Fig. 24.2.

4 Factorise $a^2b^3 - a^3b^2$.

5 Given that log 3 = 0.4771, find, without using tables, log 27.

6 An article is sold for £37.80, making a profit of $12\frac{1}{2}\%$. What was the cost price?

7 If $a = 5$ and $b = -4$, find the value of (i) $2a - 5b$; (ii) $\sqrt{(3.6a + 3.5b)}$.

8 If θ is an acute angle and $\sin\theta = \frac{4}{5}$, what is the value of $\tan\theta$?

9 Draw a Venn diagram, and shade the region $A \cup (B \cap C)$.

10 The average height of 6 boys is 1.8 m and the average height of 9 girls is 1.5 m. Calculate the average height of the group.

XVIII

1 Evaluate 31.25×0.073 (i) correct to 3 decimal places; (ii) correct to 3 significant figures.

2 If $\dfrac{1.08}{x} = 9.6$, find the value of x.

3 Find the values of (i) sin 145° (ii) cos 315°; (iii) cos 110°.

4 Factorise $\pi R^2 - \pi r^2$.

5 If A = {odd numbers less than 20}, B = {all prime numbers}, C = {all multiples of 3}, find (i) $A \cap B$; (ii) $A \cap C$; (iii) $A \cap B \cap C$.

6 What percentage is 27 of 450?

7 If $A = \begin{pmatrix} 1 & 3 \\ 2 & -1 \end{pmatrix}$, find A^2.

8 Give the co-ordinates of any two points in the region defined by $x + y > 6$.

9 Find the circumference of a circle with radius 5 cm. ($\pi = 3.14$.)

10 Prove that $a^3 - b^3 = (a+b)(a^2 - ab + b^2)$.

XIX

1 Given that $\begin{vmatrix} a & b \\ c & d \end{vmatrix}$ denotes $ad-bc$, solve the equation

$$\begin{vmatrix} x & -7 \\ x & x \end{vmatrix} = -12.$$

2 What regular solid has 4 vertices and 6 edges?

3 Find the value of x such that (i) $3^x = \frac{1}{27}$; (ii) $8^x = 16$.

4 Find the area of the triangle with sides 17 cm, 7 cm and 15 cm.

5 In triangle ABC, angle A $= 110°$, angle C $= 37°$ and AB $= 5.2$ cm. Find the other measurements of the triangle.

6 If $T = \{x : x$ is a square number less than $100\}$, list the members of the intersection of the set T with the following sets:

(i) {even numbers};

(ii) $\{x : x^3 < 200\}$.

7 Find the height of a flagstaff which is observed by a boy of height 1.7 m from a distance of 20 m at an angle of elevation of 24°.

8 Show that the matrices

$$A = \begin{pmatrix} 1 & 0 \\ 0 & 1 \end{pmatrix}, B = \begin{pmatrix} 0 & 1 \\ -1 & 0 \end{pmatrix}, C = \begin{pmatrix} -1 & 0 \\ 0 & -1 \end{pmatrix},$$

$$D = \begin{pmatrix} 0 & -1 \\ 1 & 0 \end{pmatrix}$$

form a group under matrix multiplication.

9 Draw the lines and then shade the area to show:

$x \geq 0$, $y \geq 0$, $x+2y \leq 7$, $y-3x \leq 1$.

Find the maximum value of $x+y$.

10 Draw a sketch, showing the position of the line $5-7x-3y = 0$.

XX

1 Solve the equation $\dfrac{2}{x} = \dfrac{3}{4}$.

2 Write the number 27_8 in binary.

3 A rectangle has one side three times the other and its area is 300 cm^2. If the shorter side is decreased by 10% and the longer side increased by 20 %, find the percentage change in the area of the rectangle.

4 Estimate the value of $12.13 \div 3.04$ and use your result to estimate the value of $(12.13 \times 10^4) \div (3.04 \times 10^{-5})$.

5 If $L = \{(x, y): 2x - y = 3\}$, and $M = \{(x, y): 4x + y = 9\}$, find $L \cap M$.

6 Find x if $\begin{pmatrix} 3 & 0 \\ 4 & -1 \end{pmatrix} \begin{pmatrix} x \\ x^2 \end{pmatrix} = \begin{pmatrix} 12 \\ 0 \end{pmatrix}$.

7 If $a * b$ denotes 1 less than the average of a and $2b$, find $-1 * 4$.

8 If $\mathscr{E} = \{1, 2, 3, 4, 5, 6\}$, $A = \{1, 3, 6\}$ and $B = \{2, 4, 5\}$, find $A' \cap B$.

9 If $by + c = y$, find y in terms of b and c.

10 Express $3.4 \times 10^{-3} + 5.8 \times 10^{-4}$ in standard form.

XXI

1 O is the origin of co-ordinates, A is the point (2, 2) and B is the point (3, −3). Show that $\angle AOB = 90°$ and find the radius of the circle which passes through A, O and B.

2 Evaluate $1101_2 \times 111_2$ in base 2.

3 If $\cos x = \frac{3}{5}$ and x is an acute angle, draw an appropriate right-angled triangle, and use it to find $\tan x$.

4 A man borrows £500 from a finance company. The interest on the loan is charged at 20 % and calculated at the beginning of the year. If he repays £11.50 per month, calculate how much he owes at the beginning of the second year after the interest has been added on.

5 If $a * b = a^b$, find $25 * \frac{1}{2}$.

6 Find the smallest angle of the triangle which has sides of length 8 cm, 8 cm and 10 cm.

7 The curve $y = 2ax - b$ passes through the points (1, −2) and (0, 1). Find a and b.

8 Solve the equation $x^2 - x = 0$.

9 If $A = \begin{pmatrix} 1 & 2 \\ -3 & 4 \end{pmatrix}$, find the inverse of A.

10 If $g(x) = 3x^2 - 4x + 1$, find $g(-1)$ and $g(0)$.

XXII

1 Draw on a diagram the region represented by $A = \{(x, y): x+y \leq 5, x \geq 2, y \geq -1\}$.

2 If the mean of the numbers 4, 5, a, b is 10, find the mean of the numbers 7, 8, a, b.

3 Find the co-ordinates of the image of the point $(2, -1)$ after reflection in the line $x = -2$.

4 A car travels 40 km in $\frac{3}{4}$ hour and then travels at a constant speed of 40 km/h for a further $\frac{1}{4}$ hour. What is its average speed for the whole journey?

5 If $A = \begin{pmatrix} 3 & 6 \\ 0 & -1 \end{pmatrix}$ and $B = \begin{pmatrix} 2 & 7 \\ 4 & 6 \end{pmatrix}$, find AB^2.

6 Solve the simultaneous equations $2x-3y = 7$, $x+5y = 4$.

7 A rectangle measures 8 cm by 4 cm. If the length of the longer side is increased by 10 %, and the length of the shorter side is decreased by 10 %, find the percentage change in the area of the rectangle.

8 If $\mathscr{E}$ = {quadrilaterals}, A = {squares}, B = {rhombuses} and C = {rectangles}, draw a Venn diagram illustrating the relationship between these sets.

9 If $x = p^2 - q^2$, find x when $p = 8561$ and $q = 8560$.

10 Use logarithms to evaluate $\dfrac{85.9 \times 31.7}{86.2}$.

XXIII

1 Evaluate in the octal scale $57_8 - 65_8 + 120_8$.

2 Three sets A, B and C satisfy the following conditions: $A \subset B \cup C$, $A \cap (B \cap C) \neq \emptyset$ and $A \cup (B \cap C) \subset B$. Draw three Venn diagrams to show the possible relationships between A, B and C.

3 If $x = 2\sqrt{ab^2 + c}$, find an expression for b in terms of a, c and x.

4 Solve the simultaneous equations $3x + 4y - 2 = 0$, $1 - x - y = 0$.

5 Triangles ABC and DEF are congruent. Show that it is not always possible to transform one triangle into the other by means of a single rotation.

6 If $a = 2 \times 10^3$ and $b = 3 \times 10^{-2}$, evaluate a/b^2, giving your answer in the form $A \times 10^n$ where A is correct to two decimal places and $1 \leq A < 10$.

7 Use logarithms to evaluate $3.807 \times (2.53)^2$.

8 The volume of a cone is given by the formula $V = \frac{1}{3}\pi r^2 h$, where r is the radius of the base and h is the vertical height. A block of metal of volume 100 m^3 is to be melted down and used to make a number of cones of height 0.5 m and base radius 0.21 m. What is the largest number of such cones that can be made? (Take $\pi = \frac{22}{7}$.)

9 Describe the effect that the matrix $\begin{pmatrix} 2 & 0 \\ 0 & 1 \end{pmatrix}$ will have on any figure.

10 Sketch the graph of the following function:

$$\begin{aligned} f(x) &= x^2 & -1 \le x \le 1 \\ &= 2x - 1 & x \ge 1 \\ &= x + 2 & x \le -1. \end{aligned}$$

XXIV

1 Given A = $\{a, b, c\}$ and X = {all subsets of A including A and $\emptyset$}, find the value of $n(X)$.

2 Find the image of the point $(-3, -2)$ under a rotation of $180°$.

3 If $h(x) = x^2 - x - 1$, find $h(-1) + h(1)$.

4 Evaluate the matrix product $\begin{pmatrix} 3 & 0 \\ -4 & 4 \end{pmatrix}\begin{pmatrix} 2 & -1 \\ -1 & 0 \end{pmatrix}$.

5

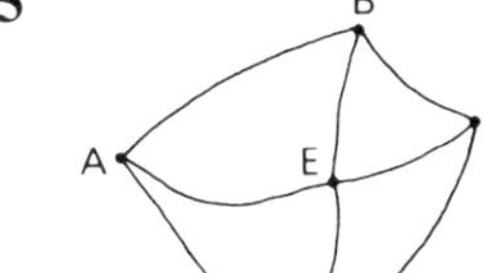

Fig. 24.3

In how many different routes can you get to A from C (in Fig. 24.3) without passing through the same point twice? What is the probability that one of these routes chosen at random passes through E?

6 If $a * b$ means $a^2 + b^2$ and $3 * x$ is a perfect square, what is x?

7

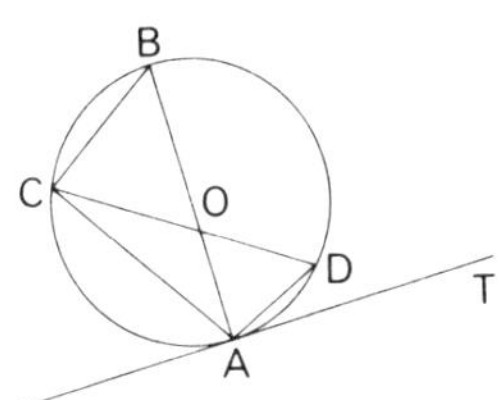

Fig. 24.4

In Fig. 24.4, O is the centre of the circle, and AT is a tangent. If $\angle BCD = 35°$, find $\angle DAT$.

8 Solve the equations $2x-5y = 8$, $x-y = 4$.

9 Use logarithms to calculate

$$\frac{23.6}{15.2 \times 0.013}.$$

Without further calculation, state the value of

$$\frac{236}{152 \times 13}.$$

10 Solve the equation $14x^2+29x = 15$.

XXV

1 If $\begin{pmatrix} 1 & 2 \\ 3 & 5 \end{pmatrix}\begin{pmatrix} a \\ 1 \end{pmatrix} = \begin{pmatrix} 6 \\ 17 \end{pmatrix}$, find a.

2 If $c^2 = a^2+b^2$, find b when $a = \sqrt{3}$ and $c = \sqrt{7}$.

3 Factorise $x^2-5x-14$.

4 ABC is a triangle, with $\angle B = 90°$. Find AC if $\angle A = 27°$ and AB = 7 cm.

5 If $\mathbf{a} = \begin{pmatrix} 3 \\ -2 \end{pmatrix}$ and $\mathbf{b} = \begin{pmatrix} -1 \\ 2 \end{pmatrix}$, find $2\mathbf{a}-3\mathbf{b}$.

6 An article costs £81. If this cost is increased by 20%, find the new cost.

7 Write down the smallest denary number which has 4 digits when expressed in the binary scale.

8 If $n(A) = 20$, $n(A \cup B) = 30$ and $n(A \cap B) = 7$, find $n(B)$.

9 Taking $\pi = \frac{22}{7}$, find the area of a circle of diameter 49 cm.

10 A television set costs £84 to produce. The breakdown of costs is represented using a pie chart. If transport costs amount to £7, what will be the angle of this sector?

Answers

1 The language of sets

Exercise 1

1 (i) $\{1, 2, 3, 4\}$ **(ii)** $\emptyset$ **(iii)** A **(iv)** $\{1, 2, 4, 5\}$ **2 (i)** $\{a, e\}$ **(ii)** $\{a, b, c, d, e, f, h\}$ **(iii)** $\{a, c, d, e, f, h\}$ **(iv)** $\{a, c, f\}$ **3 (i)** $\{3, 5, 6\}$ **(ii, iii)** $\{5, 6\}$ **(iv)** $\{6\}$ **4 (i)** $\{d, f, h\}$ **(ii)** $\{b, e\}$ **(iii)** $\{d, h\}$ **7 (i)** $\{2, 3, 5, 7\}$ **(ii)** $\{1, 3\}$ **(iii)** $\{3\}$ **(iv)** $\{4\}$ **8 (i)** 2, 10 **(ii)** 53, 97 **(iii)** 6, 111111 **(iv)** 10, 17
9 (i) $\{1, 2, 3\}$ **(ii)** $\{1, 2, 3, 4\}$
10 3 **11** 22 **13** 12 **14** 0 **15** 20, 37
16 8, 15; 15 **19** $\{a\}, \{b\}, \{c\}, \{a, b\}, \{a, c\}, \{b, c\}, \{a, b, c\}, \emptyset$
20 (iii) **21** $\subset$, $\not\supset$, $\subset$, $\supset$, $\not\supset$

Harder examples 1

1 $D' \cap R'$ **4 (i)** $\{2, 4, 6, 7\}$ **(ii)** $\{1, 3, 6\}$ **(iii)** $\{2, 5, 6, 8\}$ **(iv)** $\{2, 4\}$ **(v)** $\{2, 5, 7, 8\}$ **(vi)** $\emptyset$ **(vii)** $\{1, 2, 3\}$
5 (i) 4 **(ii)** 8 **(iii)** 16 **(iv)** 2^n
6 20 **10 (i)** 7 **(ii)** 15 **(iii)** 19
11 (a) (i) $\{5, 10, 15, 20\}$, $\{5, 10, 20\}$, $\{7, 11, 13, 17, 19\}$
(ii) $\{5, 7, 10, 11, 13, 15, 17, 19\}$, $\emptyset$, $\{5, 10, 15, 20\}$ **(b) (i)** 9 **(ii)** 24 **12** 520, 20

2 The number system

Exercise 2a

1 $\frac{1}{9}$ **2** $\frac{3}{16}$ **3** $\frac{3}{16}$ **4** $\frac{2}{7}$ **5** $\frac{1}{5}$ **6** $\frac{9}{20}$ **7** $\frac{4}{11}$ **8** $\frac{5}{9}$
9 $\frac{6}{41}$ **10** $\frac{1}{2}$ **11** $\frac{11}{90}$ **12** $\frac{5}{9}$
13 $\frac{7}{11}$ **14** $\frac{3}{7}$ **15** $\frac{5}{6}$ **16** $\frac{4}{5}$ **17** $\frac{11}{13}$ **18** $\frac{5}{6}$ **19** $\frac{3}{4}$
20 $\frac{77}{117}$ **21** $3\frac{1}{4}$ **22** $2\frac{3}{8}$ **23** $5\frac{2}{7}$
24 $5\frac{9}{11}$ **25** $8\frac{1}{2}$ **26** $3\frac{2}{5}$ **27** $9\frac{3}{4}$ **28** $11\frac{4}{7}$
29 $7\frac{4}{9}$ **30** $16\frac{1}{2}$ **31** $\frac{45}{8}$ **32** $\frac{17}{7}$ **33** $\frac{25}{6}$
34 $\frac{42}{11}$ **35** $\frac{121}{8}$

Exercise 2b

1 60 **2** 72 **3** 36 **4** 60 **5** 63 **6** 36
7 60 **8** 112 **9** 240 **10** 2310

Exercise 2c

1 $1\frac{5}{12}$ **2** $1\frac{9}{40}$ **3** $\frac{11}{36}$ **4** $\frac{1}{2}$ **5** $\frac{11}{100}$ **6** $\frac{5}{9}$
7 $\frac{1}{2}$ **8** $\frac{13}{36}$ **9** $\frac{5}{16}$ **10** $\frac{11}{192}$ **11** $4\frac{1}{4}$ **12** $6\frac{1}{6}$
13 $3\frac{16}{21}$ **14** $8\frac{13}{14}$ **15** $13\frac{19}{40}$ **16** $4\frac{5}{8}$ **17** $11\frac{7}{8}$
18 $1\frac{1}{2}$ **19** $7\frac{11}{35}$ **20** $\frac{25}{33}$ **21** $4\frac{9}{20}$
22 $16\frac{3}{8}$ **23** $8\frac{11}{40}$ **24** $5\frac{7}{8}$ **25** $21\frac{2}{15}$
26 $40\frac{3}{56}$ **27** $\frac{111}{136}$ **28** $\frac{43}{55}$ **29** $103\frac{16}{45}$ **30** $3\frac{35}{36}$

Exercise 2d

1 $1\frac{5}{21}$ **2** $1\frac{11}{36}$ **3** $1\frac{34}{63}$ **4** $\frac{3}{10}$ **5** $\frac{9}{14}$
6 $4\frac{3}{4}$ **7** $6\frac{31}{40}$ **8** $4\frac{11}{40}$ **9** $14\frac{17}{48}$ **10** $\frac{20}{29}$
11 $\frac{7}{30}$ **12** $\frac{29}{30}$ **13** $5\frac{1}{32}$ **14** $21\frac{3}{8}$ **15** $1\frac{1}{8}$
16 $\frac{39}{40}$ **17** $3\frac{1}{3}$ **18** $\frac{1}{6}$ **19** $\frac{3}{8}$ **20** $1\frac{11}{24}$
21 $4\frac{23}{28}$ **22** $\frac{3}{4}$ **23** $22\frac{1}{2}$ **24** $86\frac{5}{8}$
25 $107\frac{5}{8}$ **26** $\frac{7}{72}$ **27** $\frac{22}{63}$ **28** $\frac{2}{7}$ **29** $16\frac{1}{8}$
30 $2\frac{3}{23}$

Exercise 2e

1 -16 **2** -6 **3** 13 **4** 4 **5** -14
6 -16 **7** 29 **8** -46 **9** 2 **10** 22
11 -13 **12** -134 **13** 54 **14** 105
15 12 **16** -20 **17** -24 **18** -6
19 -192 **20** 100 **21** $-1\frac{1}{8}$ **22** -1
23 -1 **24** $-\frac{7}{8}$ **25** $1\frac{5}{12}$ **26** $-\frac{2}{3}$
27 $9\frac{1}{8}$ **28** $-1\frac{8}{35}$ **29** $-4\frac{1}{10}$ **30** $1\frac{3}{10}$
31 $7\frac{11}{30}$ **32** $18\frac{1}{2}$ **33** $-9\frac{19}{35}$ **34** $-\frac{5}{8}$ **35** $16\frac{3}{8}$
36 $6\frac{17}{56}$ **37** $22\frac{1}{4}$ 38 $-\frac{61}{66}$
39 $-13\frac{7}{8}$ **40** -7

Exercise 2f

1 $7\frac{7}{8}$ **2** $-\frac{3}{20}$ **3** $-18\frac{3}{8}$ **4** -3 **5** $-17\frac{7}{8}$
6 $-2\frac{1}{7}$ **7** $-3\frac{9}{32}$ **8** $\frac{5}{11}$ **9** $-4\frac{2}{3}$ **10** $1\frac{3}{32}$
11 $\frac{9}{16}$ **12** $-\frac{1}{32}$ **13** $-\frac{1}{42}$ **14** $-1\frac{57}{160}$
15 $-3\frac{3}{4}$ **16** $4\frac{4}{5}$ **17** -8 **18** $\frac{5}{12}$ **19** $-1\frac{1}{2}$
20 -10

Exercise 2g

1 $6\sqrt{2}$ **2** $7\sqrt{6}$ **3** $\sqrt{2}$ **4** $5\sqrt{10}$ **5** 0
6 $\sqrt{7}$ **7** $5\sqrt{5}$ **8** $-\sqrt{11}$ **9** $6\sqrt{7}$
10 $10\sqrt{2}-2\sqrt{3}$ **11** $\sqrt{2}/2$ **12** $\sqrt{2}/2$
13 $\sqrt{3}$ **14** $\sqrt{6}/6$ **15** $2\sqrt{2}$ **16** $\sqrt{5}/5$
17 $\sqrt{5}/25$ **18** $\sqrt{10}/2$ **19** $\sqrt{2}/6$
20 $-\sqrt{2}/2$

Exercise 2h

1 85 **2** 207 **3** 561 **4** 39 **5** 29
6 3240 **7** 708 **8** 6481 **9** 42 **10** 51
11 1127 **12** 823 **13** 4088 **14** 3339
15 71 **16** 1078 **17** 2248 **18** 31
19 3575 **20** 6220

Exercise 2i

1 106 **2** 135 **3** 11010 **4** 212 **5** 537
6 11110 **7** 2222 **8** 1304 **9** 1153
10 1404 **11** 14 **12** 32 **13** 100 **14** 222
15 333 **16** 10 **17** 1202 **18** 17 **19** 54
20 14

Exercise 2j

1 1067 **2** 2225 **3** 1111001 **4** 1215
5 35552 **6** 100101001 **7** 31 **8** 33
9 101 **10** 421 **11** 77 **12** 111

Exercise 2k

1 99.4 **2** 9.10 **3** 0.094 **4** 6.0 **5** 6.01
6 4.00 **7** 9330 **8** 5000000 **9** 5.87
10 0.01 **11** 4.0 **12** 100.0 **13** 0.375
14 0.5 **15** 0.571428 **16** 0.27 **17** 0.416
18 0.875 **19** 0.4375 **20** 0.40625

Harder examples 2

1 (i) 12 (ii) $2\frac{1}{2}$ **2** c **5** (i) ± 3
(ii) $\pm\sqrt{\frac{1}{2}}$ **6** 441/8 **8** (i) 6 (ii) 3
(iii) 8 **9** 7 **10** (i) 17, 18, 19 (ii) 19
11 $(9\sqrt{5}+5\sqrt{3})/4$ **12** $1\frac{2}{5}$, $\sqrt{2}$, 1.42
13 119/15 **14** 18 **15** (i) Yes, Yes
(iii) No, Yes (iii) No, No **16** $2t+1, 2t+2$

3 The metric system

Exercise 3a

1 (i) 210 (ii) 58.6 (iii) 20000 (iv) 6 (v) 0.13
(vi) 95600000 (vii) 2000 (viii) 98.5
2 (i) 0.00856 (ii) 0.09 (iii) 0.00890
(iv) 0.000865 (v) 0.000012 (vi) 0.000203
3 (i) 5000 (ii) 80 (iii) 0.560 (iv) 25000
(v) 0.0063 (vi) 850000 **4** (i) 82000
(ii) 68.7 (iii) 260000000 **5** (i) 0.8 (ii) 40000
(iii) 1000000000000 **6** (i) 0.12
(ii) 0.00005 (iii) 1

Exercise 3b

110.7, 19.32, 4, 78.9, 2.5, 80, 22.08, 0.00129, 126000

Exercise 3c

1 (i) 32.31 (ii) 11.19 (iii) 0.34 (iv) 16.67
(v) 19.05 (vi) 46.64 (vii) 1.84 (viii) 380.95
(ix) 4.68 (x) 4.48 **2** (i) 64.32 (ii) 31.5
(iii) 1023.6 (iv) 139175 (v) 80.47 (vi) 9904
3 (i) 1276.19 (ii) 775.44 (iii) 163.82
(iv) 4883.33 (v) 115.49

Exercise 3d

1 2.8×10^3 **2** 6×10^6 **3** 3.028×10^3
4 2.94×10^2 **5** 3.723×10^7 **6** 3.1×10^{-3}
7 2.7×10^{-2} **8** 1.65×10^{-1} **9** 1
10 1.2×10^{-4} **11** 2.883×10^4 **12** 4×10^{-7}
13 1.9×10^7 **14** 1.1×10^4 **15** 2.81×10^5
16 3.74×10^9 **17** 6.78×10^3
18 1.1×10^{-1} **19** 3.36×10^{-1}
20 4.4×10^{-5}

Exercise 3e

1 (i) 50 km (ii) $37\frac{1}{2}$ km/h **2** $51\frac{2}{3}$ km/h, $51\frac{2}{3}$ km **3** $3\frac{1}{3}$ s **4** £8 **5** £2.27 **6** 10
7 £106.60 **8** £117.10 **9** 187, 40
10 £110, 63p

Harder examples 3

1 £15.96 **2** 80 m **3** (i) 3 min (ii) 3 min
(iii) 120 km/h **4** £7.07 **5** 39 km/h
6 $7.44 \times 10^5, -2.08 \times 10^5$
7 £72000, £240, 50p **8** (i) £202.70
(ii) £40.80 (iii) 20.1 %
9 9.46×10^{12} km, 2.36×10^{14} km
10 −0.6 % **11** 81p **12** 7 km
(i) 9.15 a.m. (ii) 3 km

4 Indices and logarithms

Exercise 4a

1 a^{11} **2** b^{-1} **3** $12c^5$ **4** $2d^3$ **5** $27e^6$
6 $9f^2$ **7** g **8** $\frac{4}{3}h^{-1}$ **9** $\frac{5}{4}j^{-8}$ **10** k^2
11 121^{-6} **12** $-1728\,m^3$ **13** $-\frac{4}{3}n^5$
14 $1/3p^2$ **15** $3p^2$ **16** $\sqrt{2}r^4$ **17** $2s^{2/3}$
18 $16t^2$ **19** u^3 **20** $\frac{1}{4}v^{-4}$ **21** 10000 **22** $\frac{1}{100}$
23 4 **24** 0.4 **25** 3.375 **26** $\frac{1}{25}$ **27** 3 **28** $\frac{1}{2}$
29 $\frac{4}{3}$ **30** $\frac{1}{125}$ **31** 512 **32** $\frac{1}{81}$ **33** 1 **34** 8
35 64 **36** 9 **37** 7 **38** 125 **39** 7 **40** $\frac{1}{25}$

Exercise 4b

1 3.675×10 **2** 4.35×10^2
3 3.695×10^3 **4** 1×10^3
5 1.001×10 **6** 2.98×10^{-1}
7 3.75×10^{-2} **8** 2.57×10^{-4}
9 1×10^{-4} **10** 1×10^3

Exercise 4c

1 (i) 0.6 (ii) 0.7 (iii) 0.4 (iv) 0.87 (v) 0.11
(vi) 0.99 **2** (i) 1.6 (ii) 6.3 (iii) 1.9
(iv) 3.6 (v) 9.5 (vi) 4.1

Exercise 4d

1 0.5152 **2** 0.9011 **3** 0.6314 **4** 0.9373
5 1.3338 **6** 1.7745 **7** 3.0000 **8** 2.5024
9 3.6562 **10** 4.8298 **11** 4.3290
12 5.6928 **13** $\bar{1}.7580$ **14** $\bar{1}.8838$
15 $\bar{1}.0835$ **16** $\bar{3}.5054$ **17** $\bar{4}.7686$
18 $\bar{4}.0000$ **19** $\bar{4}.0000$ **20** $\bar{3}.6378$

Exercise 4e
1 3.734 **2** 2.117 **3** 8.347 **4** 2.260 **5** 37.40
6 237.6 **7** 21 180 **8** 970.7 **9** 4700
10 38.68 **11** 0.2096 **12** 0.1192
13 0.01023 **14** 0.06158 **15** 0.003475
16 0.001853 **17** 0.0139 **18** 0.9627
19 0.00003236 **20** 0.0002705

Exercise 4f
1 16.1 **2** 237 **3** 3.52 **4** 620 **5** 0.123
6 0.00234 **7** 93.7 **8** 0.147 **9** 0.0105
10 24.3 **11** 1.14 **12** 1.30 **13** 63.7 **14** 247
15 0.00541 **16** 2.20 **17** 1 800 000
18 0.000228 **19** 13.1 **20** 4290

Exercise 4g
1 54.8 **2** 62 300 **3** 39 200 **4** 0.000201
5 0.00436 **6** 4.48 **7** 3.26 **8** 1.64
9 0.756 **10** 0.418 **11** 0.885 **12** 0.838
13 3.01 **14** 0.179 **15** 24.5 **16** 2.07
17 5.59 **18** 0.111 **19** 0.125 **20** 0.562

Exercise 4h
1 0.019 **2** 0.212 **3** 2.01 **4** 0.343 **5** 0.0166
6 0.0329 **7** 0.342 **8** 0.0496 **9** 0.276
10 0.232

Harder examples 4
1 **(i)** $\frac{1}{16}$ **(ii)** 25 **(iii)** $\frac{8}{27}$ **2** 4.2 **3** 2×10^{-5}
4 3.4×10^{-3} **5** **(i)** 1.6×10^{-3} **(ii)** 4×10^{-2}
6 0.795 **7** 0.005 **8** 8596 **9** 1.34 **10** 2.18
11 $(0.75)^2$ **12** 0.0966 **13** 0.143 **14** 38.6
15 517 **16** 0 196 **17** 0.839 **18** 4.03

5 Algebra

Exercise 5a
1 $2a-6b$ **2** $5x^2-x$ **4** $9a^2-2a$
6 $5b^2-3a-3b$ **7** $7a^3-a^2$ **8** $x-4x^2$
9 $2xy+6x^2$ **10** $9at-t^2$ **11** t^2-2tm
14 $-5pq-9qp^2$ **15** $8xy+11x^2y^2$
16 $4fg+f^2g-fg^2$ **17** $2ab+ab^2+ba^2$
18 $xt+3x^2t$

Exercise 5b
1 $ab+bc+2ac$ **2** $2a^2+3ab$
3 $-2m^2-mn$ **4** $2xz-2xy-4xt$
5 $a^3+ab-ba^2-b^2$ **6** $-a^2b-a^3b$
7 $2a-2a^2-a^3$ **8** $3a-2ab$ **9** x^2+4x-6
10 $2x^3$ **11** $a^2b-3a^2-ab^2+b^2$
12 $2x^2y-2xy^2+4x^2$
13 $5mn+5mp-pm^2+pmn$ **14** ax^3-2ay
15 x^2y-xy^2

Exercise 5c
1 x^2+3x+2 **2** $2a^2+ab+2ad+db$
3 $3a^2+5ab+2b^2$ **4** $x^2+xy-2y^2$
5 $2x^2-xy-15y^2$ **6** $t^2+mt-6m^2$
7 $2a^2+3ax+x^2$ **8** $7ax-2x^2-3a^2$
9 $4y^2-a^2-3ay$ **10** $3b^2+2bc-c^2$
11 $ba+by+4ac+4cy$
12 $x^3+3x^2y-y^2x-3y^3$

Exercise 5d
1 $9a^2+6ab+b^2$ **2** $a^2-8ab+16b^2$
3 $4x^2-12xy+9y^2$ **4** $25x^2-10xy+y^2$
5 $x^2+\frac{1}{2}x+\frac{1}{16}$ **6** $25a^2-9b^2$ **7** $4x^2-y^2$
8 m^2-9n^2 **9** $a^2-\frac{1}{4}$ **10** b^2-25

Exercise 5e
1 $x(a+4x)$ **2** $3bz(z-3b)$ **3** $x(g-kx)$
4 $x(3x^2+2x+5)$ **5** $3x^2(2x^2-1)$
6 $5(x^2+2x-5)$ **7** $a^2b^2(b-a)$ **8** $4(x-2)$
9 $(a+b)(c-d)$ **10** $(x+w)(y-z)$
11 $(a+1)(4x+1)$ **12** $(a+b)(x-y)$
13 $(r-p)(q+s)$ **14** $2(y+2m)(3x-1)$
15 $(x-q)(c-d)$ **16** $(q-s)(r-t)$
17 $(a-b)(a-3)$ **18** $(x-y)(x-3)$
19 $(a+c)(b+3)$ **20** $(x-y)(a+b+c)$

Exercise 5f
1 $(x-4)(x+4)$ **2** $(2x-5)(2x+5)$
3 $2(x-1)(x+1)$ **4** $(9x-10)(9x+10)$
5 $(1-x)(1+x)$ **6** $3(9-x)(9+x)$
7 $(1-4x)(1+4x)$ **8** $(R-r)(R+r)$
10 $2(3x-7)(3x+7)$
11 $(5-6x+6y)(5+6x-6y)$ **12** $x(x-6)$
13 $4(3x-2y)(3x+2y)$ **14** $(x-2y)(x+2y)$
15 $(y-2)(y+4)$ **16** $(y+5)(y+9)$
17 $(11xy-2)(11xy+2)$
18 $8(3x-1)(3x+1)$
19 $8(x^2-2)$ **20** $(x-a-b)(x+a+b)$
21 266 **22** 5571 **23** 8.4 **24** 199 **25** 3060
26 150 000 **27** 26 600 **28** 14.4 **29** 400
30 1999

Exercise 5g
1 $(x+1)^2$ **2** $(x+4)(x+2)$ **3** $(x+5)(x+2)$
4 $(x+4)(x+1)$ **6** $(x+7)(x+3)$
8 $(x+30)(x+2)$ **9** $(x-1)^2$
10 $(x-3)(x-1)$ **11** $(x-3)(x-2)$
13 $(x-5)(x-2)$ **14** $(x-7)(x-5)$
15 $(x-20)(x-10)$ **16** $(x-8)(x-3)$
17 $(x+4)(x-1)$ **18** $(x+3)(x-1)$
19 $(x+5)(x-1)$ **20** $(x+6)(x-1)$
21 $(x+13)(x-2)$ **22** $(x+11)(x-2)$
23 $(x+72)(x-2)$ **24** $(x+24)(x-5)$
25 $(x-2)(x+1)$ **26** $(x-5)(x+1)$
27 $(x-4)(x+2)$ **28** $(x-8)(x+1)$
29 $(x-16)(x+1)$ **30** $(x-24)(x+3)$
31 $(x-6)(x+2)$ **32** $(x-8)(x+3)$

Exercise 5h

1 $(2x+3)(x+5)$ **2** $(3x+4)(x+2)$
3 $(2x+1)(2x+3)$ **4** $(2x+1)(3x+2)$
5 $(6x+5)(x+2)$ **6** $(2x-1)(4x+3)$
7 $(2x-3)(x+3)$ **8** $(2x+5)(2x-3)$
9 $(6x-5)(x+1)$ **10** $(9x-8)(x+1)$
11 $(2x-1)^2$ **12** $(2x-1)(x-3)$
13 $(5x-4)(x-1)$ **14** $(4x-1)(x-2)$
15 $(3x-1)(2x-1)$ **16** $(3x+2)(x-2)$
17 $(9x+2)(x-1)$ **18** $(2x-5)(x+1)$
19 $(4x+3)(x-2)$
20 $(4x^2+1)(x-1)(x+1)$

Exercise 5i

1 7 **2** 12 **3** $3\frac{1}{2}$ **4** $6\frac{2}{3}$ **5** -3 **6** $1\frac{1}{2}$ **7** $1\frac{2}{7}$
8 $\frac{2}{7}$ **9** 2 **10** 20 **11** 6 **12** 13 **13** 18 **14** $4\frac{1}{3}$
15 $2\frac{3}{4}$ **16** $4\frac{3}{4}$ **17** 44 **18** 33 **19** -8 **20** 6
21 5 **22** $-2\frac{2}{5}$ **23** $-\frac{4}{7}$ **24** -27 **25** $26\frac{38}{47}$
26 12

Exercise 5j

1 1, 3 **2** $-2, -3$ **3** -2, 5 **4** 3, 4
5 1, 9 **6** -4, 5 **7** $\frac{1}{2}$, 1 **8** $-\frac{1}{3}$, 2
9 1, $1\frac{1}{2}$ **10** $-\frac{1}{2}$, 3 **11** $-2\frac{4}{7}$, 1
12 1, 2 **13** -1, $1\frac{1}{3}$ **14** 1, 1
15 $\pm 2/\sqrt{3}$

Exercise 5k

1 -0.41, 2.41 **2** -4.30, -0.70
3 -12.32, -5.68 **4** 0.44, 4.56
5 -5.56, -1.44 **6** -1.83, 3.83
7 -0.53, 0.81 **8** -0.69, 0.29
9 -2.5, -0.5 **10** -0.81, 2.48
11 -3.37, 0.37 **12** -0.34, 1.09
13 -2.84, 3.34 **14** -8.36, 0.36
15 -0.11, 2.26 **16** 1, 7 **17** 0.18, 1.82
18 -2.58, 0.58 **19** 0.55, 5.45
20 -10.36, -1.64

Exercise 5l

1 $x=5$, $y=2$ **2** $x=-2$, $y=-5$
3 $x=2$, $y=1$ **4** $a=2$, $b=4$
5 $k=1\frac{1}{10}$, $l=\frac{3}{10}$ **6** $p=3$, $q=2$
7 $x=1\frac{1}{2}$, $y=-3$ **8** $x=6$, $y=-\frac{1}{2}$
9 $a=13, b=7$ **10** $m=-1\frac{2}{3}, n=1$

Exercise 5m

1 $x=\pm\sqrt{a+5}$ **2** $x=\dfrac{b+d}{c-a}$

3 $x=\dfrac{3b-a}{a-b}$ **4** $x=\dfrac{1}{1-a}$

5 $x=ab+c$ **6** $x=ab-a^2-a$

7 $x=\dfrac{1}{b-1}$ **8** $x=\dfrac{ba}{c-a}$ **9** $x=\dfrac{ab}{b-a}$

10 $x=\dfrac{5b-a}{4}$ **11** $x=\dfrac{1}{c-b}$ **12** $x=\dfrac{y-c}{m}$

13 $x=\pm\sqrt{b+a^2}$ **14** $x=\pm\sqrt{a^2-y^2}$

15 $x=\dfrac{bcy^2}{a^2}$ **16** $x=\dfrac{4}{b}$ **17** $x=\dfrac{ab^2}{1-b^2}$

18 $x=\dfrac{b^2c^2-2bc+1+ac^2}{c^2}$ **19** $x=\dfrac{ab}{a+b}$

20 $x=\dfrac{by^2-a}{y^2-1}$ **21** $T=\dfrac{100I}{PR}$

22 (i) $f=\dfrac{uv}{u+v}$ (ii) 0.3

23 (i) $r=\pm\sqrt{\dfrac{v}{\pi h}}$ (ii) ± 1.60

24 $s=\dfrac{v^2-u^2}{2g}$ **25** $u=\dfrac{2s-gt^2}{2t}$

26 $L=\dfrac{DT^2}{1-T^2}$ **27** $d=\dfrac{2(s-an)}{n(n-1)}$

28 $g=\pm\dfrac{h}{2\pi}\sqrt{t^2-4\pi^2}$

29 (i) $R=(S-\pi rh)/\pi h$ (ii) 0.82

30 $t=\dfrac{m-1}{a(m+1)}$

Exercise 5n

1 (i) Yes (ii) No (iii) Yes (iv) No (v) Yes (vi) No **2** (i) No (ii) No (iii) Yes (iv) Yes (v) No (vi) Yes (vii) Yes (viii) No
3 (i) Yes (ii) Yes (iii) No (iv) No (v) Yes (vi) Yes **4** 11, 9 **5** -9
6 6, 6 **7** $\frac{8}{3}$, $\frac{27}{20}$ (i) 12 (ii) 1
8 (i) 9 (ii) 512 **9** (i) 13 (ii) 12 (iii) 33 **10** (i) 16 (ii) 16

Exercise 5o

1 (i) 3 (ii) 2 **3** 1 **4** (i) 4 (ii) 4 **5** (i) 4 (ii) ± 5 **6** (i) 3 (ii) $1\frac{1}{2}$ (iii) 3 **7** 11, 13
8 3, 1 **10** (ii) (a) -1, 7 (b) $-1\frac{1}{2}$, 0 (iii) -2

6 Inequalities

Exercise 6a

1 $x<14$ **2** $x<-13$ **3** $x<1$ **4** $x<\frac{1}{2}$
5 $x<-1$ **6** $x<-10$ **7** $x<2\frac{2}{5}$
8 $x<-2\frac{1}{3}$ **9** $x<-5\frac{1}{2}$ **10** $x>\frac{1}{3}$
11 $x>4$ **12** $x<11$ **13** $x>-1$
14 $x>-17$ **15** $x<2\frac{1}{3}$ **16** $x>5\frac{7}{17}$
17 $x<1$ **18** $x>14$ **19** $x<-11$
20 $x<\frac{17}{18}$

Exercise 6b

1 $\{x<4\}$ **2** $\{x<1\}$ **3** $\{1, 2, 3, 4, 5, 6, 7, 8, 9\}$ **4** $\{-2, -1\}$ **5** $\{-3<x<10\}$ **6** $\{7, 8\}$ **7** $\{-3<x<4\}$ **8** $\{-4.6<x<-2.2\}$
9 $\{2, 3, 4, 5, 6, 7\}$ **10** $\emptyset$ **11** $\{0<x<\frac{1}{2}\}$
12 $\{0<x<2.5\}$ **13** $\{1<x<5.5\}$
14 $\{x<-2\}\cup\{x>2\}$

Exercise 6c

1 $\{-1<x<4\}$ **2** $\{-5<x<2\}$
3 $\{1<x<6\}$ **4** $\{x<2\}\cup\{x>9\}$
5 $\{-3<x<2\}$ **6** $\{2<x<10\}$
7 $\{2<x<3\}$ **8** $\{x<-7\}\cup\{x>1\}$
9 $\{-3<x<3\}$ **10** $\{\frac{1}{2}<x<2\}$

Harder examples 6

1 $\frac{6}{7}$ **2** 12, 2 **3** 6.04, 1.857375
4 8.4, 8.6 **5** $x<\frac{1}{30}$ **6** $x<5\frac{3}{7}$
7 $\{-1, 0, 1, 2, 3, 4\}$ **8** $\{-6<x<1\}$
9 29 **11** $\{1, 2, 3, 4, 5, 6, 7, 8\}$ **12** 6
13 3, 8 **14** -2, 0 **15** $\{1, 2, 3, 5, 7, 11, 13\}$
16 4 **18** 10 **19** 95 % **20** 25 %
21 $\{\frac{1}{2}, \frac{1}{3}, \frac{2}{3}, \frac{1}{4}, \frac{3}{4}, \frac{1}{5}, \frac{2}{5}, \frac{3}{5}, \frac{4}{5}\}$ **22** $\{5, 6, 7, 8, 9\}$, $\{7, 8\}$

7 Area and volume

Exercise 7a

1 44 cm, 154 cm^2 **2** 22 cm, $38\frac{1}{2}$ cm^2
3 132 cm, 1386 cm^2 **4** $\frac{88}{7}$ m, $\frac{88}{7}$ m^2
5 $5\frac{1}{2}$ cm, $2\frac{13}{32}$ cm^2 **6** $7\frac{1}{3}$ cm, $4\frac{5}{18}$ cm^2
7 8.8 cm, 6.16 cm^2 **8** 484 m, 18634 m^2
9 264 cm, 5544 cm^2 **10** 0.44 m, 154 cm^2
11 8.17 cm, 5.31 cm^2
12 19.79 cm, 31.18 cm^2
13 10.81 m, 9.30 m^2
14 18.32 cm, 26.70 cm^2
15 29.53 cm, 69.41 cm^2
16 36.76 m, 107.53 m^2
17 28.44 cm, 64.33 cm^2
18 124.80 cm, 1239.27 cm^2
19 50.96 m, 206.66 m^2
20 51.21 cm, 208.70 cm^2

Exercise 7b

1 384 cm^2 **2** 284 cm^2 **3** 6160 cm^2
4 238.8 cm^2 **5** 223.2 cm^2 **6** 7.96 cm

Exercise 7c

1 1848 cm^3 **2** 58.43 cm^3 **3** 33.9 cm
4 262 cm^3 **5** 65.9 cm^3 **6** 2.88 cm
7 $85\frac{1}{3}$ cm^3 **8** 17.89 cm **9** 1.26 cm
10 0.0941 m^3

Exercise 7d

1 (i) 42 cm^2 **(ii)** 10.90 cm^2 **(iii)** $66\frac{1}{2}$ cm^2
(iv) 44.13 cm^2 **2 (i)** 154 cm^2 **(ii)** $38\frac{1}{2}$ m^2
(iii) 0.0154 mm^2 **(iv)** 75.46 m^2
3 (i) 102 cm^2 **(ii)** 0.000 531 m^2
(iii) 10 600 mm^2 **(iv)** 119 cm^2
4 (i) $a(a+b)/2$ **(ii)** $4x^2$ **(iii)** $ab-\pi a^2/8$
(iv) $ab+bh/2$ **5** 960, $\frac{7}{20}$ **6** 1650 m^2
7 64.19 cm^2 **8** 30 cm^2 **9** 1.11×10^5 s
10 0.676

Harder examples 7

1 30 cm^2 **2** 24 cm^3 **3** 10:45:3
4 293 cm^2 **5 (i)** 15.6 cm^2 **6** $1437\frac{1}{3}$ cm^3, $1306\frac{2}{3}$ cm^3 **7** 2:9 **8** $15/\sqrt{2}$
9 308 cm^2, 2.71 kg **10 (i)** $20\sqrt{11}$ cm^2
(ii) $80\sqrt{11}$ cm^3 **(iii)** $50\sqrt{5}$ cm^2
(iv) 7.12 cm **11** 2:1 **12** 13 cm
(i) $7\frac{1}{5}$ cm **(ii)** $\frac{27}{123}$ **(iii)** $\frac{9}{25}$

8 Ratio and percentage

Exercise 8a

1 2:15 **2** 3:250 **3** 11:24 **4** 1:4
5 21:34 **6** 3:35 **7** 1:75 **8** 33:61
9 32:41 **10** 2:15 **11** 2:3:9 **12** 1:2:3
13 (a) 1:1.69 **(b)** 0.59:1 **14 (a)** 1:1.73
(b) 0.58:1 **15 (a)** 1:4 **(b)** 0.25:1
16 (a) 1:10 **(b)** 0.1:1 **17 (a)** 1:0.73
(b) 1.38:1 **18 (a)** 1:94.6 **(b)** 0.011:1
19 £4.50 **20** 6 kg **21** 35 **22** 34.2 m
23 3.2 **24** 9.5 **25** 8:15 **26** 28:36:27
27 3000 m., 4800 m **28** 35 cm **29** 1:4
30 73.5 g **31** 8 cm:155 cm **32** 1:64

Exercise 8b

1 20, 36 **2** £4, £8, £12 **3** 10, 6, 2
4 £$21\frac{1}{3}$, £32, £$74\frac{2}{3}$ **5** 144, 108, 72
6 6, 4, 3 **7** 26.25 cm, 35 cm, 43.75 cm
8 £4, £3 **9** £90, £50, £40
10 24°, 48°, 120°, 168° **11** 105°, 126°, 168°
12 3 km, 4.5 km, 7.5 km **13** £72, £54, £18
14 125 **15** £4.20 **16** 120°, 60°
17 £360, £450, £390 **18** £108

Exercise 8c

1 (i) $66\frac{2}{3}$ % **(ii)** 279 % **(iii)** $16\frac{2}{3}$ %
(iv) 3.75 % **(v)** 0.5 % **(vi)** 0.1 % **(vii)** 328 %
(viii) 55 % **(ix)** 150 % **(x)** 200 % **(xi)** 9.8 %
(xii) 42.86 % **2 (i)** $\frac{1}{3}$, 0.33 **(ii)** $\frac{18}{25}$, 0.72
(iii) $\frac{1}{8}$, 0.125 **(iv)** $\frac{1}{12}$, 0.83 **(v)** $\frac{3}{500}$, 0.006
(vi) $2\frac{1}{4}$, 2.25 **(vii)** $\frac{1}{400}$, 0.0025
(viii) $\frac{1}{200}$, 0.005 **(ix)** $\frac{1}{150}$, 0.006
(x) $\frac{3}{25}$, 0.12 **(xi)** $\frac{11}{40}$, 0.275 **(xii)** $\frac{73}{200}$, 0.365
3 0.0003 % **4** 8.82 % **5** 3.5 % **6** 40 %

7 8% **8** 1% **9** 3.5% **10** £1.50 **11** 51 g
12 6.3 km **13** £1.64½ **14** 2 ml **15** 5.4°
16 14.112 m **17** 40p

Exercise 8d

1 414 **2** 4.275 kg **3** 9.605 cm **4** £1.08
5 29.6 litres **6** 4.80 m **7** 44%
8 4588.5 cm^3, 87.4% **9** £235.42
10 £21.44 **11** £60 **12** £24.64
13 £20 **14** £7.20 **15** £5.04
16 £11.73 **17** £360 **18** £84

Exercise 8e

1 £67.50 **2** £40.50 **3** 14% **4** $3\frac{1}{3}$%
5 6 years **6** $1\frac{2}{3}$ years **7** £500
8 £62.50 **9** £240 **10** £59.63

Exercise 8f

1 £68.96 **2** £9.56 **3** £11.68 **4** £191.03
5 £9.66 **6** 11.41 cm^2 **7** 57.89 km/h
8 £1158.65

Harder examples 8

1 15.91% **2** £292.50 **3** 44% **4** £29.90
5 $6\frac{2}{3}$% **6** 23.9% **7** 44.00 **8** £42.86
9 20% **10** 21% **11** £17.50
12 £470, £522.50 **13** 13%
14 **(i)** £2.75 → £3.00 **(ii)** £2.20 → £2.70
15 £265, £280.90, £1685

9 Matrices

Exercise 9a

1 **(i)** $\begin{pmatrix} 8 & 6 & -2 \\ 6 & 3 & 5 \end{pmatrix}$ **(ii)** $\begin{pmatrix} 13 & 4 \\ -8 & 0 \\ 7 & 5 \end{pmatrix}$

(iii) $\begin{pmatrix} -4 & 0 \\ -8 & 4 \end{pmatrix}$ **(v)** $\begin{pmatrix} 3 & 6 & 9 \\ 0 & 3 & -6 \\ 6 & -3 & 9 \end{pmatrix}$

3 $\begin{pmatrix} -1 & 3 \\ 1 & 5 \end{pmatrix}$ **4** $\begin{pmatrix} 1\frac{1}{2} & 2\frac{1}{2} \\ 0 & -\frac{1}{2} \end{pmatrix}$

5 $\begin{pmatrix} 1 & 1 \\ -1 & -2 \end{pmatrix}$ **6** $\begin{pmatrix} 0 & 1 & -1\frac{1}{3} \\ \frac{2}{3} & -1\frac{2}{3} & 1\frac{1}{3} \end{pmatrix}$

Exercise 9b

1 (1) **2** (0) **3** (−4) **4** (0) **7** (5 −5)

8 (2 −1) **10** $\begin{pmatrix} -1 \\ 1 \end{pmatrix}$

11 $\begin{pmatrix} -3 & 2 \\ 1 & -4 \end{pmatrix}$ **12** $\begin{pmatrix} 2 & 1 & 0 \\ 1 & 3 & 0 \end{pmatrix}$

14 $\begin{pmatrix} 17 & 19 \\ 21 & 25 \end{pmatrix}$ **15** $\begin{pmatrix} 4 & -1 & 2 \\ -3 & 0 & -1 \\ 2 & 1 & 0 \end{pmatrix}$

16 $\begin{pmatrix} -21 & -18 \\ 10 & 11 \end{pmatrix}$ **18** $\begin{pmatrix} 5 & -11 & 3 \\ 1 & -1 & 0 \end{pmatrix}$

20 $\begin{pmatrix} 4 & 0 & 2 \\ 5 & 2 & -1 \\ 4 & 0 & 2 \end{pmatrix}$ **21** **(i)** $\begin{pmatrix} 0 & -2 \\ 7 & -1 \end{pmatrix}$

(ii) $\begin{pmatrix} 1 & -4 \\ 4 & -2 \end{pmatrix}$ **22** **(i, ii)** $\begin{pmatrix} -4 & -2 \\ 22 & -12 \end{pmatrix}$

Exercise 9c

4 $\begin{pmatrix} 1 & -3 \\ 1 & -1 \end{pmatrix}$ **5** $\begin{pmatrix} -1 & -\frac{1}{3} \\ -1\frac{1}{3} & -1\frac{1}{3} \end{pmatrix}$

6 $\begin{pmatrix} -6 & -10 \\ 15 & 9 \end{pmatrix}$ **7** 2, 1, 2

8 **(i)** (1 −1) **(ii)** $\begin{pmatrix} 1 & 0 \\ 0 & -1 \end{pmatrix}$

(iii) $\begin{pmatrix} 2 & 0 \\ 1 & -2 \\ -3 & 0 \end{pmatrix}$ **9** **(i)** $\begin{pmatrix} 4 & 2 \\ 2 & 10 \end{pmatrix}$

(ii) $\begin{pmatrix} 5 & 1 \\ 1 & 34 \end{pmatrix}$ **10** **(i)** $\frac{1}{2}\begin{pmatrix} 1 & -1 \\ 1 & 1 \end{pmatrix}$

(ii) $\begin{pmatrix} 2 & -1 \\ -5 & 3 \end{pmatrix}$ **(iii)** $\frac{1}{11}\begin{pmatrix} 3 & 2 \\ -1 & 3 \end{pmatrix}$

(iv) $\begin{pmatrix} 2 & 1 \\ 9 & 5 \end{pmatrix}$ **(v)** $-\frac{1}{2}\begin{pmatrix} 3 & 1 \\ -4 & -2 \end{pmatrix}$

(vi) $\frac{1}{10}\begin{pmatrix} 7 & 3 \\ -1 & 1 \end{pmatrix}$

Exercise 9d

1 4, −1 **2** $1\frac{1}{2}, \frac{1}{2}$ **3** 4, $2\frac{1}{2}$ **4** −3, −11
5 3, $\frac{2}{3}$ **6** 4, −7 **7** $-1\frac{1}{2}$, 3 **8** $\frac{1}{5}, \frac{4}{5}$
9 −2, −10, **10** $-\frac{1}{2}$, 2 **11** $2\frac{1}{2}$, 2
12 −1, −2 **13** 3, −1 **14** 3, −5

15 2, -1 **16** $\frac{1}{2}, \frac{2}{5}$ **17** 1, $1\frac{1}{2}$ **18** 2, 3
19 $\frac{2}{5}, -1\frac{1}{3}$ **20** $\frac{19}{73}, \frac{30}{73}$

Harder examples 9

1 (i) $\begin{pmatrix} 5 \\ 3 \\ -8 \end{pmatrix}$ **(ii)** $\begin{pmatrix} -3 \\ 1 \\ 2 \end{pmatrix}$

(iii) $\begin{pmatrix} -5 \\ 4 \\ 1 \end{pmatrix}$ **2** $\begin{pmatrix} 1 & 3 \\ -\frac{8}{3} & 0 \end{pmatrix}$

3 $\begin{pmatrix} -2 \\ 3 \end{pmatrix}$ **4** $\begin{pmatrix} 11 & -11 \\ 5 & -3 \end{pmatrix}$

5 $\begin{pmatrix} -1 & 4 \\ -2 & -1 \end{pmatrix}$ **6 (i)** $\begin{pmatrix} 9 & 8 \\ 4 & 9 \end{pmatrix}$

(ii) $\begin{pmatrix} 5 & 4 \\ 2 & 5 \end{pmatrix}$ **7** $a = 3, \quad b = -5, \quad c = 1$

8 $\begin{pmatrix} 10 \\ 3 \end{pmatrix}$ **9** $x = 5, \quad y = 3$

10 $x = -4\frac{1}{2}, y = -2,$ **11** $x = 5\frac{2}{3}, y = \frac{1}{3}$
12 $x = -\frac{5}{3}, \quad y = \frac{5}{3}$ **13** $a = -9, \quad b = 4$
14 $a = 5, b = 1, c = 8$
15 $a = 3, b = 11, c = 4$
16 $x = \frac{1}{3}, y = -\frac{2}{3}$

17 $\begin{pmatrix} x+2y \\ -x+3y \end{pmatrix}, \begin{pmatrix} 5 & 0 \\ 0 & 5 \end{pmatrix}, x = 2, y = -1$

18 $-\frac{1}{7}\begin{pmatrix} -1 & 1 \\ 5 & 2 \end{pmatrix}, \quad x = 3\frac{2}{7}, \quad y = -\frac{3}{7}$

19 1, $\begin{pmatrix} 3 & -2 \\ -4 & 3 \end{pmatrix}, \quad x = 3, \quad y = -\frac{1}{2}$

20 $\begin{pmatrix} 1 & 0 \\ 0 & 1 \end{pmatrix}, \begin{pmatrix} -1 & 0 \\ 0 & -1 \end{pmatrix},$

$\begin{pmatrix} 2 & 1 \\ -2 & -1 \end{pmatrix}, \begin{pmatrix} -1 & 1 \\ -2 & 2 \end{pmatrix}$

10 Mappings and functions

Exercise 10a

1 (i) 3 **(ii)** 6 **(iii)** 0 **2 (i)** 3 **(ii)** 8 **(iii)** -102 **3 (i)** 7 **(ii)** -3 **(iii)** 17 **4 (i)** 10 **(ii)** -5 **(iii)** -11 **5 (i)** 4 **(ii)** 4 **(iii)** 0 **6 (i)** 5 **(ii)** 0 **(iii)** 32 **7 (i)** 33 **(ii)** 9 **(iii)** 19 **8 (i)** $7\frac{1}{7}$ **(ii)** $2\frac{1}{2}$ **(iii)** $-2\frac{1}{2}$ **9 (i)** 8 **(ii)** 3 **(iii)** 3
10 (i) 7 **(ii)** 7 **(iii)** 1

Exercise 10b

1 (i) $1\frac{1}{3}$ **(ii)** 3 **(iii)** $\frac{3}{4}$ **2 (i)** 30 **(ii)** 55 **3** $g^{-1}: y \to (y-1)/3$ **4** 0 **5** $-\frac{3}{2}$ **6 (i)** 21 **(ii)** 168 **7 (i)** 3, 11 **(ii)** $3 < x < 11$ **8 (i)** 2 **(ii)** 0, 1 or 2 **(iii)** 0 **9 (i)** -3 **(ii)** 4, 0.5
10 (i) $f^{-1}: y \to 1-y$
(ii) $g^{-1}: y \to 1/y$ **(iii)** $h^{-1}: y \to \sqrt{y}$
(iv) $m^{-1}: y \to (1-y)/3$
(v) $n^{-1}: y \to (1-y)/y$
(vi) $q^{-1}: y \to (b-dy)/(cy-a)$

Harder examples 10

1 $a = 1, b = 3$ or $a = -1, b = -3$
2 $a = 2, b = 5$ or $a = -2$, b = 7
5 $f^{-1}: y \to (y-1)/4$, $g^{-1}: y \to y+1$, $\frac{3}{2}$
6 (i) 20, 5, Thursday **(ii)** 24, 11
7 (i) 9, 12 **(ii)** $\{10, 11, 12\}$ **(iii) (a)** never **(b)** always **(c)** sometimes, $\{x : x \neq 3k+1\}$
8 (i) (a) $x \to 2$ **(b)** $x \to 4$ **(ii)** gh
(iii) $m : x \to 3x-3$ **9** (i), (iv) true
10 (i) 12

11 Co-ordinates, graphs and linear programming

Exercise 11b

12 -4 **13 (i)** -1.79, 2.79
(ii) 3.45, -1.45 **(iii)** -2.08, 2.41
14 1.71, 0.29 **15** 4.8

Exercise 11d

1 (i) none; none; (10, 0); (10, -8); none; (none, 0); (12, 0); none; (3, none); (9, -1) **(ii)** (none, -3); (-1, none); (14, -1); (15, -8); (1, none); (none, 7); (8, 0); none; none; (3, -7) **2** £72 **3 (i)** 4, 8 **(ii)** 15, 0

Harder examples 11

1 (i) (a) -5 **(b)** 16 **(c)** $1\frac{1}{3}$ **(ii)** 5
2 (a) false **(b)** true **(c)** false
3 (i) max. = 16, min. = 0; no solution
(ii) max. = 29, min. = -3; 1.28, -0.78
(iii) no max. or min.; 1.14, -0.14
(iv) max. = 64, min. = -64; -1.33
(v) max. = 17, min. = 1; no solution
4 (i) 6.05, 5.9375, less **(ii)** 2.45
5 -2.8, 0, 2.8
6 $y \leq 4, \quad 5y \geq 4x, \quad 2x+y \geq 4$
7 (a) (7.2, 15.2) **(b)** (8, 15) **8** 16
9 (3, 3) (4, 2)
12 (i) -12 **(ii)** 3 **(iii)** 26

12 Right-angled triangles and beyond

Exercise 12
1 67°, 16.5 cm, 17.9 cm
2 67° 23′, 22° 37′, 12 cm
3 26° 34′, 63° 26′, 6.71 cm
4 66°, 1.20 m, 2.96 m
5 66° 45′, 60.15 cm, 140 cm
6 48° 36′, 41° 24′, 4.74 cm
7 22°, 68°, 1295 m
8 41° 24′, 48° 36′, 0.16 m
9 54° 41′, 8.86 km, 10.86 km
10 56° 51′, 33° 9′, 0.059 cm
11 2.92 s **12** 58 m
13 186.3 km **14** 0.9 m/s
15 6

Harder examples 12
1 (i) 2.83 m, 5.66 m (ii) 45° (iii) 1.414 (iv) 54° 44′ **3** (i) 30°, 150° (ii) 150°, 210° **4** 257 m **5** 9.08 cm, 17.82 cm (i) 72° (ii) 1.2 cm **6** 2.41 cm **7** 7.44 cm **8** 13 cm **9** 8 cm, 73° 44′ **10** (i) 54° 44′ (ii) 70° 32′ (iii) 109° 28′ **12** 15.9 km, $199\frac{1}{2}$°

13 Variation and proportion

Exercise 13a
1 (i) $6\frac{2}{3}$ (ii) 8 (iii) 20 (iv) 0.1 (v) $1\frac{1}{2}$ (vi) $3a$ (vii) 60 (viii) 0.15 (ix) 0.035 (x) $(a+b)^2$ **2** $y = 5x$, 0.3

Exercise 13b
1 (i) 11.1 (ii) 6.4 (iii) 16 (iv) 0.002 (v) 9 (vi) $9a$ (vii) 120 (viii) 2.25 (ix) 0.008 (x) $(a+b)^3$ **2** 40
3 21.6 **4** $\frac{22}{9}$, (i) $\frac{22}{3}$ (ii) $\frac{44}{9}$
5 ± 10.14

Harder examples 13
1 $y = 1\frac{1}{2}$, $z = \pm 6$ **2** $1\frac{4}{5}$
3 (i) incorrect (ii) incorrect (iii) correct (iv) correct **4** (i) 2 (ii) 1 (iii) -1
5 $a = \frac{1}{2}$, $n = 3$ **6** 1:12 **7** 3.37
8 4:9 **9** $20b^2/a^2$ **10** 937.5 m^2

14 Straight lines and circles

Exercise 14a
1 40°, 40°, 30° **2** 90° 55°, 65°, 35°
4 50°, 60°, 50° **5** 50°, 50° **6** 120°, 30°
7 80° **8** 110°, 30°, 30° **9** 36°, 36°, 72°
10 130° **11** $49\frac{6}{11}$° **12** 8

Exercise 14b
1 70° **2** 40°, 70° **3** 25° **4** 70°
5 130°, 130° **6** 45°, 50°, 45°
7 50°, 50° **8** 15°, 10°, 150°

Exercise 14c
1 2 straight lines, cone
2 2 parallel lines, cylinder
3 1 parallel line, plane
4 ellipse, ellipsoid **5** disc, solid sphere
6 infinite solid 4 cm thick
7 cylindrical hole in space
8 circle, sphere **9** circle, sphere
10 half plane, half space

Harder examples 14
1 55°, 100° **2** 110°, 70° **3** $2x$, 18°
4 80°, 55° **5** 100° **7** AQO, TOP, QTC
9 (i) $(180-x)$°, (ii) (iii), (iv), $360°/x$
10 130°, 30°, 50° **11** $(1, \pm\sqrt{3})$

15 Vectors

Exercise 15a
1 $\begin{pmatrix}4\\1\end{pmatrix}$ **2** $\begin{pmatrix}5\\2\end{pmatrix}$ **3** $\begin{pmatrix}7\\-1\end{pmatrix}$ **4** $\begin{pmatrix}-1\\-1\end{pmatrix}$

5 $\begin{pmatrix}3\\-2\end{pmatrix}$ **6** $\begin{pmatrix}-2\\3\end{pmatrix}$ **7** $\begin{pmatrix}8\\1\end{pmatrix}$ **8** $\begin{pmatrix}-1\\5\end{pmatrix}$

9 $\begin{pmatrix}-11\\2\end{pmatrix}$ **10** $\begin{pmatrix}-3\\8\end{pmatrix}$ **11** $\begin{pmatrix}-3\\14\end{pmatrix}$

12 $\begin{pmatrix}-18\\7\end{pmatrix}$ **13** $\begin{pmatrix}11\\10\end{pmatrix}$ **14** $\begin{pmatrix}2\\3\end{pmatrix}$ **15** $\begin{pmatrix}24\\7\end{pmatrix}$

16 $\begin{pmatrix}5\\-7\end{pmatrix}$ **17** $\begin{pmatrix}17\\7\end{pmatrix}$ **18** $\begin{pmatrix}-4\\-7\end{pmatrix}$

19 $\begin{pmatrix}35\\-7\end{pmatrix}$ **20** $\begin{pmatrix}-16\\9\end{pmatrix}$

Exercise 15b
1 $\sqrt{17}$ **2** $\sqrt{29}$ **3** $5\sqrt{2}$ **4** $\sqrt{2}$ **5** $\sqrt{13}$
6 $\sqrt{13}$ **7** $\sqrt{65}$ **8** $\sqrt{26}$ **9** $5\sqrt{5}$
10 $\sqrt{73}$ **11** $\sqrt{205}$ **12** $\sqrt{373}$
13 $\sqrt{221}$ **14** $\sqrt{13}$ **15** 25 **16** $\sqrt{74}$
17 $\sqrt{338}$ **18** $\sqrt{65}$ **19** $\sqrt{1274}$ **20** $\sqrt{337}$

Exercise 15d
1 (i) $\mathbf{b}+\mathbf{c}$ (ii) $\mathbf{a}+\mathbf{b}+\mathbf{c}$ (iii) $\mathbf{a}+\mathbf{b}$
2 (i) $\mathbf{b}-\mathbf{a}$ (ii) $-\mathbf{a}-\mathbf{b}$ (iii) $2\mathbf{b}$ **3** (i) $\mathbf{a}$
(ii) $\mathbf{a}+\mathbf{b}$ (iii) $-2\mathbf{a}-\mathbf{b}$ **4** (i) $\begin{pmatrix}3\\3\end{pmatrix}$

(ii) $\begin{pmatrix} -1 \\ 3 \end{pmatrix}$ **(iii)** $\begin{pmatrix} -1 \\ 1 \end{pmatrix}$ **(iv)** $\begin{pmatrix} 7 \\ -1 \end{pmatrix}$ **(v)** $\begin{pmatrix} 6 \\ 2 \end{pmatrix}$

(vi) $\begin{pmatrix} -7 \\ 5 \end{pmatrix}$ **(vii)** $\begin{pmatrix} -5 \\ 9 \end{pmatrix}$ **(viii)** $\begin{pmatrix} -1\frac{3}{4} \\ \frac{3}{4} \end{pmatrix}$

(ix) $\begin{pmatrix} -\frac{3}{2} \\ \frac{3}{2} \end{pmatrix}$ **(x)** $\begin{pmatrix} \frac{1}{4} \\ 6\frac{1}{4} \end{pmatrix}$ **5 (i)** $\begin{pmatrix} -6 \\ -2 \end{pmatrix}$

(ii) $\begin{pmatrix} 7 \\ -2 \end{pmatrix}$ **(iii)** $\begin{pmatrix} 1 \\ -4 \end{pmatrix}$ **6** (5, −3), $\begin{pmatrix} 2 \\ -4 \end{pmatrix}$

7 2, −4 **9 (i)** $\mathbf{r}-\mathbf{p}$ **(ii)** $\mathbf{q}-\mathbf{r}$ **(iii)** $\mathbf{r}-\mathbf{p}-\mathbf{q}$
10 $-\frac{1}{3}$, $-\frac{2}{3}$ **11** Circle centre A, radius 3
12 (6, 4), (18, 8), $\sqrt{61}$

Harder examples 15
1 (−7, 18)

4 $x\begin{pmatrix} 1 \\ -1 \end{pmatrix}+y\begin{pmatrix} -4 \\ 2 \end{pmatrix}=\begin{pmatrix} 2m \\ n \end{pmatrix}$

5 (ii) 16 km due west **(iii)** 3.2, 13.84 km **6** (7, 4) **7** 45° **8 (i) (a)** 0 **(b)** 90° **(ii)** $2\mathbf{b}$, $\mathbf{b}-\mathbf{a}$, 5.5 **9 (i)** $h\mathbf{b}-\mathbf{a}$, $\frac{1}{2}(h\mathbf{b}+\mathbf{a})$ **(ii)** $(1-k)\mathbf{a}+k\mathbf{b}$, $\frac{3}{4}(1-k)\mathbf{a}+k\mathbf{b}$; $h=\frac{1}{2}, k=\frac{1}{3}$; OM/MB = 1; AL/LB = $\frac{1}{2}$ **10** $a=-1$ **11** $(8, 7\frac{1}{2})$, $(-8, -4\frac{1}{2})$

16 Transformation geometry

Exercise 16a
1 (i) reflection in $y = x$ **(ii)** rotation of 90° **(iii)** rotation of −90° **(iv)** enlargement and shear **(v)** dilatation and shear **2** rotations **4** 6

Harder examples 16
3 9 **5** $\frac{1}{5}$, $y=\frac{1}{3}x$ **6** $\begin{pmatrix} -1 & 0 \\ 0 & -1 \end{pmatrix}$

7 (b) B, B, I (c) CB, CBA, BC
12 (i) (a) (3, 5) **(b)** 6, 31.5, 7.5 **(c)** $2x$

17 Statistics

Exercise 17a
3 14.6% **4** 351

Exercise 17b
1 (i) 7 **(ii)** 8.5 **(iii)** 2.02 **(iv)** 0.2 **(v)** −0.267 **(vi)** $\frac{5}{8}$ **2** 5.5, 55.5 **3** $-1\frac{2}{3}$, $218\frac{1}{3}$ **4 (i)** 81.86 **(ii)** 52.17 **(iii)** 5248.6 **(iv)** −0.9017 **5 (i)** 41 **(ii)** 64 **(iii)** 70 **(iv)** $\frac{1}{2}$ **6 (a) (i)** 6 **(ii)** 5 **(iii)** 4 **(b) (i)** 15.5 **(ii)** 15 **(iii)** 19 **7** 1 month **9 (i)** 3.79 **(ii)** 6 **(iii)** 3.5 **10 (i)** 4 **(ii)** 3.9

Exercise 17c
2 30–39 **4** 53.1 **5** 99.5

Exercise 17d
1 (ii) 2.06 **2** 2.47 **3** 51, 13.6

Harder examples 17
4 72 **5** 747 **6** 5 **7 (i)** 2.5 **(ii)** 30° **8** 17 **9** 8 **10 (i)** $6\frac{3}{4}$ **(ii)** 7 **11** 6.27 **12** 5.1 **13 (ii)** 7.01 **14** 9.2, 60.5%, 2–9 **15** 185, 57.2, 51.25

18 Probability

Exercise 18
1 {HHHH, HHHT, HHTH, HTHH, THHH, HHTT, HTHT, THHT, HTTH, THTH, TTHH, HTTT, THTT, TTHT, TTTH, TTTT} **4 (i)** $\frac{1}{13}$ **(ii)** $\frac{1}{3}$ **(iii)** $\frac{4}{11}$ **(iv)** $\frac{9}{25}$ **(v)** $\frac{3}{5}$ **5 (i)** $\frac{8}{25}$ **(ii)** $\frac{3}{50}$ **(iii)** $\frac{1}{10}$

Harder examples 18
1 (a) false **(b)** true **(c)** true **(d)** false
2 $\frac{1}{2}$ **3** $\frac{7}{10}$ **5** $\frac{1}{2}$, $\frac{1}{5}$, $\frac{1}{10}$ **6** $1-\pi/4$ **7 (i)** $\frac{3}{5}$ **(ii)** $\frac{7}{20}$ **8 (i)** $\frac{1}{2}$ **(ii)** $\frac{2}{9}$ **9 (i)** $\frac{2}{9}$ **(ii)** $\frac{1}{3}$ **10 (i)** $\frac{25}{81}$ **(ii)** $\frac{5}{18}$ **11** $\frac{2}{5}$ **12** $\frac{1}{5525}$ **13 (i)** $\frac{8}{27}$ **(ii)** $\frac{20}{27}$ **14** {(1, 1), (1, 2), (1, 3), (1, 4), (2, 1), (2, 2), (2, 3), (2, 4), (3, 1), (3, 2), (3, 3), (3, 4), (4, 1), (4, 2), (4, 3), (4, 4)} **(i)** $\frac{3}{8}$ **(ii)** $\frac{13}{16}$ **15** $1-p$, $(1-p)^2$ **16 (i)** 0.141 **(ii)** 0.53 **17 (i)** $\frac{19}{34}$ **(ii)** $\frac{15}{34}$ **18 (i) (a)** $\frac{1}{4}$ **(b)** $\frac{1}{2}$ **(ii) (a)** $\frac{1}{5}$ **(b)** $\frac{1}{2}$ **19 (i)** $\frac{1}{3}$ **(ii)** $\frac{1}{216}$ **(iii)** $\frac{1}{36}$ **(iv)** $\frac{1}{18}$ **20 (a)** $\frac{1}{4}$ **(b)** 27 **21** $4x = 3y, x = 15, y = 20$

19 The solution of the general triangle

Exercise 19a
1 (i) 9.335 **(ii)** 12.08 **(iii)** 10.30 **(iv)** 4.355 **(v)** 8.902 **(vi)** 11.26 **(vii)** 74.6 **(viii)** 50.14 **2 (i)** 34° 50′ **(ii)** 45° 51′ **(iii)** 41° 22′ **(iv)** 78° 24′ **3 (i)** 3.32 **(ii)** 29.6 **(iii)** 2.94 **(iv)** 13.82 **(v)** 27.3 **(vi)** 119 **(vii)** 176 **4 (i)** 55° 47′ **(ii)** 113° 46′ **(iii)** 23° 15′

Exercise 19b
1 (i) 15.7 mm^2 **(ii)** 0.0712 km^2 **(iii)** 1.03 cm^2 **(iv)** 0.106 cm^2 **(v)** 43.5 cm^2 **2 (i)** 45° 35′ **(ii)** 63° 14′ **(iii)** 23° 57′ **(iv)** 39° 48′ **(v)** 2.71 cm **(vi)** 3.62 cm **(vii)** 4.08 cm **(viii)** 75 mm

Exercise 19c
1 **(i)** 5.48 km/h; 32° 34′
(ii) 4.90 km/h; 44° 25′
(iii) 19.85 km/h; 49° 5′
(iv) 3.08 km/h; 8.46 km/h
(v) 5.69 km/h; 1.47 km/h
(vi) 23.3 km/h; 22.2 km/h
2 **(i)** 077° 36′; 410 km/h
(ii) 115° 37′; 491 km/h
(iii) 280° 35′; 287 knots
(iv) 019° 26′; 34.9 km/h
(v) 285° 28′; 110 knots
(vi) 090° 17′; 32 km/h
(vii) 71° 59′; 398 km/h
(viii) 170° 34′; 246 knots
(ix) 193° 48′; 183 km/h
(x) 162° 22′; 395 km/h
(xi) 186° 2′; 318 knots
(xii) 211° 42′; 326 km/h

Harder examples 19
1 10.44 **2** 44°25′ **3** 7.03, 5.22 **4** 21°44′
5 11.6 cm **6** **(i)** 2.59 cm **(ii)** 8.70 cm
7 143 km **8** 2087 m, 247° 30′
9 **(i)** 6.71 cm **(ii)** 36° 52′
(iii) 13.5 cm^2 **10** 8.72; 10 cm **11** 256 m^2
12 **(i)** 800 cm^2 **(ii)** 774 cm^2
13 14° 48′; 064° 48′
14 276; 326 knots
15 324° 36′; 360 km/h; 15.10 h
16 018°–019°; 101°–102° **(i)** $v = 3$
(ii) $v = 1.5$ **17** **(i)** 73° 44′ **(ii)** 3.93; 9.93
18 18.34 km; 1.92 km

20 Techniques of drawing and construction

Exercise 20a
1 **(i)** 7.8 cm **(ii)** 2.1 cm **(iii)** 9.1 cm
3 2.5 cm **7** 2.1 cm **8** 1.7 cm

Exercise 20b
3 22.4 cm

Harder examples 20
1 3.5 cm **4** 12.7 m, 15 200 m^2
5 7 cm **7** 200 m^2, 305 m^2

21 Latitude and longitude

Exercise 21a
1 **(i)** $\frac{1}{12}$ **(ii)** $\frac{1}{24}$ **(iii)** $\frac{1}{30}$ · **(iv)** $\frac{7}{360}$ **(v)** $\frac{1}{80}$
(vi) $\frac{1}{240}$ **(vii)** $\frac{5}{32}$ **(viii)** $\frac{29}{900}$ **(ix)** $\frac{131}{1440}$
(x) $\frac{7}{16}$ **2** **(i)** 8.56 cm **(ii)** 14 400 km
(iii) 0.447 m **(iv)** 2510 km
3 **(i)** 86° **(ii)** 57° **(iii)** 5° **(iv)** 1°
4 **(i)** 14.3 m **(ii)** 32.2 cm **(iii)** 631 km
(iv) 204 km **5** **(i)** 0.523 **(ii)** 1.256 **(iii)** 3.14
(iv) 2.355 **(v)** 0.419 **(vi)** 0.131 **6** **(i)** 11° 28′
(ii) 114° 39′ **(iii)** 40° 8′ **(iv)** 60° **(v)** 120°
(vi) 180°

Exercise 21b
1 40000 km **2** 111.1 km **3** 12 330 km
4 4° 30′ **5** 2° 42′ **6** 1334 km **7** 1.26 cm
8 4504 km **9** 6181 km **10** 6 h

Harder examples 21
1 **(i)** **(a)** 3562 km **(b)** 22 370 km
(ii) **(a)** 5281 km **(b)** 33 160 km
(iii) **(a)** 6153 km **(b)** 38 640 km
2 15 920 km **3** 2610 km
4 11 500 km **5** 89.9 km **6** 889 km
7 12 220 km **8** 1925 km
9 572 km **10** 180° W, 51° S
11 0° **12** 10.9 h **13** 8703 km
14 **(i)** 63 **(ii)** 72 **15** **(i)** 70° 32′ **(ii)** 30
16 2290 km, 14 h 4 min

22 Group theory

Exercise 22a
1, 4, 5(ii), 8, 12, 14 **18** yes **19** yes **20** yes
22 20, 21

Exercise 22b
1 $\{e, a\}$, $\{e, b\}$, $\{e, c\}$
2 $\{e, \ p^2\}$ **3** $\left\{\begin{pmatrix}1 & 0\\ 0 & 1\end{pmatrix}, \begin{pmatrix}-1 & 0\\ 0 & -1\end{pmatrix}\right\}$
4 $\{1, 3\}$, $\{1, 5\}$, $\{1, 7\}$ **5** $\{0, 3\}$, $\{0, 2, 4\}$
6 $\{e, p, p^2, p^3\}$, $\{e, p^2\}$, $\{e, a\}$, $\{e, b\}$, $\{e, c\}$, $\{e, d\}$

Harder examples 22
1 1, 14 **3** **(i)** c **(ii)** b, d **(iii)** yes
4 **(ii)** symmetry group of rectangle
(iii) **(a)** S
(b) Inverses of P, Q, R, S are P, R, Q, S
5 **(ii)** 2, 4 **(iii)** $x = 1$, $y = 2$ or $y = 3$
6 $I \leftrightarrow 6$, $A \leftrightarrow 2$, $B \leftrightarrow 4$, $C \leftrightarrow 8$
7 $\{0, 2, 6, 8\}$

23 The calculus

Exercise 23a

1 $4x^3$ **2** $15x^4$ **3** $-\dfrac{8}{x^3}$ **4** $2x$ **5** $3x^2-7$
6 $2x+3$ **7** $8x^3-2x$ **8** $24x^2-3$
9 $42x^5+10x$ **10** $6+\dfrac{4}{x^3}$ **11** $\frac{1}{2}x^{-1/2}$
12 $2.4x^{-0.6}$ **13** $\frac{2}{3}x^{-1/3}+\frac{1}{3}x^{-2/3}$
14 $-\frac{1}{2}x^{-3/2}$ **15** $2x+2$ **16** $2x$
17 $4x^3+4x$ **18** $1-\dfrac{1}{x^2}$ **19** $\frac{1}{2}x^{-1/2}$
20 $1-\dfrac{1}{x^2}$ **21** $4x^3$ **22** $\dfrac{1}{x^2}+\dfrac{4}{x^3}$
23 $2x-\dfrac{2}{x^3}$ **24** $4x^3+3x^2+1$
25 $36x^3+48x$

Exercise 23b

1 MIN (0, 0) **2** MIN (0, 5) **3** MIN (0, −7)
4 MIN $(-\frac{5}{2}, -\frac{1}{4})$ **5** MAX (0, 2)
6 MAX $\left(\dfrac{1}{\sqrt{3}}, \dfrac{2}{3\sqrt{3}}\right)$, MIN $\left(-\dfrac{1}{\sqrt{3}}, \dfrac{-2}{3\sqrt{3}}\right)$
7 None **8** MIN (1, 2), MAX (−1, −2)
9 MIN $(-\frac{5}{4}, -\frac{121}{8})$
10 MAX (0, 1), MIN (±1, 0)

Exercise 23c

1 2 s, 11 m **2** $10-10t$, 10 m/s, 5 m, 2 s
3 6, 36 m, 6 m/s, -6 m/s^2, 3 s
4 2, $213\frac{1}{3}$ m
5 $\frac{1}{2}$, $1\frac{1}{2}$; 9 cm/s, 24 cm/s^2, 3 cm, $1\frac{1}{2}$ s

Exercise 23d

1 $\dfrac{x^5}{5}+C, \dfrac{x^6}{2}+C, \dfrac{-4}{x}+C, \dfrac{x^3}{3}+9x+C,$
$\dfrac{x^4}{4}-\dfrac{7x^2}{2}+x+C, \dfrac{x^3}{3}+\dfrac{3x^2}{2}+5x+C,$
$\dfrac{2x^5}{5}-\dfrac{x^3}{3}+x+C, 2x^4-\dfrac{3x^2}{2}+C,$
$x^7+\dfrac{5}{3}x^3-4x+C, 3x^2+\dfrac{2}{x}+C, \dfrac{2}{3}x^{3/2}+C,$
$\dfrac{30x^{1.4}}{7}+C, \dfrac{3}{5}x^{5/3}+\dfrac{3}{4}x^{4/3}+C, 2x^{1/2}+C,$
$\dfrac{x^3}{3}+x^2+C, \dfrac{x^3}{3}-x+C, \dfrac{x^5}{5}+\dfrac{2x^3}{3}+x+C$
2 **(i)** $4\frac{1}{3}$ **(ii)** $-3\frac{3}{4}$ **(iii)** $2\frac{2}{3}$ **(iv)** 0 **(v)** $5\frac{1}{3}$
(vi) $\frac{5}{6}$ **(vii)** $55\frac{13}{15}$ **(viii)** −3 **3** **(i)** 3
(ii) $13\frac{1}{3}$ **(iii)** $10\frac{2}{3}$ **(iv)** 1.9 **(v)** 2, sq. units.
4 **(i)** 6.6π **(ii)** $91\frac{11}{15}\pi$ **(iii)** $34\frac{2}{15}\pi$ **(iv)** $\frac{7999}{3000}\pi$
(v) $3\frac{1}{15}\pi$ cubic units

Harder examples 23

1 **(a)** $y=4x-12$
(b) MAX (0, 13) MIN (4, −19)
2 **(a)** **(i)** −1, 4 **(ii)** 9 **(b)** **(ii)** 1 **(iii)** $\dfrac{80}{3}$ m
3 **(i)** 3 cm **(ii)** −3 cm/s **(iii)** 2 cm/s^2
(iv) $\dfrac{4}{3}$ s **4** 4 s **5** **(i)** 1, 3 s
(ii) 9 cm/s **(iv)** 0 **(v)** 2 cm **(vi)** 3 s
6 **(i)** 10 cm/s **(ii)** −5 cm/s^2 **(iii)** 5 s
(iv) $11\frac{1}{6}$ cm **7** **(a)** $1.2x^8-\frac{3}{2}x^4+5x+C$
(b) 0, 2 **8** (2, ±16), 28 **9** **(ii)** $3\frac{1}{6}$
(iii) 1.78 **10** **(c)** $(\frac{1}{2}, \frac{3}{4})$ **(d)** $\frac{16}{15}\pi$

24 Test papers

1 6.5163 **2** 6.50 **3** 7 **4** $\begin{pmatrix} 4 \\ 13 \\ 1 \end{pmatrix}$
5 $\frac{3}{13}$ **6** 34 **7** $\frac{4}{11}, \frac{3}{7}, \frac{5}{9}$ **8** 12 **9** **(i)** −1
(ii) 11 **10** $12\frac{1}{2}$

II

1 $4\frac{1}{10}$ **2** $\frac{1}{2}$ **3** −1 **4** $\begin{pmatrix} 11 & -3 \\ 7 & 2 \end{pmatrix}$
5 $\frac{3}{5}, \frac{3}{4}, \frac{7}{9}$ **6** 1.54 **7** {3, 6} **8** £3.60
9 6 **10** $4b-2a$

III

1 $4\frac{2}{5}$ **2** $\frac{1}{8}$ **3** 3 **4** $\begin{pmatrix} 7 & -1 \\ 8 & -4 \end{pmatrix}$ **5** $5\frac{3}{4}$
6 7 **7** 4 **8** 60 **9** −12 **10** 87.5%

IV

1 62 **2** no **3** $3y-2x$ **4** 30° 58′
7 $\frac{1}{2}$ **8** 3 **9** £8, £12 **10** 5

V

1 384 m^2 **2** 8 **3** 3 **4** £60 **5** 750
6 $r=C/2\pi$ **7** 13 cm **8** 2 **9** 0.277 **10** **(ii)**

VI

1 $1\frac{1}{6}$ **2** $2x^2-x+7$ **3** 17 **4** $x<-4$ **5** 920
6 ±4 **7** 3.2×10^4 **8** 200 **9** 17 **10** 0.0431

VII

1 15 **2** $\begin{pmatrix} -4 & -2 \\ -7 & -3 \end{pmatrix}$ **3** **(i)** 1 **(ii)** 1 **4** 1.75
5 3 **6** 2 cm^2 **7** 10111 **8** $\frac{3}{4}$ **9** $2x^2+2x-5$
10 7

VIII

1 16 **2** $\begin{pmatrix} -6 & 6 \\ -20 & 4 \end{pmatrix}$ **3** (i) 2 (ii) -10
4 13 **5** -3 **6** $33\frac{3}{4}$ cm^3 **7** 1 110 011 **8** $\frac{1}{6}$
9 $3x^2-x-5$ **10** $\{31, 37\}$

IX

1 342 **2** 5.48 **3** 2580 **4** $\begin{pmatrix} 6 & 5 \\ 10 & 0 \end{pmatrix}$
5 (i) 0 (ii) -8 **6** 16 **7** 9 **9** -9
10 $2a^2+b^2$

X

1 7.6659 **2** 1.33 **3** 16 **4** $\begin{pmatrix} 5 \\ 0 \end{pmatrix}$ **5** $\frac{2}{3}$ **6** 12
8 5 **9** $8u^2$

XI

1 £23.44 **2** $2\times3\times5^2\times7$ **3** 1.22×10^6
4 $h=(A-2\pi r^2)/2\pi r$
5 $(x-2)(x-5)$, $x<2$, $x>5$
6 $3\frac{13}{14}$ **7** $\frac{12}{13}, \frac{12}{5}$ **8** $-\frac{3}{2}$ **9** $6.32, -0.32$ **10** $8\frac{1}{3}$

XII

1 -2 **2** $-3.2\le x\le14.8$ **3** (i) 16.1
(ii) -3.5 **4** 44% **5** (i) 36 (ii) 17, 18
6 2, -1 **7** 106° 16′ **8** $d=1/(c-1)$
9 $\pm\frac{5}{13}, \pm\frac{12}{13}$ **10** 50.5

XIII

1 $\frac{1}{8}$ **2** 1 **3** 0.376 **4** $7\frac{11}{12}$ **5** 5 **6** 5, 2
7 $25.8\le \text{MN}\le41.4$ **8** $(x+10)(x-2)$
9 14 **10** $-1/(x+2)$

XIV

1 65% **2** 6 **3** $\frac{9}{10}$ **4** 1.8 **5** $x=(pt-y)/pz$
6 37° **7** $y=13-x$ **8** $3\frac{3}{4}$, $15\frac{15}{16}$ **9** -1, 7
10 (i) 4 (ii) 3

XV

1 £2.87 **2** 2.52 **3** 5^6 **4** $\begin{pmatrix} 1 & 2 \\ 0 & 1 \end{pmatrix}$,
shear in x-direction **5** 1.38×10^{-2}
6 $\varnothing$, A, {1}, {2}, {3}, {4}, {1, 2}, {1, 3},
{1, 4}, {2, 3}, {2, 4}, {3, 4}, {1, 2, 3},
{1, 2, 4}, {2, 3, 4}, {1, 3, 4}
7 (i) 7 (ii) 1 (iii) 1 **8** 9, -1 **9** (i) $\frac{1}{4}$
(ii) $\frac{1}{4}$

XVI

1 $a(a-2b)(a+2b)$ **2** $97\frac{1}{2}$p **3** (iii), (iv), (v)
4 (i) 2^{14} (ii) 2^5
5 $a(3a+2)$ **6** $\begin{pmatrix} -21 \\ -14 \end{pmatrix}$
7 18, 1p **8** 29 **9** 2.83 **10** 4

XVII

1 0.825 **2** 27 **3** 8, 10 **4** $a^2b^2(b-a)$
5 1.4313 **6** £33.60 **7** (i) 30 (ii) ±2 **8** $\frac{4}{3}$
10 1.62 m

XVIII

1 (i) 2.281 (ii) 2.28 **2** $\frac{9}{80}$ **3** (i) 0.5736
(ii) 0.7071 (iii) -0.3420 **4** $\pi(R-r)(R+r)$
5 (i) {1, 3, 5, 7, 11, 13, 17, 19} (ii) {3, 9, 15}
(iii) {3} **6** 6% **7** $\begin{pmatrix} 7 & 0 \\ 0 & 7 \end{pmatrix}$ **8** (3, 5), (3, 6)
9 31.4 cm

XIX

1 $-3, -4$ **2** tetrahedron **3** (i) -3 (ii) $\frac{4}{3}$
4 40.6 cm^2 **5** 8 cm, 4.5 cm
6 (i) {4, 16, 36, 64} (ii) {1, 4}
7 10.6 m **9** 7

XX

1 $2\frac{2}{3}$ **2** 10111 **3** 8% **4** 4, 4×10^9
5 {(2, 1)} **6** 4 **7** 2.5 **8** {2, 4, 5}
9 $y=c/(1-b)$ **10** 3.98×10^{-3}

XXI

1 $\frac{1}{2}\sqrt{26}$ **2** 1011011 **3** $\frac{4}{3}$ **4** 554.40 **5** 5
6 51° 19′ **7** $-\frac{2}{3}, -1$ **8** 0, 1
9 $\frac{1}{10}\begin{pmatrix} 4 & -2 \\ 3 & 1 \end{pmatrix}$ **10** 8, 1

XXII

2 11.5 **3** $(-6, -1)$ **4** 50 km/h
5 $\begin{pmatrix} 288 & 552 \\ -32 & -64 \end{pmatrix}$ **6** $3\frac{8}{13}, \frac{1}{13}$ **7** 1%
9 17121 **10** 31.6

XXIII

1 112 **3** $b=\pm(x^2-4c)/4a$ **4** 2, -1
6 2.22×10^6 **7** 24.4 **8** 4329
9 enlarges in horizontal direction

XXIV

1 8 **2** (3, 2) **3** 0 **4** $\begin{pmatrix} 6 & -3 \\ -12 & 4 \end{pmatrix}$ **5** 9, $\frac{7}{9}$
6 ±4 **7** 55° **8** 4, 0 **9** 120, 0.12
10 $\frac{3}{7}, -2\frac{1}{2}$

XXV

1 4 **2** ±2 **3** $(x-7)(x+2)$ **4** 7.86 cm
5 $\begin{pmatrix} 9 \\ -10 \end{pmatrix}$
6 £97.20 **7** 8 **8** 17 **9** 1886.5 cm^2 **10** 30°

LOGARITHMS

	0	1	2	3	4	5	6	7	8	9	1	2	3	4	5	6	7	8	9
10	0000	0043	0086	0128	0170						4	9	13	17	21	26	30	34	38
						0212	0253	0294	0334	0374	4	8	12	16	20	24	28	32	36
11	0414	0453	0492	0531	0569						4	8	12	15	19	23	27	31	35
						0607	0645	0682	0719	0755	4	7	11	15	19	22	26	30	33
12	0792	0828	0864	0899	0934						3	7	11	14	18	21	25	28	32
						0969	1004	1038	1072	1106	3	7	10	14	17	20	24	27	31
13	1139	1173	1206	1239	1271						3	7	10	13	16	20	23	26	30
						1303	1335	1367	1399	1430	3	7	10	13	16	19	22	25	29
14	1461	1492	1523	1553	1584						3	6	9	12	15	19	22	25	28
						1614	1644	1673	1703	1732	3	6	9	12	15	17	20	23	26
15	1761	1790	1818	1847	1875						3	6	9	11	14	17	20	23	26
						1903	1931	1959	1987	2014	3	6	8	11	14	17	19	22	25
16	2041	2068	2095	2122	2148						3	5	8	11	14	16	19	22	24
						2175	2201	2227	2253	2279	3	5	8	10	13	16	18	21	23
17	2304	2330	2355	2380	2405						3	5	8	10	13	15	18	20	23
						2430	2455	2480	2504	2529	2	5	7	10	12	15	17	20	22
18	2553	2577	2601	2625	2648						2	5	7	9	12	14	16	19	21
						2672	2695	2718	2742	2765	2	5	7	9	11	14	16	18	21
19	2788	2810	2833	2856	2878						2	4	7	9	11	13	16	18	20
						2900	2923	2945	2967	2989	2	4	6	8	11	13	15	17	19
20	3010	3032	3054	3075	3096	3118	3139	3160	3181	3201	2	4	6	8	11	13	15	17	19
21	3222	3243	3263	3284	3304	3324	3345	3365	3385	3404	2	4	6	8	10	12	14	16	18
22	3424	3444	3464	3483	3502	3522	3541	3560	3579	3598	2	4	6	8	10	12	14	15	17
23	3617	3636	3655	3674	3692	3711	3729	3747	3766	3784	2	4	6	7	9	11	13	15	17
24	3802	3820	3838	3856	3874	3892	3909	3927	3945	3962	2	4	5	7	9	11	12	14	16
25	3979	3997	4014	4031	4048	4065	4082	4099	4116	4133	2	3	5	7	9	10	12	14	15
26	4150	4166	4183	4200	4216	4232	4249	4265	4281	4298	2	3	5	7	8	10	11	13	15
27	4314	4330	4346	4362	4378	4393	4409	4425	4440	4456	2	3	5	6	8	9	11	13	14
28	4472	4487	4502	4518	4533	4548	4564	4579	4594	4609	2	3	5	6	8	9	11	12	14
29	4624	4639	4654	4669	4683	4698	4713	4728	4742	4757	1	3	4	6	7	9	10	12	13
30	4771	4786	4800	4814	4829	4843	4857	4871	4886	4900	1	3	4	6	7	9	10	11	13
31	4914	4928	4942	4955	4969	4983	4997	5011	5024	5038	1	3	4	6	7	8	10	11	12
32	5051	5065	5079	5092	5105	5119	5132	5145	5159	5172	1	3	4	5	7	8	9	11	12
33	5185	5198	5211	5224	5237	5250	5263	5276	5289	5302	1	3	4	5	6	8	9	10	12
34	5315	5328	5340	5353	5366	5378	5391	5403	5416	5428	1	3	4	5	6	8	9	10	11
35	5441	5453	5465	5478	5490	5502	5514	5527	5539	5551	1	2	4	5	6	7	9	10	11
36	5563	5575	5587	5599	5611	5623	5635	5647	5658	5670	1	2	4	5	6	7	8	10	11
37	5682	5694	5705	5717	5729	5740	5752	5763	5775	5786	1	2	3	5	6	7	8	9	10
38	5798	5809	5821	5832	5843	5855	5866	5877	5888	5899	1	2	3	5	6	7	8	9	10
39	5911	5922	5933	5944	5955	5966	5977	5988	5999	6010	1	2	3	4	5	7	8	9	10
40	6021	6031	6042	6053	6064	6075	6085	6096	6107	6117	1	2	3	4	5	6	8	9	10
41	6128	6138	6149	6160	6170	6180	6191	6201	6212	6222	1	2	3	4	5	6	7	8	9
42	6232	6243	6253	6263	6274	6284	6294	6304	6314	6325	1	2	3	4	5	6	7	8	9
43	6335	6345	6355	6365	6375	6385	6395	6405	6415	6425	1	2	3	4	5	6	7	8	9
44	6435	6444	6454	6464	6474	6484	6493	6503	6513	6522	1	2	3	4	5	6	7	8	9
45	6532	6542	6551	6561	6571	6580	6590	6599	6609	6618	1	2	3	4	5	6	7	8	9
46	6628	6637	6646	6656	6665	6675	6684	6693	6702	6712	1	2	3	4	5	6	7	7	8
47	6721	6730	6739	6749	6758	6767	6776	6785	6794	6803	1	2	3	4	5	5	6	7	8
48	6812	6821	6830	6839	6848	6857	6866	6875	6884	6893	1	2	3	4	4	5	6	7	8
49	6902	6911	6920	6928	6937	6946	6955	6964	6972	6981	1	2	3	4	4	5	6	7	8

LOGARITHMS

	0	1	2	3	4	5	6	7	8	9	1 2 3	4 5 6	7 8 9
50	6990	6998	7007	7016	7024	7033	7042	7050	7059	7067	1 2 3	3 4 5	6 7 8
51	7076	7084	7093	7101	7110	7118	7126	7135	7143	7152	1 2 3	3 4 5	6 7 8
52	7160	7168	7177	7185	7193	7202	7210	7218	7226	7235	1 2 2	3 4 5	6 7 7
53	7243	7251	7259	7267	7275	7284	7292	7300	7308	7316	1 2 2	3 4 5	6 6 7
54	7324	7332	7340	7348	7356	7364	7372	7380	7388	7396	1 2 2	3 4 5	6 6 7
55	7404	7412	7419	7427	7435	7443	7451	7459	7466	7474	1 2 2	3 4 5	5 6 7
56	7482	7490	7497	7505	7513	7520	7528	7536	7543	7551	1 2 2	3 4 5	5 6 7
57	7559	7566	7574	7582	7589	7597	7604	7612	7619	7627	1 2 2	3 4 5	5 6 7
58	7634	7642	7649	7657	7664	7672	7679	7686	7694	7701	1 1 2	3 4 4	5 6 7
59	7709	7716	7723	7731	7738	7745	7752	7760	7767	7774	1 1 2	3 4 4	5 6 7
60	7782	7789	7796	7803	7810	7818	7825	7832	7839	7846	1 1 2	3 4 4	5 6 6
61	7853	7860	7868	7875	7882	7889	7896	7903	7910	7917	1 1 2	3 4 4	5 6 6
62	7924	7931	7938	7945	7952	7959	7966	7973	7980	7987	1 1 2	3 3 4	5 6 6
63	7993	8000	8007	8014	8021	8028	8035	8041	8048	8055	1 1 2	3 3 4	5 5 6
64	8062	8069	8075	8082	8089	8096	8102	8109	8116	8122	1 1 2	3 3 4	5 5 6
65	8129	8136	8142	8149	8156	8162	8169	8176	8182	8189	1 1 2	3 3 4	5 5 6
66	8195	8202	8209	8215	8222	8228	8235	8241	8248	8254	1 1 2	3 3 4	5 5 6
67	8261	8267	8274	8280	8287	8293	8299	8306	8312	8319	1 1 2	3 3 4	5 5 6
68	8325	8331	8338	8344	8351	8357	8363	8370	8376	8382	1 1 2	3 3 4	4 5 6
69	8388	8395	8401	8407	8414	8420	8426	8432	8439	8445	1 1 2	2 3 4	4 5 6
70	8451	8457	8463	8470	8476	8482	8488	8494	8500	8506	1 1 2	2 3 4	4 5 6
71	8513	8519	8525	8531	8537	8543	8549	8555	8561	8567	1 1 2	2 3 4	4 5 5
72	8573	8579	8585	8591	8597	8603	8609	8615	8621	8627	1 1 2	2 3 4	4 5 5
73	8633	8639	8645	8651	8657	8663	8669	8675	8681	8686	1 1 2	2 3 4	4 5 5
74	8692	8698	8704	8710	8716	8722	8727	8733	8739	8745	1 1 2	2 3 4	4 5 5
75	8751	8756	8762	8768	8774	8779	8785	8791	8797	8802	1 1 2	2 3 3	4 5 5
76	8808	8814	8820	8825	8831	8837	8842	8848	8854	8859	1 1 2	2 3 3	4 5 5
77	8865	8871	8876	8882	8887	8893	8899	8904	8910	8915	1 1 2	2 3 3	4 4 5
78	8921	8927	8932	8938	8943	8949	8954	8960	8965	8971	1 1 2	2 3 3	4 4 5
79	8976	8982	8987	8993	8998	9004	9009	9015	9020	9025	1 1 2	2 3 3	4 4 5
80	9031	9036	9042	9047	9053	9058	9063	9069	9074	9079	1 1 2	2 3 3	4 4 5
81	9085	9090	9096	9101	9106	9112	9117	9122	9128	9133	1 1 2	2 3 3	4 4 5
82	9138	9143	9149	9154	9159	9165	9170	9175	9180	9186	1 1 2	2 3 3	4 4 5
83	9191	9196	9201	9206	9212	9217	9222	9227	9232	9238	1 1 2	2 3 3	4 4 5
84	9243	9248	9253	9258	9263	9269	9274	9279	9284	9289	1 1 2	2 3 3	4 4 5
85	9294	9299	9304	9309	9315	9320	9325	9330	9335	9340	1 1 2	2 3 3	4 4 5
86	9345	9350	9355	9360	9365	9370	9375	9380	9385	9390	1 1 2	2 3 3	4 4 5
87	9395	9400	9405	9410	9415	9420	9425	9430	9435	9440	0 1 1	2 2 3	3 4 4
88	9445	9450	9455	9460	9465	9469	9474	9479	9484	9489	0 1 1	2 2 3	3 4 4
89	9494	9499	9504	9509	9513	9518	9523	9528	9533	9538	0 1 1	2 2 3	3 4 4
90	9542	9547	9552	9557	9562	9566	9571	9576	9581	9586	0 1 1	2 2 3	3 4 4
91	9590	9595	9600	9605	9609	9614	9619	9624	9628	9633	0 1 1	2 2 3	3 4 4
92	9638	9643	9647	9652	9657	9661	9666	9671	9675	9680	0 1 1	2 2 3	3 4 4
93	9685	9689	9694	9699	9703	9708	9713	9717	9722	9727	0 1 1	2 2 3	3 4 4
94	9731	9736	9741	9745	9750	9754	9759	9763	9768	9773	0 1 1	2 2 3	3 4 4
95	9777	9782	9786	9791	9795	9800	9805	9809	9814	9818	0 1 1	2 2 3	3 4 4
96	9823	9827	9832	9836	9841	9845	9850	9854	9859	9863	0 1 1	2 2 3	3 4 4
97	9868	9872	9877	9881	9886	9890	9894	9899	9903	9908	0 1 1	2 2 3	3 4 4
98	9912	9917	9921	9926	9930	9934	9939	9943	9948	9952	0 1 1	2 2 3	3 4 4
99	9956	9961	9965	9969	9974	9978	9983	9987	9991	9996	0 1 1	2 2 3	3 3 4